Process Instrumentation

Second Edition

Technical Editor

John L. Sluder
River Parishes Community College
Gonzales, LA

Managing Director, Career Development and Employability: Leah Jewell
Specialist, Alliance/Partnership Management: Greg Oliver
Director, Employability Solutions: Kelly Trakalo
Assistant Content Producer: Alma Dabral
Director, Digital Studio and Content Production: Brian Hyland
Media Producer: Jose Carchi
Development Editor: Rachel Bedard, Editorial Consultants
Instructor and Student Supplement Development: Linda Malcak, Perci LLC and Malcak, Tyler Associates
Cover Design: SPi Global U.S., Inc.
Cover Image Credit: KhanunHaHa/Shutterstock
Chapter Opening Image: KhanunHaHa/Shutterstock
Project Management, Editorial and Full-Service Production and Composition Services: SPi Global
Printer/Bindery: LSC Communications
Cover Printer: Phoenix Color

Credits and acknowledgments borrowed from other sources and reproduced, with permission, in this textbook appear on appropriate page within text.

Credits not appearing on page are from NAPTA, Introduction to Process Technology, 2nd Ed., © 2018. Reprinted and electronically produced with permission by Pearson Education, Inc., New York, NY. These assets appear on pages: 4, 6, 8, 9, 10, 30, 34, 46 (2), 52, 67, 77, 89, 90, 130, 131, 132 (2), 228.

Credits not appearing on page are from NAPTA, Process Technology Equipment, 2nd Ed., © 2019. Reprinted and electronically produced with permission by Pearson Education, Inc., New York, NY. These assets appear on pages: 3, 69, 133, 134, 138 (2), 254, 278.

Credits not appearing on page are from NAPTA, Safety, Health, and Environment, 2nd Ed., © 2019. Reprinted and electronically produced with permission by Pearson Education, Inc., New York, NY. These assets appear on pages: 332 (2), 334, 336.

Copyright © 2020, 2010 Pearson Education, Inc., All rights reserved. Printed in the United States of America. This publication is protected by Copyright and permission should be obtained from the publisher prior to any prohibited reproduction, storage in a retrieval system, or transmission in any form or by any means, electronic, mechanical, photocopying, recording, or likewise. To obtain permission(s) to use material from this work, please visit http://www.pearsoned.com/permissions/.

Library of Congress Cataloging-in-Publication Data

Names: North American Process Technology Alliance.
Title: Process instrumentation: North American Process Technology Alliance
  / technical editor, Sluder John, River Parishes Community College,
  Gonzales, LA.
Description: Second edition. | Hoboken: Pearson, 2020. | Includes index.
Identifiers: LCCN 2019046615 | ISBN 9780135213926 (paperback) | ISBN
  9780135984642 (epub) | ISBN 9780135213957
Subjects: LCSH: Process Technology Instrumentation.
Classification: LCC TP155.7 .I598 2020 | DDC 660/.28--dc23
LC record available at https://lccn.loc.gov/2019046615

11 2022

www.pearsonhighered.com

ISBN-10:   0-13-521392-4
ISBN-13: 978-0-13-521392-6

Preface

The Process Industries Challenge

In the early 1990s, the process industries recognized that they would face a major staffing shortage because of the large number of "baby boomer" employees who would be retiring. Industry partnered with community colleges, technical colleges, and universities to remedy this situation. Together, they developed this series, which provides consistent curriculum content and exit competencies for process technology graduates to ensure a knowledgeable and competent staff that is ready to take over the demands of the field. The collaborators in education and industry also recognized that training for process technicians would benefit industry by reducing the costs associated with training and traditional hiring methods. This was how the NAPTA series on Process Technology was born.

To achieve consistency of exit competencies among graduates from different schools and regions, the North American Process Technology Alliance identified a core technical curriculum for the Associate Degree in Process Technology. This core consists of eight technical courses and is taught in alliance member institutions throughout the United States.

Instructors who teach the process technology core curriculum, and who are recognized in industry for their years of experience and depth of subject matter expertise, requested that a textbook be developed to match the standardized curriculum. Reviewers from a broad range of process industries and educational institutions participated in the production of these materials so that the presentation of content would address the widest audience possible. This textbook is intended to provide a common national standard reference for the *Instrumentation* course in the Process Technology degree program.

This textbook is intended for use in high schools, community colleges, technical colleges, universities, and any corporate setting in which process technology is taught. Current and future process technicians will use the information within this textbook as their foundation for work in the process industries. This knowledge will make them better prepared to meet the ever-changing roles and responsibilities within their specific process industry.

What's New!

The second edition has been thoroughly updated and revised.

- **New** Learning Outcome alignment with NAPTA core objectives, with links from objective to text page provided.
- **All New** Full Color Art Program with more than 350 fully revised drawings and photos, including all new visuals.
- **New** Key term definitions on the text page where content appears, as well as at the beginning of each chapter.
- **Content Organization** to enhance learning-process variables (Chapters 1–6); symbology, hardware, and instrumentation communication (Chapters 7–9); control loops, controllers, control schemes, digital control, and PLC (Chapters 10–18); DCS, power supply, ESD, malfunctions, and troubleshooting (Chapters 19–23).
- **New Content** on intrinsically safe display devices/tablets, wireless technologies, and open Internet (intranet) protocol transmission/addressing.
- **Expanded** Additional coverage of the roles of a process technician related to cybersecurity, industry 4.0, and analytic data trends.
- **Updated Review and New Answers Appendix!** Checking Your Knowledge questions have been updated to meet new chapter objectives and now include an Appendix with answers.
- **ALL NEW Instructor Resource Package** including lesson plans, test banks, review questions, PowerPoints, and a correlation guide to NAPTA curriculum.

Organization of the Textbook

This book has been divided into four parts with 23 chapters. Part 1 (Chapters 1–6) provides an overview of instrumentation and describes the key process variables. Part 2 (Chapters 7–9) describes symbology, hardware, and instrumentation communication. Part 3 (Chapters 10–18) focuses on control loops, controllers, and programmable logic. Part 4 (Chapters 19–23) addresses DCS, power supply, interlocks, shutdowns, malfunctions, and troubleshooting.

Each chapter has the same organization.

- **Objectives** for each chapter are aligned with the revised NAPTA curriculum and can cover one or more sessions in a course.
- **Key Terms** list important words or phrases and their respective definitions, which students should know and understand before proceeding to the next chapter.
- The **Introduction** might be a simple introductory paragraph or might introduce concepts necessary to the development of the chapter's content.
- **Key Topics** and subtopics address the objectives stated at the beginning of each chapter.
- The **Summary** is a restatement of important points in the chapter.
- **Checking Your Knowledge** questions are designed to help students do self-testing on essential information in the chapter.

- **Student Activities** provide opportunities for individual students or small groups to apply some of the knowledge they have gained from the chapter. These activities generally should be performed with instructor involvement.

Acknowledgments

The second edition of *Process Instrumentation* was a huge undertaking. It would not have been possible without the support of the NAPTA Board and the continued leadership of Executive Director Eric Newby. This edition relied heavily on input and guidance from faithful members of the NAPTA Curriculum Committee and the interim Curriculum Committee Co-chair Martha McKinley.

A particular and special thank you goes to Technical Editor John Sluder for his key role in the revision of this title. His vision for the book, organizational insight, and collaborative approach gave this edition its structure and are greatly appreciated.

Finally, the production of *Process Instrumentation* 2nd edition owes a huge debt of gratitude to industry partners Emerson Automation Solutions, Simulation Solutions, and The Dow Chemical Company, who provided images that bring the world of process technology to life. Individuals from industry and education who volunteered their time and skills are recognized below in the Contributor section.

Contributors

Sheri Bankston, Alliance Safety Council
Baton Rouge, LA (Instructor supplements)

Mike Brouse, Emerson Automation, Texas (Art program)

Gayle Cannon, Middlesex County College
Millstone Township, NJ (Text revisions)

Matthew Garvey, Simulation Solutions, Inc., New Jersey (Art program)

Thomas Germusa, FirstEnergy Corp
Shippingport, PA (Text revisions)

Don Glaser, Simulation Solutions, Inc., New Jersey (Art program)

Gary Hicks, Brazosport Community College, Texas (Art program)

Michael Kean, Los Medanos College
Pittsburg, CA (Instructor supplements)

Martha McKinley, McKinley Consulting
Longview, TX (Text revisions)

Alisha Nash, The Dow Chemical Company, Louisiana (Art program)

Michele Ramshur, Emerson Automation Solutions, Texas (Art program)

Terry Rings, SOWELA Technical Community College
Jennings, LA (Text revisions)

John Sluder, River Parishes Community College
Gonzales, LA (Text revisions)

Steve Wethington, College of the Mainland
Texas City, TX (Text revisions)

Reviewers

The following organizations and their dedicated personnel supported the development of the first edition of this textbook. Their contributions set the foundation for this revision and continue to be greatly appreciated.

Industry Content Developers and Reviewers

Lisa Arnold, Marathon Ashland, Texas
Ted Borel, Equistar Chemical, Texas
Curtis Briggs, INEOS, Texas
Gerald Canady, The Dow Chemical Company, Louisiana
Tim Carroll, BASF Corporation, Louisiana
Lester Chin, Hovensa, LLC, U.S. Virgin Islands
Sheldon Cooley, Westlake Group, Louisiana
Regina Cooper, Marathon Petroleum, Texas
Ed Couvillion, Sterling Chemicals, Texas
Greg Curry, The Dow Chemical Company, Texas
Lloyd Davis, Halliburton, Oklahoma
Steve Erickson, GCPTA, Texas
Steve Ernest, ConocoPhillips, Oklahoma
Billy Fridelle, ExxonMobil, Texas
Richard Honea, The Dow Chemical Company, Texas
Lee Hughes, BASF Corporation, Texas
Jennifer Martin, OG&E Horseshoe/Mustang Plant, Oklahoma
Mike McBride, BASF, Texas
Martha McKinley, Eastman Chemical, Texas
Jason Oxley, OG&E Horseshoe/Mustang Plant, Oklahoma
Aracelli Palomo, The Dow Chemical Company, Texas
Allen Parker, Crompton Corporation, Louisiana
Shawn Parker, Marathon Ashland, Texas
Clemon Prevost, BP, Texas
Robert Rabalaise, The Dow Chemical Company, Louisiana
Raymond Robertson, PPG, Louisiana
Matt Scully, BP, Alaska
Roy St. Romain, Eastman Chemical, Texas
Paul Summers, The Dow Chemical Company, Texas
Dennis Thibodeaux, Citgo, Louisiana
Robert Toups, Lyondell Chemical, Louisiana

Education Content Developers and Reviewers

Sazar Ali, Houston Community College, Texas
Louis Babin, ITI Technical College, Louisiana
Allen Bargar, Brazosport College, Texas
Carrie Braud, Baton Rouge Community College, Louisiana
Tommie Ann Broome, Mississippi Gulf Coast Community College, Mississippi
Larry Callaway, Bellingham Technical College, Washington
Chuck Carter, Lee College, Texas
Lou Caserta, Alvin Community College, Texas
Ken Clark, Perry Technical Institute, Washington
Michael Connella III, McNeese State University, Louisiana
Richard Cox, Baton Rouge Community College, Louisiana
Mark Demark, Alvin Community College, Texas
Ned Duffey, South Arkansas Community College, Arkansas
Jerry Duncan, College of the Mainland, Texas
Tommy Edgar, Texas State Technical College, Texas
Steve Ernst, Northern Oklahoma College, Oklahoma
Charles Gaffen, Elizabeth High School, New Jersey
Gary Hicks, Brazosport Community College, Texas
Kevin Holmstrom, Bismarck State College, North Dakota
Raymond Johns, Northern Oklahoma College, Oklahoma
Larry Johnson, Nova Scotia Community College, Nova Scotia
Mike Kukuk, College of the Mainland, Texas
Tony Kuphaldt, Bellingham Technical College, Washington
Linton LeCompte, Sowela Technical Community College, Louisiana
Michael Murray, Copiah-Lincoln Community College, Mississippi
Richard Ortloff, Bellingham Technical College, Washington
Anthony Pringle, Remington College, Texas
Denise Rector, Del Mar College, Texas
Robert Robertus, Montana State University - Billings, Montana
Paul Rodriguez, Lamar Institute of Technology, Texas
Dean Schwarz, Southwestern Illinois College, Illinois
Mike Speegle, San Jacinto College, Texas
Wayne Stephens, Wharton County Junior College, Texas
Mark Stoltenberg, Brazosport Community College, Texas
Robert Walls, Del Mar College, Texas
Scott Wells, Brazosport Community College, Texas

Contents

Part 1 Introduction and Process Variables

1 Introduction to Instrumentation — 1

Introduction — 3
Evolution of Process Instrumentation — 3
Monitoring Process Variables: Importance of Monitoring Process Variables — 5
Importance of Instrumentation — 5
Role of Process Operators in Safety — 5
Process Variables — 6
Pressure — 6
Temperature — 7
Flow — 7
Level — 8
Analytics — 8
Instrument Component Types — 9
Instruments Categorized by Location — 9
Instruments Categorized by Function — 9
Control Instruments Categorized by Signal Type — 11
Types of Signals Produced — 12
Process Variable Relationships — 14
What Is the Effect of Temperature Change on Pressure in a Closed Container? — 14
What Is the Effect of the Height of Liquid on Pressure at the Bottom of a Vessel? — 15
What Is the Effect of Pressure Change on the Boiling Point of a Material? — 17
What Is the Effect of Temperature Change on the Volume of a Material? — 17
What Is the Effect of Temperature Change on the Density of a Material? — 18
What Is the Effect of Variable Change on Accuracy of Measurements? — 18
Summary — 19
Checking Your Knowledge — 20
Student Activities — 20

2 Process Variables, Elements, and Instruments: PRESSURE — 21

Introduction — 23
Pressure Defined — 23
Components That Affect the Force Exerted by Molecules — 24
Pressure Measurement — 26
Differential Pressure (Delta P, ΔP, D/P) — 26
Pressure Measurement Scales — 27
Unit Equivalency — 29
Pressure Sensing and Measurement Instruments — 29
Manometers — 30
Pressure Gauges — 31
Differential Pressure Cells — 32
Strain Gauge Transducers — 33
Capacitance and Piezoelectric Transducers — 34
Pressure Conversions — 34
Pressure Units — 35
Pressure Conversion Calculations — 36
Summary — 37
Checking Your Knowledge — 38
Student Activities — 39

3 Process Variables, Elements, and Instruments: TEMPERATURE — 41

Introduction — 43
Temperature and Effects of Heat — 43
Effects of Heat Energy on Molecular Movement — 44
Temperature Measurement, Differential, and Scales — 45
Temperature Measurement Scales — 46
Boiling and Freezing Points of Water — 47
Effects of Pressure and Contaminants on the Boiling Point of Water — 48
Temperature Sensing and Measurement Instruments — 49
Thermometer — 49
Bimetallic Strip — 51
Resistance Temperature Detector (RTD) — 52
Thermocouples — 53
Thermistors — 54
Thermowell — 55
Temperature Gauge — 56
Temperature Conversion — 57
Converting Fahrenheit to Celsius — 58
Converting Celsius to Fahrenheit — 58
Converting Celsius to Kelvin — 58
Converting Kelvin to Celsius — 58
Converting Kelvin to Rankine — 58
Summary — 59
Checking Your Knowledge — 60
Student Activities — 60

4 Process Variables, Elements, and Instruments: LEVEL — 63

Introduction to Level and Terms — 65
What Is Level? — 65
Innage and Ullage (Outage) — 66
Level Measurement Categories and Methods — 67

Interface Level	68	Chromatography	111
Meniscus	69	Combustion	111
Calculating Head Pressure	70	Analytical Sensing or Measuring Instruments	111
Calculating the Height of a Liquid	70	Purpose of Analytical Devices	111
Level, Temperature, Density, and Volume	71	Common Types of Analyzers	112
Density and Hydrostatic Head Pressure	72	Lower Explosive Limit (LEL)–Upper Explosive Limit (UEL)	117
Level Sensing and Measurement Instruments	73	Chromatography	117
Gauge or Sight Glass	74	Mass Spectrometers	118
Floats and Float Gauges	75	Total Carbon Analyzers	119
Tapes and Tape Gauges	76	Miscellaneous Monitors	120
Differential Pressure (D/P) Cells	77	Miscellaneous Devices	121
Bubblers	78	Vibration	121
Displacer and Transmitter	80	Speed	123
Distance Measuring Devices	81	Analytical Instruments and the Role of the Process Technician	124
Nuclear Device	82	Summary	124
Load Cells	83	Checking Your Knowledge	125
Summary	84	Student Activities	126
Checking Your Knowledge	85		
Student Activities	86		

5 Process Variables, Elements, and Instruments: FLOW 87

Part 2 Symbology, Hardware, and Instrumentation Communication

Introduction to Flow	89		
Reynolds Number	90		
Direct and Indirect Flow Measurement	91		
Positive Displacement Flow Measurement	91		
Percentage Flow Rate	91		
Volumetric Flow Units	91		
Mass Flow Units	91		

7 Process Diagrams and Instrumentation Symbology 129

Introduction to Process Diagrams	130
Introduction to Symbology	133
ISA S5.1	133
ISA Instrument Tag Number	134
Instrumentation Symbols	136
Instrumentation Symbol Interpretation	137
Summary	146
Checking Your Knowledge	147
Student Activities	148

Types of Flow Measurement Devices	92
Orifice Plates	92
Venturi Tubes	93
Flow Nozzles	94
Pitot Tubes	94
Multiport Pitot Tubes	95
Rotameters	95
Magmeters (Electromagnetic Flow Meters)	96
Turbine Meters	97
Mass Flow Meters	98
Differential Pressure (D/P) Transmitters	98
Miscellaneous Flow Meters	100
Flow Rate, Total Volume Flow, and Volumetric Flow	101
Summary	102
Checking Your Knowledge	103
Student Activities	104

8 Switches, Relays, and Alarms 150

Introduction	151
Switch Categories	151
Switch Nomenclature and Identification	152
Toggle or Hand-Off-Auto (HOA) Switches	153
Switches for Alarm Application	153
Switches for Shutdown Application	153
Switches for Autostart Application	154
Switches for Bypass Application	154
Limit Switches	155
Proximity Switches	155
Vibration Sensors	156
Process Variable Switches	156
Switch Symbols	158
Relays	158
Relay Functionality	159
Relay Applications	159

6 Process Variables, Elements, and Instruments: ANALYTICS 106

Introduction to Analytical Instruments and Terms	109
pH	110
Oxidation Reduction Potential (ORP)	110
Conductivity	110
Optical Measurement	111

Alarm Systems	159
Summary	161
Checking Your Knowledge	162
Student Activities	162

9 Signal Transmission and Conversion — 163

Introduction	164
Signal Transmission	165
Transmitter Purpose and Operation	165
Protecting Integrity and Reliability of Signal Transmission	166
Grounding and Shielding	166
Insulation	167
Error Correction	167
Signal Conditioning	168
Construction Material Selection	168
Other Safety Concerns	168
Common Types of Transmissions	168
Analog Transmission	168
Digital Transmission	169
Optical Transmission	169
Signal Conversion	169
Purpose and Operation of Signal Converter Equipment	170
"Zeroing Out" Wet Leg Instruments	170
Scaling Calculations	171
Instrument Scale Definitions	171
Common Linear Signal Conversions	172
Nonlinear Signal Conversion	172
Wireless Technologies—Industry 4.0	173
Open Internet Protocol Transmission/Addressing	173
Summary	174
Checking Your Knowledge	175
Student Activities	176

Part 3 Control Loops, Controllers, Control Schemes, and Programmable Logic

10 Introduction to Control Loops: Simple Loop Theory — 177

Introduction	179
Process Control and Control Loop Elements	179
Closed and Open Control Loops	180
Closed Loop Control (Feedback)	181
Open Loop Control	182
Control Loop Functions and Signals	182
Sensor: Sensing Element	183
Transmitter: Converting and Transmitting Element	183
Controller: Comparing, Calculating, and Correcting Element	184
Transducer: Converting Element	185
Final Control Element: Manipulating Element	186
Signal Transmissions, Loop Error, Live Zero	186
Pneumatic (Analog)	187
Electronic (Analog)	187
Electronic (Digital)	188
Mechanical (Link or Linkage)	188
Loop Error	189
Live Zero	189
Summary	190
Checking Your Knowledge	191
Student Activities	192

11 Control Loops: Primary Sensors, Transmitters, and Transducers — 193

Introduction	194
Component Purpose and Operation	194
Sensors	195
Transmitters	197
Transmitter Input and Output Signals and Scaling	198
Transducers and Signals	200
Pneumatic and Electrical Signals	201
Control Schemes and Control Loops	201
Summary	202
Checking Your Knowledge	202
Student Activities	203

12 Control Loops: Controllers and Final Control Element Overview — 205

Introduction	207
Controller Characteristics	208
Front Panel Controllers	208
Side Panel	209
Tuning	210
Controller Switching—Bumpless Transfer	211
Controller Switching—Mode Changes	212
Automatic to Manual Mode	212
Manual to Automatic Mode	212
Controller Applications	212
Local Controllers	212
Remote Controllers	213
Controllers with Split-Range Valves	213
Cascade/Remote Set Point (RSP) Controllers	214
Ratio Controllers	215
Final Control Element Overview	216
Control Valves	217
Louver and Damper	218
Electric Motors	218
Summary	219
Checking Your Knowledge	221
Student Activities	222

13 Control Loops: Control Valves and Regulators — 225

Introduction	227
The Control Valve	227

Control Valve Failure Conditions	229
Fail Open	229
Fail Closed	229
Fail in Place or Fail Last	230
Control Valve Actuators	230
Actuator and Valve Actions for Sliding Stem Valves	230
Pneumatically Driven Actuators	231
Valve Positioners	232
Valve Positioner Operation	233
Output Signals	234
Valves for Control Configurations	235
Globe Control Valves	235
Three-Way Control Valves	235
Butterfly Control Valves	236
Ball or Segmented Ball Control Valves	237
Instrument Air Regulators	238
Back-Pressure Regulators	239
Pressure-Reducing Regulators	239
Summary	240
Checking Your Knowledge	241
Student Activities	242

14 Controllers 243

Introduction	245
Process Dynamics and Control	245
Process Equilibrium	245
Load Change	245
Lag Time	246
Dead Time	246
Over Range	246
Spanning	247
Controllers and Control Actions or Modes	247
Feedback Control	248
Feedforward Control	249
Direct or Reverse Acting Controllers	249
On/Off Control	250
Other Control Terms	251
Process Recording	252
Offset	253
Trend	253
Hunting	254
Steady State	254
Tuning Modes (Gain, Reset, and Rate)	255
Gain (Proportional Band) Mode	255
Reset (Integral) Mode	258
Rate (Derivative or Preact) Mode	259
PID	260
Combined Control Action	261
Troubleshooting	262
Cycling	262
Offset	262
Summary	263
Checking Your Knowledge	263
Student Activities	264

15 Control Schemes 265

Introduction	266
Types of Control Schemes	266
On/Off Control	266
Lead/Lag Control	266
Feedback Control	267
Feedforward Control	268
Controller Modes	268
Local Control (Field-Mounted)	268
Local Manual	268
Local Automatic	269
Remote Control	269
Remote Manual	270
Remote Automatic Control	270
Cascading of the Remote Automatic Control	270
Summary	271
Checking Your Knowledge	271
Student Activities	272

16 Advanced Control Schemes 273

Introduction	274
Cascade Control	274
Primary versus Secondary Controller	274
Purpose of Cascade Loops	274
Examples of Cascade Control	276
Ratio Control	278
Examples of Ratio Control	279
Split-Range Control	282
Multivariable Control	283
Override Controllers	285
Summary	286
Checking Your Knowledge	286
Student Activities	287

17 Introduction to Digital Control 288

Introduction	289
Brief History of Controllers	290
Analog versus Digital Control	290
Analog	290
Digital	291
Other Components of a Digital Control System	291
Remote Processing Units (RPUs)	291
Programmable Logic Controllers (PLCs)	292
Discrete Controllers	292
Multiplexers/Demultiplexers (MUX/DEMUX)	292
Signal Conversion and Transmission	292
Digital to Analog (D/A)	293
Analog to Digital (A/D)	293
Stand-Alone Digital Controllers	294

Smart Transmitters	294
Summary	294
Checking Your Knowledge	295
Student Activities	296

18 Programmable Logic Controls — 297

Introduction	298
Historical Background of PLCs	298
Reading Ladder Diagrams	299
Practical Tips	300
PLCs Today	300
Logic in PLCs	301
Programming a PLC	301
Summary	302
Checking Your Knowledge	303
Student Activities	303

Part 4 DCS, Power Supply, Shutdowns, Malfunctions, and Troubleshooting

19 Distributed Control Systems (DCSs) — 304

Introduction	305
Historical Background	305
DCS Operation	306
Function of a Typical DCS Node	307
DCS Architecture	307
Input/Output (I/O) Devices	307
Data Highway	308
Operator Stations or Work Stations	310
Redundancy	311
Configuration Station	312
Accessory Devices	312
Plant Network	312
DCS Advantages	312
Repeatability and Accuracy	312
Speed	312
Cost-Effectiveness	313
Improved Data Management	313
Space Efficiency	313
Safety and Reliability	313
Bumpless Transfer	313
Processes, Equipment, and Safety for DCSs	313
Analytics/Trending	313
Intrinsically Safe Display Devices/Tablets	314
Cybersecurity	314
Summary	315
Checking Your Knowledge	315
Student Activities	316

20 Instrumentation Power Supply — 317

Introduction	318
Plant Power Supply Overview	318
Uninterruptible Power Supply (UPS) Systems	321
UPS Components	322
Troubleshooting Tips and Warnings	323
Summary	323
Checking Your Knowledge	323
Student Activities	324

21 Emergency Shutdown (ESD), Interlocks, and Protective Devices — 327

Introduction	329
Permissives and Interlocks	329
Startup Permissives	329
Interlocks	330
Emergency Shutdown (ESD)	330
Interlocks in Emergency Shutdown	331
Breaker Switches in Emergency Shutdown	332
Electric Eye Sensors in Emergency Shutdown	333
Alarms	333
High Alarm (H)	333
High High Alarm Shutdown (HHSD)	333
Low Alarm (L)	334
Low Low Alarm Shutdown (LLSD)	335
Deviation Alarm	335
Vibration Alarm and Shutdown	335
Gas Detection Alarm and Suppression Systems	336
Avoiding Unnecessary Shutdowns	336
Redundancy	337
Voting Logic	337
Emergency Shutdown (ESD) Resetting and Testing	338
Simple Reset	338
Multistage Reset	338
Testing an ESD	338
Full Testing	339
Dry Testing	339
Practical Tips	340
ESD Bypassing	340
ESD Bypass Management	340
Summary	341
Checking Your Knowledge	341
Student Activities	342

22 Instrumentation Malfunctions — 343

Introduction	344
Identifying Instrument Malfunction	344
Isolating the Problem Source	344
Calibration of Control Valves	345
Calibration of Sensors	346

Failure Types and Possible Causes	346
Instrument Failure Modes	**349**
Temperature Elements	349
Flow Elements	350
Pressure Elements	351
Analytical Elements	351
Loop Response to Instrumentation Failure	**352**
Primary Sensing Elements	352
Transmitters	352
Controllers	352
Control Valve	354
Summary	354
Checking Your Knowledge	354
Student Activities	355

23 Instrumentation Troubleshooting — 356

Introduction	357
Typical Malfunctions	358
Troubleshooting Methods	358
Dividing the Loop in Half	360
Checking the Controller	361
Process to Controller (Before the Controller)	361
Controller to Process (Beyond the Controller)	362
Communication during Troubleshooting	**362**
Troubleshooting Equipment	**362**
Calibration and Troubleshooting	363
Troubleshooting "Smart" Instruments	364
Troubleshooting Safety and Environmental Issues	**364**
Summary	364
Checking Your Knowledge	365
Student Activities	366

Appendix Answers to Checking Your Knowledge Questions — 367

Glossary — 371

Index — 383

PART 1 Introduction and Process Variables

Chapter 1
Introduction to Instrumentation

Objectives

After completing this chapter, you will be able to:

1.1 Discuss the evolution and importance of process instrumentation to the process industries. (NAPTA Introduction to Instrumentation 1*) p. 3

1.2 Explain the importance of monitoring process variables, the importance of instrumentation to the process technician, and the operator's role in safety. (NAPTA Introduction to Instrumentation 2, 3, 4; Monitoring Process Variables 1) p. 5

1.3 Categorize the major process variables controlled in the process industries and define their units of measurement:

pressure

temperature

flow

level

analytics. (NAPTA Introduction to Instrumentation 6) p. 6

1.4 Describe instrumentation with regard to location, function, type, and signal produced:

local/field

remote

indicating

transmitting

recording

controlling

control loop

*North American Process Technology Alliance (NAPTA) developed curriculum to ensure that Process Technology courses will produce knowledgeable graduates to become entry-level employees in process technology. Objectives from that curriculum are named here in abbreviated form. For example, "(NAPTA Introduction to Instrumentation 1)" means that this chapter's objective relates to objective 1 of NAPTA's course content introducing instrumentation.

manipulating

pneumatic

electronic

analog

digital

   distributed control system (DCS)

   programmable logic controller (PLC)

analog–digital hybrid

split range. (NAPTA Introduction to Instrumentation 5; Monitoring Process Variables 1) p. 9

1.5 Examine the relationship between common process variables:

pressure in a closed container when temperature increases or decreases

temperature in a closed container when pressure increases or decreases

vessel bottom pressure when height of liquid increases or decreases

boiling point of a material when pressure increases or decreases

volume of a material when temperature increases or decreases

density of a material when temperature increases or decreases

differential pressure when the flow increases or decreases

effect of changes in variables on accuracy. (NAPTA Introduction to Instrumentation 7) p. 14

Key Terms

Analog—a continuous signal that changes to mirror the actual process variable, **p. 12.**

Analytics—use of a logical technique to perform an analysis; in instrumentation, a measurement of a physical or chemical property, **p. 8.**

Control loop—the series of instruments that work together to monitor and control a single process variable, **p. 9.**

Controlling—keeping a variable at a specific quantity, **p. 10.**

Differential pressure (Δp, D/P)—the difference between two related pressure measurements; usually used in measurements of pressure, temperature, level, and flow, **p. 6.**

Digital—a transmission method that employs discrete electrical signals as opposed to continuous signals or *waves*, **p. 12.**

Electronic—powered by electricity, **p. 11.**

Flow—the movement of fluids, **p. 7.**

Flow rate—the specific amount of fluid moving past a given point per unit of time, **p. 7.**

Indicating—the showing of a current process variable condition via a readout (e.g., digital or analog), **p. 9.**

Level—a position of height or depth along a vertical axis; in the process industries, level usually means the height of the surface of a material compared to a zero reference point, **p. 8.**

Local—related to instrumentation in close physical proximity to the process, generally found outside the control room, **p. 9.**

Pneumatic—powered by a gas, **p. 11.**

Pressure—the amount of force a substance or object exerts over a specific area, **p. 6.**

Process variable (PV)—measurements of the chemical or physical properties of a process stream, generally relating to operating parameters or product specifications, **p. 6.**

Proportional, integral, derivative (PID) control—tuning parameters utilized in a controller to ensure appropriate response to process variable changes in comparison to set point, **p. 4.**

Recording—the keeping of historical data by making a documented record, **p. 10.**

Remote—relating to instrumentation located away from the process (e.g., in a control room) and controlled by the distributed control system (DCS) or other panel board mounted controller, **p. 9.**

Temperature—the measure of the thermal energy of a substance, **p. 7.**

Total flow—the quantity of fluid moving past a given point per day (e.g., 4 GPM × 60 min/hr × 24 hr = 240 × 24 = 5760 gal/day), **p. 7.**

Transmitting—communicating via electronic or pneumatic signal from one control loop component to another, **p. 10.**

1.1 Introduction

This chapter describes the evolution of instrumentation in the process industries, from the earlier versions through the many technological advances that provide the automated control systems of today. An introduction to various common terms is discussed with a preview of coming in-depth material on instrumentation that will be addressed in later chapters of the book. In addition, various relationships among process variables are introduced through the posing of questions and answers.

Evolution of Process Instrumentation

In prehistory, as well as in modern times, humankind has attempted to control processes and machinery. Early process control was mainly completed by visually watching a process and making a change, or by controlling that process by opening or closing a manual valve. Vessel levels, process flows, temperatures, and pressures were all controlled in this manner. One of the first types of industrial speed controls was the application of the flyball governor. Its design provided one of the first examples of feedback control (Figure 1.1) in 1775. Although there had been previous attempts, James Watt was the first to apply it successfully to the steam engine as a speed control mechanism.

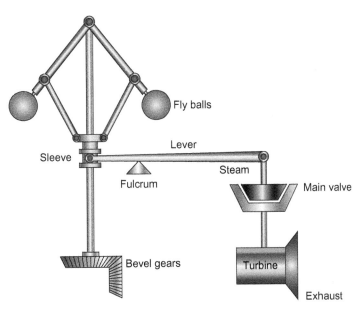

Figure 1.1 Flyball governor.

Rudimentary forms of automatic control (e.g., pneumatic) had begun to appear in the mid to late 1920s. However, the vast majority of processes were controlled manually. In manual control, a process technician looks at a gauge or sight glass (Figure 1.2) and then makes an appropriate adjustment to a piece of equipment that directly affects a process variable. Throughout the workday, technicians had to control each variable in every piece of equipment using this time- and energy-consuming manner.

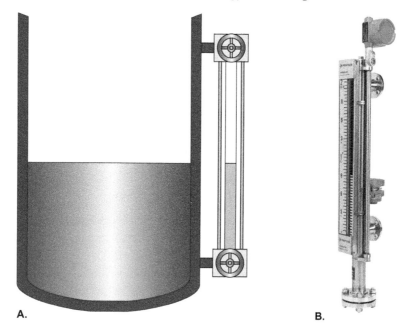

Figure 1.2 **A.** Manual control (tubular sight glass). **B.** Photo of modern magnetic liquid level gauge.

CREDIT: B. © Emerson 2019.

Local control systems, using these newly found forms of control, began to appear in the 1930s. In 1932, Harry Nyquist published the first theoretical paper on automatic process control. Zeigler and Nichols introduced **proportional, integral, derivative (PID) control** in the early 1940s.

Proportional, integral, derivative (PID) control tuning parameters utilized in a controller to ensure appropriate response to process variable changes in comparison to set point.

The implementation of modern control theory was slow but steady until 1941 when the United States entered World War II. In those days, the workforces in plants consisted almost entirely of men who were then leaving these jobs to join the war effort. This left a huge need for technicians to fill these positions. Concurrently, the war was generating an even greater need for the products that these plants were producing. Something had to be done. The answers to the problem were to hire and train women (Figure 1.3) and to develop greater automatic control.

Figure 1.3 Image of Rosie the Riveter.

CREDIT: TinaImages/Shutterstock.

Where new ideas and new hardware are concerned, there is always a development period followed by an implementation period. Although analog electronic instrumentation (Figure 1.4A) was in limited use during the 1940s, its use escalated in the 1960s.

The first use of computers in industry dates back to 1958, but computers were large and expensive, so computer use did not develop rapidly at that time. This began to change in 1971 when the Intel Corporation introduced the Model 4004 microprocessor, the world's first processor on a chip (Figure 1.4B). The microprocessor paved the way for development of the modern digital control that is in use today. Distributed control systems (DCSs) are the latest evolution of digital control and the most advanced control application in the process industries. They are largely used in process plants where continuous control is needed.

Figure 1.4 A. Analog electronic instrumentation. **B.** Computer chips in the early 1970s.

CREDIT: **A.** © Emerson 2019. **B.** Graphego/Shutterstock.

A.

B.

1.2 Monitoring Process Variables: Importance of Monitoring Process Variables

Why is process monitoring important? The answers may seem obvious. Industries work with equipment, fuels, and materials that are potentially deadly. Vessels in process industries are often under pressure. Leaks, explosions, or runaway reactions could cause widespread damage to people, the unit, and the surrounding communities if a problem went unrecognized or was left untended.

Processes themselves require monitoring. For example, if a material will not undergo the intended physical or chemical change unless it reaches a particular temperature, both time and money can be wasted if temperature is unmonitored and remains too low.

Process monitoring is critically important to ensure the correct product specifications. These specifications may be set by the customer or by the company, depending on the type of product. Process monitoring is also very important in regard to business economics to ensure the product is produced in a manner that is cost effective without sacrificing quality.

Importance of Instrumentation

Process operators benefit from the extensive use of instrumentation monitoring systems in industry. Without instrumentation, industries would need individuals to stand physically next to each piece or set of equipment to look, listen, and check on its operation whenever machines were in operation. With instrumentation, the instruments themselves function somewhat like those "eyes and ears." As a tool, they identify aspects of a process and whether it is functioning within expected parameters. If process variables are not within normal limits, instruments can communicate to the process technician that there may be a need for troubleshooting or for repair.

Role of Process Operators in Safety

Process operators rely upon instrumentation daily, but they have a major role in ensuring safety. An important aspect of their work is responding appropriately and in a timely way to instrument alarms or signals. When needed, process operators also help identify the source of the alarm and communicate with managers and/or maintenance if repairs are needed. Even in a situation without sounding alarms, it is the operator's responsibility to observe the array of instruments and be on the lookout for problems. The process operator is the person who can identify and help interrupt a failed or failing process most quickly.

1.3 Process Variables

Instrumentation is used in industry to *measure* and *control* a **process variable (PV)** such as pressure, temperature, level, flow, and analytics. Instrumentation is the primary overseer and decision making element of a process condition. Some of these PVs respond to changes in valve position and other controllable process instruments. A controller receives instrument readings as inputs, compares the process value readings to operator entered set points, and then adjusts the controllable outputs (such as valves) to keep process values at set point. The following information provides a brief description for each of these common process variables and also includes various terminologies used when describing process variables.

Process variables include the following:

- Pressure
- Temperature
- Flow
- Level
- Analytics

Process variable (PV) a measurement of the chemical or physical properties of a process stream, generally relating to operating parameters or product specifications.

Pressure

Pressure can be defined as a force applied to a unit of area. Pressure is one of the most critical measurements taken in industry. Nearly all industrial processes use or produce liquids, gases, or both. Gases and vapors apply force uniformly over all surfaces. In contrast, liquids apply force in accordance with their depth and density. One example of a local device used to measure pressure is a pressure gauge (Figure 1.5).

Pressure is measured in pounds per square inch gauge (PSIG), pounds per square inch atmospheric or absolute (PSIA), and pounds per square inch of mercury vacuum (in. Hg Vac). (The different pressure scales will be discussed in Chapter 2, *Process Variables, Elements, and Instruments: Pressure.*)

Pressure the amount of force a substance or object exerts over a specific area.

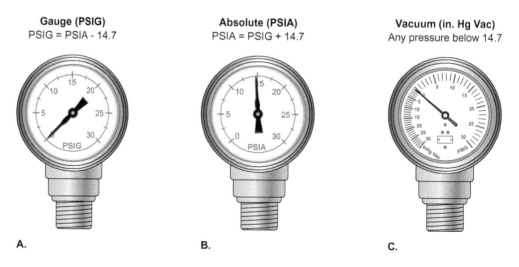

Figure 1.5 A. PSIG scale. B. PSIA gauge. C. Vacuum gauge.

DIFFERENTIAL PRESSURE (ΔP, D/P) Simply put, differential pressure is the difference between two related pressure measurements taken from two separate points. The designation of delta (Δ) is commonly used in **differential pressure** measurement (Figure 1.6) in industry. Two commonly used ways to express differential are with the Greek letter delta (Δp) and the letter combination D/P.

Differential pressure (Δp, D/P) the difference between two related pressure measurements; usually used in measurements of pressure, temperature, level, and flow.

Figure 1.6 Process variables–integral differential pressure.

CREDIT: © Emerson 2019.

Temperature

Temperature is defined as a measure of the average kinetic energy (hotness or coldness) of a substance as indicated on a reference scale (Figure 1.7). The two most common temperature measurement scales are Fahrenheit (degrees F) and Celsius or Centigrade (degrees C). Process plants control temperature in almost every major process unit.

Temperature the measure of the thermal energy of a substance.

Flow

In process industries, three important terms are **flow**, **flow rate**, and **total flow**. The word flow is used interchangeably with the term flow rate. Flow rate can be defined as the quantity of fluid that moves through a pipe (Figure 1.8) or channel within a given period of time. Flow rate is usually expressed in volume or mass units per unit of time, such as gallons per minute (GPM) or pounds per hour (PPH). Total flow is the quantity of fluid moving past a given point per day. Total daily flow is expressed as volume (gallons) or mass (lb) and is assumed to be for a 24-hour day. Total flow for a batch operation such as filling a vessel, truck, or rail car is also in volume or mass units.

Flow the movement of fluids.

Flow rate the specific amount of fluid moving past a given point per unit of time.

Total flow the quantity of fluid moving past a given point per day (e.g., 4 GPM × 60 min/hr × 24 hr = 240 × 24 = 5760 gal/day).

Figure 1.7 Thermometer.

CREDIT: Tomas Ragina/Shutterstock.

Figure 1.8 Flow.

CREDIT: Interior Design/Shutterstock.

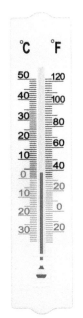

Level

Level a position of height or depth along a vertical axis; in the process industries, level usually means the height of the surface of a material compared to a zero reference point.

Level is defined as the position of either height or depth (Figure 1.9) along a vertical axis. In industry, the term level specifically means the surface position of a material in a vessel. Checking levels is very important in all types of equipment including tanks, railcars, towers, and other vessels.

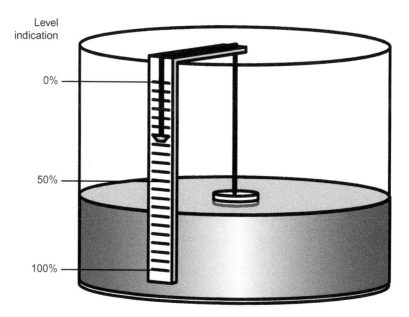

Figure 1.9 Level measurement. The level indicates 0% when the pulley sensor reaches the bottom of the tank.

Analytics

Analytics use of a logical technique to perform an analysis; in instrumentation, a measurement of a physical or chemical property.

Analytics involves instruments (Figure 1.10) that measure the chemical and/or physical properties of a process stream. Some common analytical instruments include chromatographs, pH meters, and viscosity meters. Proximity switches are used to detect the nearness of an object without physical contact. Ammeters measure electrical current in amperes. Analytical instruments can be located within the process equipment as well as in a laboratory.

There are many other process variables measured and controlled in the process industry besides the types listed above. These are discussed later in this textbook in greater detail but some examples are briefly mentioned here:

- Weight: weighing materials in tanks, drums, or even on a conveyor belt
- Speed: measuring and controlling the rotational speeds of motors and turbines

Figure 1.10 Analytical instrument.

CREDIT: © Emerson 2019.

- Vibration: detecting vibration of rotating machinery
- Acceleration: generally measuring the rate of change of velocity per unit of time

1.4 Instrument Component Types

Process instrumentation components may be categorized by location, function, and signal type. Owing to the natural overlapping of these categories, a single instrument may fall into more than one category. For example, a pneumatic pressure transmitter is a local instrument because it resides in the processing area and its instrument signal is powered by air. The following category designations explore the differences in location, function, and signal types to clarify the categorization of instruments.

Instruments Categorized by Location

An instrument may be categorized by location with respect to its proximity to the actual process it is measuring and/or controlling.

LOCAL OR FIELD An instrument located at or near the process is called a **local** instrument (Figure 1.11A and 11B). Another name for a local instrument is a field instrument.

Local related to instrumentation in close physical proximity to the process, generally found outside the control room.

Figure 1.11 A and B. Local level instrument (sight or level glass). **C.** Remote instrument data in a control room.
CREDIT: C. © Emerson 2019.

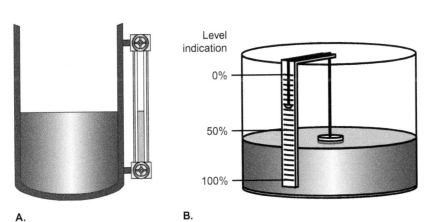

A.   B.   C.

REMOTE An instrument located away from the process is called a **remote** instrument (Figure 1.11C). Control room instruments are examples of the most common remote instruments. The term remote designates an instrument that receives a signal from and/or transmits a signal to another location. Generally speaking, all instrumentation not located in the immediate vicinity of the process equipment is considered to be remote.

Remote relating to instrumentation located away from the process (e.g., in a control room) and controlled by the distributed control system (DCS) or other panel board mounted controller.

Instruments Categorized by Function

Instruments are also categorized by their function such as sensing, measuring, indicating, transmitting, comparing and/or controlling, and manipulating. These function designations are easier to understand by relating them to a **control loop** (discussed in Chapter 10, *Introduction to Control Loops: Simple Loop Theory*). A control loop must consist of all the parts necessary to control a process variable. Typical control systems have a measurement sensor, a transmitter or transmitters, a controller, and a final control element, which is usually a valve.

Control loop the series of instruments that work together to monitor and control a single process variable.

INDICATING Instrumentation showing the current conditions, such as temperature, flow, and so on., is categorized as **indicating**. Any instrument that has a digital readout or graduated scale with a pointing device is called an indicating instrument. Transmitters, recorders, and controllers can be indicating instruments.

Indicating the showing of a current process variable condition via a readout (e.g., digital or analog).

Transmitting communicating via electronic or pneumatic signal from one control loop component to another.

TRANSMITTING Instrumentation that sends a signal (measurement represented in a standardized form such as 3–15 PSI or 4–20 mA) (Figure 1.12) to a remote location is categorized as **transmitting**. Generally speaking, transmitters are both a measuring and a transmitting device built into a single unit. For example, a unit may perform a pressure sensing activity, convert this measurement to a pneumatic (3–15 PSI) or electronic (4–20 mA) signal, and then transmit the signal to a controller.

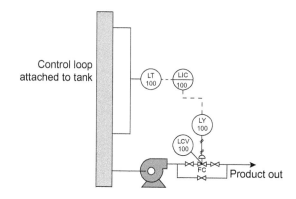

Figure 1.12 Types of transmitter signals.
CREDIT: Eastman Chemicial Company.

Recording the keeping of historical data by making a documented record.

RECORDING Instrumentation that records or registers a process variable over time (Figure 1.13) is categorized as **recording**. Recorders keep historical data on process variables. Sometimes these records are kept for extended periods of time due to strict directives from regulatory agencies such as the Environmental Protection Agency (EPA). Recorded information may be very helpful when trying to troubleshoot a process or an instrument related problem.

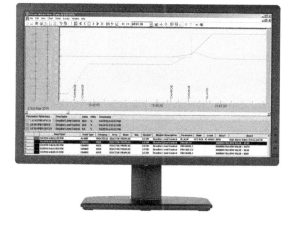

Figure 1.13 Pressure recorder with digital strip chart.
CREDIT: © Emerson 2019.

Controlling keeping a variable at a specific quantity.

CONTROLLING Instrumentation that maintains a process variable at a specific quantity is categorized as **controlling**. A controller is a device that performs the control function. It may exist as either a physical entity (local controller) or an algorithm (subroutine or program) in a remote location (Figure 1.14), such as a controller in a computer or control room rack or cabinet.

SPLIT RANGE Control loops may operate with an output signal from the controller to the final control element, typically a control valve, in which the transmission signal may be split into two separate actions. Example: The vent pressure controller on a storage tank may be operated with an output signal of 3–9 PSI (4–12 mA) to open a control valve to allow entry of a gas to maintain the tank pressure at a specific setting amount. That same controller may

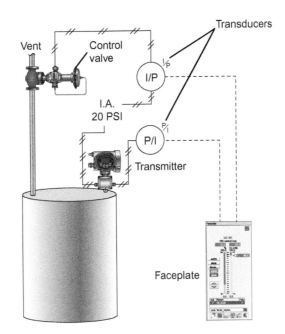

Figure 1.14 Remote pressure controller.

operate a second control valve that opens from 9–15 PSI (13–20 mA) to vent excess pressure from the tank. This controller is said to have split range operation.

MANIPULATING The final element in a control loop system is the component that manipulates a control valve, damper, fan, or other device in response to the controller's output signal.

Control Instruments Categorized by Signal Type

PNEUMATIC Instruments powered by air or other gases such as nitrogen are called **pneumatic** instruments. Pneumatic instrumentation (often supplied by instrument air; see I.A. in Figure 1.14; note that in some locations, instrument air is designated as A.S. for air supply) was implemented during the early part of the 1900s and was the first generation of control systems. It is still in use today but, with the exception of control valves, its use has steadily declined. Pneumatic instruments are still a viable option in many industrial plants and will likely maintain a niche market in the process industry for the foreseeable future for two reasons:

Pneumatic powered by a gas.

- They are self purging, which keeps them clean and free of debris, and they are intrinsically safe, which makes them desirable in highly corrosive environments or where any electrical potential is unacceptable.
- Since most control valves are still pneumatically actuated, it would be cost prohibitive to convert pneumatic instrumentation in a plant to an electrical power source.

ELECTRONIC Instruments powered by electricity are called **electronic** instruments. The umbrella of electronic instrumentation is subdivided by the type of signal. Electronic instrumentation signals can be either analog (continuous) or digital (discrete) in design. Analog instrumentation signals operate in a continuous pattern that can vary or change. In contrast, digital instrumentation operates discrete signals that are binary in nature; these signals have set parameters or steps. Generally speaking, analog signals are used between the components in a control loop and are then converted to digital signals upon entry to a distributed control system (DCS).

Electronic powered by electricity.

When combination power sources are used within a loop, categorization must be done for individual instruments within the loop. These categories are pneumatic and electronic. Figure 1.15 shows an illustration of the various types of incoming signal power.

Figure 1.15 Types of signals.

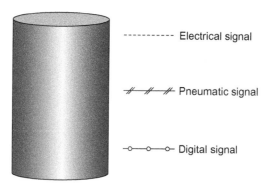

Types of Signals Produced

As mentioned, the umbrella of electronic instrumentation is subdivided by the type of signal.

Analog a continuous signal that changes to mirror the actual process variable.

ANALOG SIGNALS By definition, the word **analog** means analogous, or similar to, something else. In the case of a measured variable, that something else may be an indication of a variable such as pressure or temperature. In an analog system, pressure and temperature measurement changes are measured in a continuous manner that mirrors the actual process variable. An analog signal is a continuously variable representation of a process variable. Analog signals follow in the same manner as an analog measurement. An analog signal contains all the values from one point to another. For example, the typical analog *pneumatic* signal is 3–15 PSI and the typical analog *electronic* signal is 4–20 mA (Figure 1.16).

Figure 1.16 Analog pressure indicators.

CREDIT: © Emerson 2019.

Analog electronic instruments are those instruments that use continuously variable electrical quantities to measure, amplify, and/or produce a standard output signal such as 4–20 milliampere (mA).

Digital a transmission method that employs discrete electrical signals as opposed to continuous signals or *waves*.

DIGITAL SIGNALS **Digital** is a term applied to a device that uses binary numbers to represent continuous values or discrete (individual or distinct) states. Measurements taken from a process are continuously variable quantities (analog) that must be converted into binary numbers to enter into a computer for processing.

Digital instruments are microprocessor based and can produce a digital as well as an analog output signal. A digital signal is presented in a coded form by packets of ones and zeros. Generally, they use a serial communication format similar to a computer network in an office or classroom.

Example of an 8-bit binary packet produced by a transmitter scaled for 0–100% and reading 67%:

1000011

ANALOG/DIGITAL HYBRID Electronic instruments such as smart transmitters may be microprocessor based but produce an analog output such as a 4–20 mA signal. The microprocessor in these instruments can be communicated via a frequency signal that shares the same wires as the analog signal.

A binary number representing an analog quantity can be produced by means of a device called an analog to digital (A/D) converter.

As stated above for analog signals, there is an infinite number of values between any two points. That is, the pressure in a vessel does not jump from 1 PSI to 2 PSI without going through all values between the two. A digital readout device (Figure 1.17) must jump from one numerical value to the next, whereas an analog pressure gauge with an indicating pen moves smoothly across all the values between 1 PSI and 2 PSI. When a digital indicator changes from 1 PSI to 2 PSI, it is only the readout device that is skipping numerical values rather than the actual pressure jumping from one value to another.

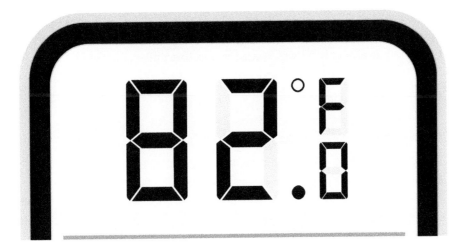

Figure 1.17 Digital temperature indicator.

CREDIT: Aydngvn/Shutterstock.

Transmitters called *smart transmitters* (microprocessor-based instruments) are used in industry. These instruments have an analog to digital (A/D) converter built into them so that they can process information within the onboard microprocessor. There are many advantages to using microprocessor-based instruments, including programmability, self-diagnostics, and the ability to recalibrate from a remote location. The output signal from a smart transmitter may or may not be digital.

Microprocessor-based transmitters are available as analog–digital hybrids and as pure digital instruments. A transmitter with a digital output signal would have a typical computer communication output that transmits data in a serial manner just as computers would if networked together in an office.

In short, industry uses both analog and digital systems because they both have qualities desirable in specific situations. The true power of a digital system lies in its ability to manage information far beyond the capabilities of an analog system. Digital instrumentation has become the norm, and analog systems are on the decline. However, analog equipment continues to retain a significant role in the process industry.

1.5 Process Variable Relationships

Common process variables affect one another continuously in industry. The physics of each variable dictates its response within a process unit or system. Processes are designed to account for these reactions. To help illustrate these relationships, a discussion through the use of questions and answers is provided below.

What Is the Effect of Temperature Change on Pressure in a Closed Container?

A temperature increase causes a pressure increase; a temperature decrease causes a pressure decrease. Temperature and pressure are directly proportional to one another. For example, the pressure in the radiator of a car rises as its temperature increases. You would never uncap a hot radiator because you know it is pressurized.

RELATIONSHIP AMONG PRESSURE, VOLUME, AND TEMPERATURE IN A CLOSED CONTAINER Consider the measurement conditions inside of a typical tank. Process plants are like large chemistry labs with different types of vessels that are used to store, blend, or even react chemicals in a controlled environment. Individual processes have their own unique characteristics, but physical attributes such as pressure and temperature follow strict laws of physics. Pressure and volume are inversely proportional to one another.

The following equation shows the relationship among pressure (P), volume (V), and temperature (T) as applied to gases. The pressure, volume, and temperature on the left side of the equation represent condition (1), that is, the condition found prior to a change in conditions. The conditions on the right represent the changed condition (2). Pressure and temperature values must be expressed in absolute units.

$$\frac{P_1 V_1}{T_1} = \frac{P_2 V_2}{T_2}$$

In this equation, pressure and volume are directly proportional to each other and temperature is inversely proportional to both of them.

RELATIONSHIP BETWEEN PRESSURE AND TEMPERATURE The previous equation explains that there is a specific relationship among the volume, temperature, and pressure of gases. Common sense dictates that if the temperature in a closed container is increased, its pressure will rise. A simple and very common example of this phenomenon is the pressure change that occurs in the tires on a car between the time it is at rest and after it has been driven a while.

Here is an industrial example: A vessel, somewhat like a tire, is a rigid structure and its volume therefore remains constant. Because the volume stays constant, it can be effectively removed from the previous equation, leaving a simpler relationship between temperature and pressure (Charles' Law).

$$\frac{P_1}{T_1} = \frac{P_2}{T_2}$$

Consider an actual example of how temperature affects pressure in a fixed volume. If a tank were blocked in (all inlets and outlets are closed) late in the evening with a relatively mild temperature of 68 degrees Fahrenheit (528 degrees Rankine), a pressure gauge attached to it would indicate an ambient (surrounding) pressure (14.7 PSIA). The next day, as the sun begins to shine directly on the tank, its inside temperature would begin to rise and so would its pressure. If the temperature inside the tank rises to 158 degrees Fahrenheit (618 degrees Rankine), how much pressure would now be in the tank?

NOTE: Pressure (PSI, PSIA, PSIG, etc.) is covered in Chapter 2, *Process Variables, Elements, and Instruments: Pressure*. Temperature scales and scale conversions are covered in Chapter 3, *Process Variables, Elements, and Instruments: Temperature*.

$$P_2 = \frac{P_1 T_2}{T_1}$$

$$P_2 = \frac{(14.7 \text{ PSIA})(618°R)}{528°R}$$

$$P_2 = 17.2 \text{ PSIA}$$

or

If converted to PSIG, then $(17.2 - 14.7) = 2.5$ PSIG

Since tanks are designed to withstand pressure from the inside out, this increase in pressure may not create a problem. But, what would happen if the opposite were the case?

WHAT IS THE EFFECT OF PRESSURE CHANGE ON TEMPERATURE IN A CLOSED CONTAINER? A pressure increase causes a temperature increase; a pressure decrease causes a temperature decrease. For example, heat causes gases to become more active, thus taking up more space. So, these expanded molecules are compressed and their total energy is concentrated, resulting in a measurable rise in temperature. Notice in the following equation how the pressure (P_2) increases as the temperature (T_2) increases.

$$T_2 = \frac{P_2 T_1}{P_1}$$

$$T_2 = \frac{(17.2 \text{ PSIA})(528°R)}{14.7 \text{ PSIA}} = 618°R$$

$$T_2 = 618°R - 460°R = 158°F$$

Again, the inverse of this situation is that the pressure of a gas decreases as the temperature decreases.

RELATIONSHIP BETWEEN TEMPERATURE AND PRESSURE INSIDE A CLOSED CONTAINER Using a situation similar to the previous example, what would happen to the pressure in a tank if it were blocked in during the heat of the day (158 degrees Fahrenheit) and then a cold wet storm swept across the area later that evening (68 degrees Fahrenheit)?

$$P_1 = 14.7 \text{ PSIA}$$

$$T_1 = 158°F + 460 = 618°R$$

$$T_2 = 68°F + 460 = 528°R$$

$$P_2 = \frac{P_1 T_2}{T_1}$$

$$P_2 = \frac{(14.7 \text{ PSIA})(528°R)}{618°R} = 12.6 \text{ PSIA}$$

This is 2.1 PSIA below atmospheric pressure.

Since vessels are usually built to withstand internal pressure, the vessel walls could collapse under vacuum conditions. Fortunately, most storage vessels are equipped with vacuum breakers that guard against too much negative pressure placed on a tank.

What Is the Effect of the Height of Liquid on Pressure at the Bottom of a Vessel?

A level increase causes the vessel bottom pressure to increase; a level decrease lowers the pressure. For example, consider a tank containing 10 feet of water that is vented to the atmosphere. Problems of a pressurized tank are taken out of the scenario, leaving head pressure

(pressure exerted by a liquid) as the only type of pressure considered. The pressure at the bottom of the tank is affected by the height of the liquid in the tank and by its density. The following equation expresses this concept:

Where:
Head = height of the fluid (h)
For conversion to PSIG:

As h increase, p increases

h = liquid height

p = pressure at the bottom

$$d = \text{density} = \frac{62.4 \text{ lb}}{\text{cubic foot}} = \frac{62.4 \text{ lb}}{\text{ft}^3}$$

$$p = \left(\frac{62.4 \text{ lb}}{\text{ft}^3}\right)(10 \text{ ft})\left(\frac{\text{ft}^2}{144 \text{ in.}^2}\right) = \frac{(62.4 \text{ lb})(10 \text{ ft})}{144 \text{ in.}^2 - \text{ft}}$$

$$= \left(\frac{0.433 \text{ lb}}{\text{in.}^2 - \text{ft}}\right)(10 \text{ ft}) = 4.3 \text{ PSI}$$

RELATIONSHIP BETWEEN BOTTOM PRESSURE AND HEIGHT OF LIQUID As mentioned, pressure exerted by a liquid is called *head pressure*. Atmospheric head pressure is a result of the height of the liquid times its density. As the liquid height increases, the head pressure also increases. Consider a tank containing a fluid with the equivalent of three-quarters of the density of water, or a specific gravity (SG) of 0.75.

Review Table 1.1 to see the pressure readings at each of the various levels within the tank.

Table 1.1 Comparison of PSIG and PSIA Pressure Readings at Different Heights

Pressure Reading Level Points	Pressure (PSIG)	Pressure (PSIA)
C	0.0	14.7
B	0.6	15.3
A	1.2	15.9

The following equation expresses this concept:

$$P_B = (h)(SG)$$

Where:
P_B = bottom pressure
h = height of liquid (in feet)
SG = specific gravity of liquid
D_w = density of water $(0.433 \text{ lb/in.}^2)/(\text{ft of water})$
As h increases, P increases.

Example:
$P_B = (16 \text{ ft})(0.433 \text{ lb/in.}^2)$
$P_B = (16 \text{ ft})(0.433 \text{lb/in.}^2)$
$P_B = 6.9 \text{ lb/in.}^2$ or 6.9 PSI

What Is the Effect of Pressure Change on the Boiling Point of a Material?

The boiling point increases when the material pressure increases; the boiling point decreases when pressure is lowered. For example:

- The boiling point of a liquid is directly proportional to applied pressure. There is less atmospheric pressure at the top of the mountain than at sea level; atmospheric pressure is the highest at the surface of the ocean (sea level). This is why water boils at a lower temperature on a mountain as compared to sea level. Pressure, such as atmospheric pressure, on the surface of a liquid holds the molecules in the liquid state. The greater the pressure, the less likely that the liquid molecules can vaporize and break the surface (boil).
- In a pressure cooker, the pressure is allowed to increase: As the pressure above a liquid increases, the boiling point of the liquid also increases. This allows the food being cooked to heat up faster and cook more rapidly and in a shorter time period.

RELATIONSHIP BETWEEN BOILING POINT AND PRESSURE Pressure applied to the surface of a liquid directly affects its boiling point. To understand how the boiling point of a liquid is affected by pressure, an understanding of boiling point is necessary. An easy way of identifying when a liquid is boiling is by the bubbles being produced and through the release of vapor. This is a good observation, but what is really happening?

Simply put, the *normal* boiling point of a liquid is the temperature at which the liquid boils at atmospheric pressure. All liquids have a vapor pressure. If the vapor pressure exceeds atmospheric pressure, the liquid evaporates.

To repeat, as the pressure applied to the surface of a liquid increases, the boiling point of that liquid also increases. Conversely, as the pressure applied to the surface of a liquid decreases, the boiling point of that liquid also decreases.

What Is the Effect of Temperature Change on the Volume of a Material?

The volume increases when the temperature increases and decreases when the temperature decreases. The coolant system in a car is a good example of this concept. Look at the coolant level in the overflow/fill reservoir of a car when it is cool and then look at it again when it is hot. The reservoir should show a higher level when it is hotter. This is why there are level markings (look for a COOL line and a HOT line) to use when filling the reservoir. The extra space in the reservoir is needed for the expanding material. If the expansion allowance space were not calculated into the total filling volume, the expanding volume would drain out of the reservoir.

RELATIONSHIP BETWEEN VOLUME AND TEMPERATURE The volume of a substance is directly affected by temperature. Almost all materials expand when heated and contract when cooled. With any given temperature change, the change in volume experienced by gases is much greater than that experienced by liquids and solids.

The volume changes of liquids and solids are based more upon the characteristics of each individual material than upon a general volume versus temperature relationship that applies to gases. In the following formula, the volume (V) of a specific material responds directly (ΔV) to a temperature change (ΔT) by the factor (β), which is its coefficient of volume expansion.

$$\text{Liquids: } \Delta V = (\beta)(\Delta T)$$

Where:
ΔV = change in volume
β = coefficient of volume expansion
ΔT = change in temperature

To illustrate this concept, consider again the coolant level in the overflow/fill reservoir of a car. When the car is cool, there is less coolant volume in it as compared to it when it is at operating temperature (hot).

An exception to this is water. Water that is cooled contracts until reaching four degrees Celsius (39 degrees Fahrenheit). At that point it expands until it reaches the freezing point. At freezing point it expands approximately nine percent in volume.

What Is the Effect of Temperature Change on the Density of a Material?

Density (mass per unit volume) decreases when the temperature increases; the density increases when the temperature decreases. In the radiator example above, the cooling system contains a certain amount of liquid (e.g., 22 pounds of coolant). As the temperature of this system increases, its volume increases. If the volume expands and the weight (mass) remains the same, then the density of the fluid will have changed.

Where:

$$\text{Density} = \frac{\text{mass}}{\text{volume}}$$

At a lower temperature,

$$\text{Density} = \frac{22 \text{ lb}}{2.5 \text{ gallons}}$$

$$\text{Density} = 8.8 \text{ lb/gal}$$

At a higher temperature:

$$\text{Density} = \frac{22 \text{ lb}}{2.625 \text{ gallons}}$$

$$\text{Density} = 8.38 \text{ lb/gal}$$

If density is defined as mass per unit volume, and the material is expanding, its density is decreasing.

What Is the Effect of Variable Change on Accuracy of Measurements?

Because of the interrelationship of one variable to another, a change in one variable in a process can have a profound effect on another measurement variable. For example, if the temperature in a tank increases, then the volume increases, causing a decrease in density. Since a tank has rigid sides, extra volume increases the height of the level. If a pressure transmitter is being used to read a tank level, then the measurement is no longer accurate. This is because the pressure at the bottom of the tank remains the same, but the actual height of the liquid (level) has increased.

$$P = \left(\frac{0.433}{(\text{in.}^2)(\text{ft})}\right)(\text{height})(\text{SG})$$

NOTE: Not all variables in a process are measured, but they may still have to be kept constant so that an accurate measurement of the controlled variable (the variable that is meant to be controlled) may be obtained.

RELATIONSHIP BETWEEN VARIABLE CHANGES AND ACCURATE MEASUREMENT
Design engineers should ensure that all reasonable conditions are accounted for before the plant ever starts up. Even with all the thought and detail put into the design, chemical consistencies can change. In a plant, process technicians are part of the first line of defense against process problems, and they need to be cognizant of the potential ramifications of these changes.

Summary

Instrumentation used to control processes date back to 1775 when a flyball governor was used to control the speed of a steam engine. Manual control has been around much longer than automatic control which did not appear until the late 1920s. Modern control theory began with Harry Nyquist in the early 1930s and was slow in implementation until World War II. Zeigler and Nichols introduced proportional, integral, derivative (PID) control in the early 1940s. Analog electronic devices were developed as early as the 1940s and are still common in many plants today. Computer control began to appear in the late 1950s, and usage multiplied with the invention of the microprocessor (chip) in the early 1970s. From the development of microprocessors came distributed control systems (DCSs) used widely today.

Instrumentation is used to measure and control process variables such as pressure, temperature, level, flow, and various analytical properties.

Instruments are categorized by various methods to include the following:

- *Location* (local [or field] or remote): Local instruments are located close to the process equipment and remote are located away from the process.
- *Function* (e.g., sensing, indicating, transmitting, comparing, and/or controlling): These functions are usually related to a particular control loop that has all the parts necessary to control a process. These parts include a measurement sensor, a transmitter, a controller, and a final control element.
- *Power source* (e.g., pneumatic or electronic): These power sources categorizations depend on how all the instruments within a control loop function together or how each individual instrument functions as an entity. For an entire loop to be categorized as pneumatic, each instrument must use air (or other gases) as the power source. Instruments may also be electronic (analog, digital, and analog/digital hybrid) if their power source is electricity.
- *Type of signal produced* (e.g., analog or digital): This designation is a stronger than that of power source. Analog signals are continuous as compared to digital signals that jump from one value to the next. Analog signals can be converted to digital by an analog to digital (A/D) converter or smart transmitters.

The many relationships between process variables include the following:

- Between pressure and volume when temperature is increased: In a closed vessel, when temperature increases, pressure increases. In an open vessel, when temperature increases, fluid volume increases. When temperature decreases in a closed vessel, the pressure decreases.
- Temperature in a closed container when pressure increases or decreases: A pressure increase causes a temperature increase and a pressure decrease causes a temperature decrease. Opening vents on closed containers prevents overpressurization or negative pressure (vacuum) from occurring on temperature changes.
- Pressure at the bottom of a vessel when the height of the liquid increases or decreases: A level increase causes the vessel bottom pressure to increase, whereas a level decrease lowers the pressure. The pressure at the bottom of the tank is affected by the height of the liquid in the tank and by its density. Pressure exerted by a liquid is called head pressure.
- Between the boiling point of a material when pressure increases or decreases: The boiling point increases when the material pressure increases, and the boiling point decreases when pressure is lowered. Pressure applied to the surface of a liquid directly affects its boiling point.
- Volume of a material when temperature increases or decreases: The volume increases when the temperature increases and decreases when temperature decreases. Almost all materials expand when heated and contract when cooled, but the change in volume of gases is much greater than that experienced by liquids and solids.
- Density of a material when temperature increases or decreases: Density (mass per unit volume) decreases when the temperature increases, and density increases when temperature decreases. If a material is expanding, then its density must be decreasing.
- Variable changes and accurate measuring: A change in one variable in a process can have a profound effect on another measurement variable.

Checking Your Knowledge

1. Place the following historical events in chronological order from the earliest (first) to the most current (fourth).
   a. Analog electronic instrumentation becomes commonplace in the industry.
   b. James Watt successfully applies feedback control to the steam engine.
   c. Digital process control emerges into the form of Distributed Control Systems.
   d. Nyquist publishes the first theoretical paper on automatic process control.

2. Match the following terms to their appropriate definition:

Term	Definition
I. Flow	a. Force applied to a specific area
II. Level	b. The movement of fluids
III. Differential pressure	c. Height of the surface of a material compared to a zero reference point
IV. Temperature	d. The measure of thermal energy of a substance
V. Pressure	e. Difference between pressure measurements taken from two separate points

3. Which of the following is an important role of a process technician to ensure safety?
   a. Repair instruments when they malfunction.
   b. Respond to instrument alarms or signals.
   c. Shut down operations when an instrumentation alarm sounds.
   d. Revise the process flow when an instrument fails.

4. Which process variable generally measures the rate of change of velocity per unit of time?
   a. Ammeter
   b. Proximity
   c. Acceleration
   d. Analytics

5. Why will pneumatic instruments be used for the foreseeable future in the process industries?
   a. They are inexpensive.
   b. They are easy to use.
   c. They are self-purging.
   d. They are safer than electrical components.

6. When the pressure applied to a liquid increases, the boiling point _____.
   a. increases
   b. decreases
   c. stays the same
   d. fluctuates

7. When the temperature increases, volume _____.
   a. increases
   b. decreases
   c. stays the same
   d. fluctuates

NOTE: Answers to Checking Your Knowledge questions are in the Appendix.

Student Activities

1. Given your earlier process training or experience with instrumentation, think of at least five scenarios in which you can create a discussion about the effects of various process variables in plant situations. Record your comments and share with other classmates to determine if you have discovered all the possibilities for each scenario.

2. Tour local plant facilities with an experienced process technician and have the technician point out various types of instrumentation within their working environment. If this type of tour is not available, then your instructor should perform for you a similar demonstration of available instrumentation within the classroom or lab environment.

Chapter 2
Process Variables, Elements, and Instruments: PRESSURE

Objectives

After completing this chapter, you will be able to:

2.1 Identify the concept of pressure and the three components that affect the force exerted by molecules:

speed (temperature)

mass (atomic weight)

density (number of molecules per volume). (NAPTA Process Variables . . . : Pressure 2) p. 23

2.2 Manipulate units of measurement associated with pressure and pressure instruments:

PSI

PSIA

PSIG

atmospheres

bars

inches H₂O

inches Hg (mercury) or metric equivalent mmHg

inches Hg Vac (NAPTA Process Variables . . . : Pressure 1) p. 26

*North American Process Technology Alliance (NAPTA) developed curriculum to ensure that Process Technology courses will produce knowledgeable graduates to become entry-level employees in process technology. Objectives from that curriculum are named here in abbreviated form. For example, "(NAPTA Process Variables . . . : Pressure 1)" means that this chapter's objective relates to objective 2 of NAPTA's course content on instrumentation about pressure measurement.

2.3 Illustrate common types, purpose, and operation of pressure sensing or pressure measuring instruments used in the process industries:

manometers

pressure gauges (internals)

differential pressure (D/P) cells

strain gauge transducers. (NAPTA Process Variables . . . : Pressure 3, 4) p. 29

2.4 Using a standard calculator and conversion formulas, convert between the following pressure scales:

pounds per square inch gauge (PSIG) and pounds per square inch absolute (PSIA)

inches of mercury (in. Hg) and inches of water (in. H_2O)

pounds per square inch (PSI) and inches of water column (in. H_2O). (NAPTA Process Variables . . . : Pressure 5) p. 34

Key Terms

Atmosphere(s)—the pressure at any point in the atmosphere due solely to the weight of the atmospheric gases above the point concerned; 14.7 PSIA, which equals one atmosphere, is the pressure at sea level, **p. 28.**

Atmospheric pressure—the pressure at the surface of the earth (14.7 PSIA at sea level), **p. 28.**

Bar—measurement of pressure equal to 0.987 atmospheres, **p. 28.**

Capacitance transducer—an electrical device that contains a measurement diaphragm and capacitor plates; it converts an applied pressure to an electronic signal measurement that is proportional to the process being measured, **p. 34.**

Density—mass per unit volume, **p. 25.**

Differential pressure (D/P)—the difference between two related pressure measurements; usually used in measurements of pressure, temperature, level, and flow, **p. 26.**

Differential pressure cell (D/P cell)—an instrument that measures the difference between two related pressure points and produces a corresponding output signal that is proportional to the process being measured, **p. 32.**

Force—energy that causes a change in the motion of an object, involving strength or direction of push or pull, **p. 24.**

Inches of mercury (in. Hg)—most common measurement scale for a manometer; measures in pounds per square inch; 1 inch Hg = 13.0687 inches H_2O, **p. 28.**

Inches of water (in. H_2O)—a very small measurement of pressure equal to 0.036 pounds per square inch at 4 degrees Celsius (39.2 degrees Fahrenheit), **p. 28.**

Local indicators—instruments placed on equipment in the field only to be read in the field; may be used for comparison with transmitted instrumentation readings or used on noncritical processes to indicate process values, **p. 31.**

Mass—the amount of matter in a body or object measured by its resistance to a change in motion, **p. 25.**

Molecular speed—the rate of motion of molecules in a substance; it may change as a result of factors such as heat or collision with other molecules (e.g., the walls of a vessel), **p. 25.**

Pressure—the amount of force a substance or object exerts over a specific area, **p. 23.**

Pressure gauge—the most common local instrument type for measuring pressure; the three primary types of pressure gauge measuring scales are absolute, gauge, and vacuum, **p. 29.**

PSIA—pounds per square inch absolute; zero on the absolute scale would represent the absence of any pressure, **p. 27.**

PSIG—pounds per square inch gauge; pressure measurement that references 14.7 PSIA as its zero point, **p. 28.**

Specific gravity—the ratio of the density of a liquid or solid to the density of pure water *or* the density of a gas to the density of air at standard temperature and pressure (STP), **p. 26.**

Strain gauge transducer—a pressure sensing and measuring device consisting of a group of wires that stretch when pressure is applied, creating resistance, thereby changing the process pressure into an electronic signal, **p. 33.**

U-tube manometer—a gravity balanced pressure measuring device with two fluid chamber tube gauges connected by a U-shaped tube so fluid (a liquid or mercury) flows freely between the chambers, **p. 30.**

Vacuum—a condition in which the pressure measured is less than atmospheric pressure; usually measured in inches of mercury (in. Hg), **p. 29.**

2.1 Introduction

This chapter reviews the concept and definition of pressure and, in general terms, how to calculate the amount of pressure for various applications. Also discussed are the various types of measurement scales used to determine the amount of pressure in a given situation and how to convert the pressure reading from one scale to another. Different types of pressure measuring devices are identified as to their purpose and operation, including the respective locations where they may be used in a processing environment.

Pressure Defined

One of the most common process variables encountered is pressure. **Pressure** (P) is defined as the amount of force (F) per unit of area (A). If force is expressed in pounds and area in square inches, then pressure would be expressed in pounds per square inch (PSI). Pressure can be defined mathematically in the following equation:

$$P = F/A$$

Where:
P = pressure (PSI)
F = force (pounds)
A = area (square inches)

To illustrate how both force and area affect pressure, think of a solid object resting on a table (Figure 2.1). If the object weighs 1 pound and is resting on 1 square inch of surface,

Pressure the amount of force a substance or object exerts over a specific area.

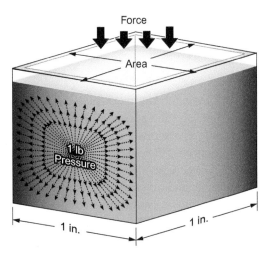

Figure 2.1 Pressure.
CREDIT: Fouad A. Saad/Shutterstock.

then it is applying 1 PSI of pressure. If the same weight is instead resting on one-half square inch of area then it is now applying 2 PSI of pressure.

Pressure directly affects the boiling point of a substance. The greater the amount of pressure on a substance, the greater the amount of heat required to bring the substance to its boiling point. Because of the effects of this principle on the properties of a substance, process industries must account for pressure in the mixing, creation, and/or separation of chemicals. A process technician is responsible for both monitoring and controlling pressure in pipes and equipment. Failure to control pressure adequately can have disastrous consequences.

Components That Affect the Force Exerted by Molecules

As stated earlier, pressure can be defined as a force applied to a unit of area. For a given force, if the area is smaller, then the pressure is greater.

Force is a push or pull exerted on an object (solid) that causes the object to change direction. Liquids and gases are both considered to be fluids in processing, but they behave differently when force is exerted on them in a static or stationary environment. The total height of a liquid column in a given vessel (e.g., storage tank), the surface pressure (atmospheric pressure), and the density (compactness) of the material determine the amount of force at the bottom of the liquid column for each unit of area. In the example in Figure 2.2, h represents the height of the liquid. As the column of liquid rises in the vessel, the amount of pressure on the bottom also increases. The height of a liquid column (e.g., product in a storage tank) and the amount of force (head pressure) it exerts at the bottom are often used to move (push) the liquid to another destination. The pressure is not released until a valve is opened, allowing the liquid to flow out from the vessel. The vessel itself is designed to resist the internal forces caused by the liquid. All vessels and equipment are designed to handle a certain amount of total pressure.

Force energy that causes a change in the motion of an object, involving strength or direction of push or pull.

Figure 2.2 Pressure versus height. Note that pressure is high at the bottom of the tank and low at the top.

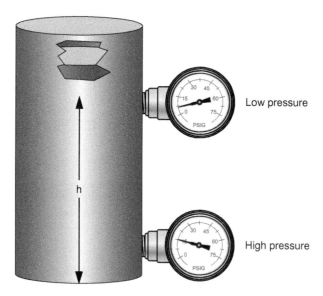

In contrast to liquids and solids, gases and vapors apply pressure equally to all surfaces in a container. The pressure produced by this state of matter is primarily caused by the extraordinary energy of the individual molecules and, to a lesser degree, by gravity. Gases and vapors are different from liquids and solids because each molecule exists apart from the others. They are highly energized and react accordingly. Imagine billiard balls bouncing off one another and off any surface they may encounter. The more energized they become, the faster they move, producing more pressure.

Factors affecting the force that molecules contribute to pressure include the following:

- Speed (temperature)
- Mass (atomic weight)
- Density (number of molecules per volume)

SPEED The temperature of a substance is usually the greatest factor affecting **molecular speed**. The greater the amount of heat applied, the more the molecules become excited and move around (Figure 2.3). The more they move, the greater the distance the molecules try to put between themselves. The greater the amount of heat applied to a liquid, the more the molecules increase in energy and the greater their ability to overcome surface tension and become a vapor. If confined, the increase in speed also increases the amount of pressure exerted on the container (vessel) walls. The vessel itself confines the movement of the molecules and promotes pressure increase as the molecules increase in speed.

Molecular speed the rate of motion of molecules in a substance; it may change as a result of factors such as heat or collision with other molecules (e.g., the walls of a vessel).

MASS **Mass** is the amount of matter (atomic or molecular weight) in a body or object. Mass has to do with the amount of matter in a molecule rather than the molecule's size. The greater the collective weight of the individual atoms that make up a molecule, the more the molecule weighs. For example, an atom of iron (Fe) weighs more than an atom of either hydrogen (H) or oxygen (O). Water is the standard used to measure the mass of other liquids and solids.

Mass the amount of matter in a body or object measured by its resistance to a change in motion.

The example of a solid object resting on a table (see Figure 2.1) is only one way of describing pressure. A liquid applies pressure that varies with its depth to that portion (walls) of the vessel that contains it. The force that produces the pressure created by solids and liquids is primarily derived from its mass being drawn toward a center of gravity. The more mass that is applied to a unit of area, the greater the pressure. Additional mass could be in the form of more molecules or heavier molecules.

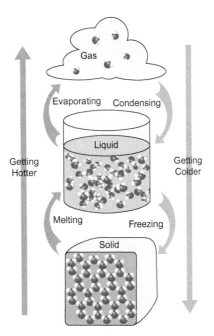

Figure 2.3 Molecular motion.
CREDIT: OSweetNature/Shutterstock.

DENSITY Another factor affecting the pressure of gases and vapors is its **density**. Density is the total number of molecules per unit volume at a given temperature. Unlike liquids, which are essentially not compressible, the density of gases can change. Density (the total number of molecules per unit volume at a given temperature) is another factor that affects the pressure of gases and vapors. With gases and vapors, the number of molecules per unit of volume defines density. The greater the number of molecules, the greater the number

Density mass per unit volume.

of collisions that occur between molecules. What this implies is that a greater number of molecules colliding within the container walls of a vessel creates a higher pressure. These collisions create more force within the material.

Using a container of water (more molecules) versus the same type of container of paraffin (fewer molecules), and applying heat to both, the result would show that the container of water would begin to vaporize more readily than the paraffin.

The density of fluid gases is dependent on the allowable space that the molecules occupy. The greater the space, the less dense the gas would become since the molecules would spread out until they filled the entire space. If the container volume size is reduced for the same number of gas molecules, then internal pressure on the walls of the container increases. This is why substances that are gases at ambient temperatures have specially designed pressure vessels.

Specific Gravity. **Specific gravity** is the density of a substance relative to the density of water, defined as 1.0 at 39 degrees Fahrenheit (4 degrees Celsius). Liquids or solids have a specific gravity that is either greater or less than water. Specific gravity of a liquid is calculated as the weight of a volume of liquid divided by the weight of an equal volume of water at a standard temperature.

Specific gravity of a gas is not associated with water. Instead, the specific gravity of a gas is the weight of a volume of gas divided by an equal weight of air at standard temperature and pressure.

Specific gravity the ratio of the density of a liquid or solid to the density of pure water *or* the density of a gas to the density of air at standard temperature and pressure (STP).

2.2 Pressure Measurement

Pressure measurement is often taken at unique points in a process operation. An example of this is a single local pressure gauge located on a pump suction or discharge line. When this measurement is vital to controlling the process, the measurement taken is converted to a pneumatic or electronic signal and transmitted to another location. This measurement signal may be used as a single numeric value indicating a monitoring process, or it may be used to signal a process change in a control loop arrangement. This will be discussed in depth in later chapters.

Measuring and controlling pressure in the process industry allows for a safer and more productive plant. Controlling the pressure inside a reactor helps optimize the chemical reaction. This results in higher yields and increased profits. From a safety standpoint, controlling pressure is a matter of common sense. Pipes or vessels exposed to pressures that exceed their pressure ratings can rupture, causing death, bodily injury, equipment destruction, and environmental consequences.

Knowing how pressure is measured and the appropriate scale to use for each application is critical. The following definitions and descriptions of pressure measurements are given to help process technicians make more informed decisions during daily operations.

Differential Pressure (Delta P, ΔP, D/P)

Differential pressure (D/P) is the difference between two related pressure measurements. This is one of the most common measurements taken in the process industry. Differential pressure is used to infer liquid level and flow rate as well as pressure differences between two points. Two commonly used ways to express differential pressure are the Greek letter delta (Δ) and the abbreviation D/P.

Differential pressure (D/P) the difference between two related pressure measurements; usually used in measurements of pressure, temperature, level, and flow.

Knowing the pressure difference (delta) between two points in a process is important. A D/P cell (differential pressure measuring device) may be located so that it measures the difference in pressure, such as across an orifice plate (Figure 2.4) that is installed in a pipeline to measure flow.

$$\Delta P = P_{upstream} - P_{downstream}$$

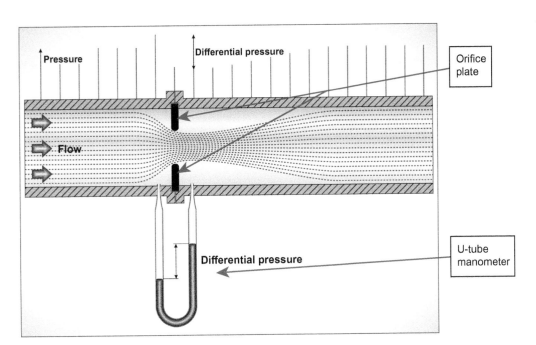

Figure 2.4 Cutaway of an orifice plate installed in process piping.

CREDIT: Fouad A. Saad/Shutterstock.

Pressure Measurement Scales

Whether from a single local pressure indicator (e.g., pressure gauge), or a more sophisticated D/P cell, pressure is still measured against a particular scale. Typical pressure measurement scales include the following:

- Pounds per square inch (PSI)
- Pounds per square inch absolute (PSIA)
- Pounds per square inch gauge (PSIG)
- Atmospheres (atm)
- Bars
- Inches of water (in. H_2O)
- Inches of mercury (in. Hg) or metric equivalent (mmHg)
- Inches Hg Vac (negative in. Hg)

PSI (POUNDS PER SQUARE INCH) Pounds per square inch is the basic formula for pressure: force (expressed in pounds) divided by the affected area (expressed in square inches). This PSI scale originates at true zero pressure, that is, a total vacuum.

PSIA (POUNDS PER SQUARE INCH ABSOLUTE) Pounds per square inch absolute (**PSIA**) is the basic formula for pressure caused by the specific gravity of air (a combination of various gases). *At sea level*, 14.7 PSI is the standardized (averaged) pressure measurement, and this is the number we will use to explain the concepts in this chapter. Keep in mind that at other elevations, this number changes. The PSIA is equivalent to one atmosphere (1 atm). Zero on an absolute scale would represent no pressure at all. The higher the inches of mercury, the lower the absolute pressure.

PSIA pounds per square inch absolute; zero on the absolute scale would represent the absence of any pressure.

High in the mountains, the atmospheric pressure is much lower than at sea level. Below sea level, atmospheric pressure is much higher. This principle of atmospheric pressure must be taken into account if the processing facilities are significantly above or below sea level. For example, liquids boil at much lower temperatures when they are not affected by as much atmospheric pressure, so liquids in a plant located at high altitude will boil more rapidly than those in a plant at sea level.

PSIG (POUNDS PER SQUARE INCH GAUGE) This PSI scale originates at atmospheric pressure. That means that regardless of where you are, the ambient (surrounding) pressure establishes zero on this scale.

PSIG pounds per square inch gauge; pressure measurement that references 14.7 PSIA as its zero point.

The pounds per square inch gauge (**PSIG**) starts the scale at zero (0). The 14.7 atmospheric pressure (or the atmospheric pressure where the measurement is being taken) is already taken into account in the scale. This is the most common pressure measuring scale used in modern processing facilities.

ATMOSPHERE(S) One **atmosphere** (atm) of pressure is equal to 14.7 PSIA. Two (2) atmospheres of pressure are equal to 29.4 PSIA.

Atmosphere the pressure at any point in the atmosphere due solely to the weight of the atmospheric gases above the point concerned; 14.7 PSIA, which equals one atmosphere, is the pressure at sea level.

Atmospheric pressure is defined as the force per unit area exerted against a surface by the weight of the air molecules above that surface. As stated earlier, one atmosphere is equivalent to 14.7 PSIA. To state this differently, for every square inch of the earth's surface at sea level, there are 14.7 pounds of force weighing down on it. This 14.7 pounds of pressure is comparable to the weight of the atmospheric gases above the point where the measurement is taken.

Atmospheric pressure the pressure at the surface of the earth (14.7 PSIA at sea level).

Atmospheric pressure is measured with an instrument called a barometer. Atmospheric pressure is often referred to as barometric pressure. In process operations, atmospheric pressure may be factored into the calibration of a pressure measuring device. Table 2.1 provides pressure measurement scales and conversions.

Table 2.1 Pressure Unit Conversion Chart*

Pressure Units	PSIA	bar	mbar	in. Hg	in. H_2O	mmHg	mmH_2O	Atm
PSIA	1	14.504	0.014504	0.49118	0.036127	0.019337	0.0014223	14.7
bar	0.068946	1	0.001	0.033865	0.0024908	0.0013332	9.8068×10^{-5}	0.98692
mbar	68.946	1000	1	33.865	2.4908	1.3332	0.098068	1013.25
in. Hg	2.0359	29.529	0.029529	1	0.073552	0.039368	0.0028959	29.92
in. H_2O	27.68	401.47	0.40147	13.596	1	0.53525	0.039372	407
mmHg	51.714	750.06	0.75006	25.401	1.8683	1	0.073558	760
mmH_2O	703.05	0.10197	10197	345.32	25.339	13.595	1	10332
atm	0.068045	0.98692	0.00098692	0.033422	0.0024583	0.0013158	9.6788×10^{-5}	1

This table is read in columns rather than in rows. Values in the top row are converted into pressure units indicated in the left column.

Note: The consistency or explanation of rounding is utilized in the field on the inches of water column for pressure conversions. In other words, 27.68 in. H_2O might be rounded to 27.7 in. H_2O, the equivalent to 1 PSI. However, use of this rounding, which is standard in the field, can yield inaccuracies on very tall vessels. For example, a 0.02-inch difference in a 180-foot vessel would result in a difference of 3.6 inches.

Bar measurement of pressure equal to 0.987 atmospheres.

BAR(S) A **bar** is a metric unit primarily used to measure atmospheric pressure.

$$1 \text{ bar} = 14.5 \text{ PSIA}$$

One bar of pressure is equal to 0.987 atmospheres. The term *bar* is the equivalent to 1,000 millibars. A millibar is the unit of measurement used on barometers to determine atmospheric pressure. The scale is based on inches of Hg. Most weather reporters use the phrase *atmospheric pressure* or *barometric pressure* rather than saying millibars or bars.

Inches of water (in. H_2O) a very small measurement of pressure equal to 0.036 pounds per square inch at 4 degrees Celsius (39.2 degrees Fahrenheit).

INCHES H_2O Inches of water (in. H_2O) is a pressure unit equal to the pressure exerted by the height of 1 inch of water.

$$27.7 \text{ in. } H_2O = 1 \text{ PSI}$$

Inches of mercury (in. Hg) most common measurement scale for a manometer; measures in pounds per square inch; 1 inch Hg = 13.0687 inches H_2O.

INCHES Hg Inches of mercury (in. Hg) is a pressure unit equal to the pressure exerted by the height of 1 inch of mercury.

$$2.04 \text{ in. Hg} = 1 \text{ PSI}$$

760 mmHg equals 14.7 PSIA. The unit of measurement mmHg is used to measure barometric pressure. Anything less than 760 mmHg is considered to be vacuum pressure. Pressure expressed in mmHg is the metric equivalent to inches of Hg.

INCHES Hg Vac Zero on an absolute scale would represent no pressure at all. A total vacuum is considered to be 14.7 PSI below atmospheric pressure (0 PSIA). On an absolute scale, a total vacuum would be −14.7 PSIG. Vacuum is often measured in inches of mercury (in. Hg).

Unit Equivalency

$$1 \text{ atm} = 1.013 \text{ bars} = 14.7 \text{ PSIA} = 407 \text{ in. H}_2\text{O} = 29.92 \text{ in. Hg.} = 14.7 \text{ PSI}$$

$$1 \text{ PSI} = 27.7 \text{ in. H}_2\text{O} = 2.04 \text{ in. Hg}$$

In common usage, 1 bar is equated to 1 atmosphere.

One problem encountered by new technicians is misunderstanding of pressure units and their magnitudes. The consequences can be catastrophic. The difference between thinking PSIA (absolute pressure) and reacting with PSIG (gauge pressure) could create a high pressure situation that could rupture a tank. The difference between PSI and inches water column can cause damage and pose a danger.

2.3 Pressure Sensing and Measurement Instruments

Many devices are used to indicate the pressure of the process, one of the most common being the **pressure gauge**. In process industries, three types of pressure are measured: absolute, gauge, and vacuum. All pressure is actually the weight exerted on a certain area. For example (Figure 2.5), if we were at sea level and examined a one square inch area at the surface, we would find that the pressure or force would be 14.7 lb. This means that a column of air one square inch and extending from the surface to the space beyond our atmosphere would weigh 14.7 lb. This is referred to as absolute pressure (weight of the air comprising the atmosphere) and is considered to be 14.7 lb. per square inch. Atmospheric pressure is used as the basic reference point for gauge pressure.

Pressure gauge the most common local instrument type for measuring pressure; the three primary types of pressure gauge measuring scales are absolute, gauge, and vacuum.

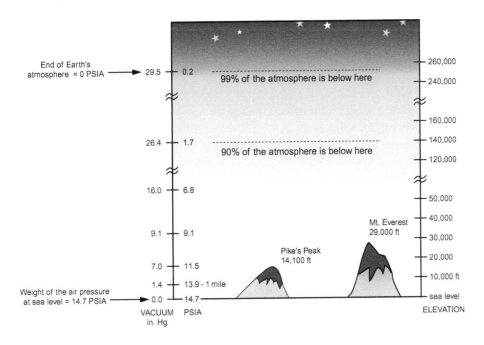

Figure 2.5 Atmospheric pressure.

A pressure gauge reads zero at sea level even though there is 14.7 lb per square inch because its mechanism is calibrated to include 14.7 PSI as the zero point. If the gauge reading increased to 50 PSIG, the PSIA pressure would now be 50 PSI above 14.7 PSI or 64.7 PSIA.

If it were possible to remove all the pressure from a line or vessel so that it is actually 14.7 PSI below atmospheric pressure, or zero pressure, we would say our pressure is now *absolute zero*. Then, any pressure measured above this absolute zero point would be called absolute pressure.

In industry, we are sometimes interested in maintaining pressures less than atmospheric; this is referred to as **vacuum**. Vacuum is measured from the 14.7 PSI atmospheric pressure point in units of inches or millimeters of mercury vacuum or inches of water (column).

Vacuum a condition in which the pressure measured is less than atmospheric pressure; usually measured in inches of mercury (in. Hg).

Common types of pressure sensing/measuring instruments used in the process industry include the following:

- Manometers
- Pressure gauges
- Differential pressure (D/P) cells
- Strain gauge transducers
- Capacitance transducers

Manometers

Visual type liquid manometers have been used to measure pressure for hundreds of years. They are among the most reliable measuring devices in industry, provided they remain clean and dry. However, from a practical standpoint, manometers are indicating instruments only.

The following are the common manometer types (Figure 2.6):

- **U-tube manometer**
- Well (reservoir) manometer
- Inclined manometer
- Barometer

U-tube manometer a gravity balanced pressure measuring device with two fluid chamber tube gauges connected by a U-shaped tube so fluid (a liquid or mercury) flows freely between the chambers.

Figure 2.6 Manometers.

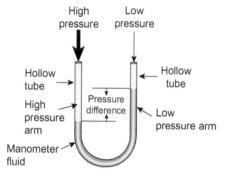

The accuracy of manometers is limited by how well the reading is taken by the technician. For example, when reading a manometer filled with a liquid that produces a convex column, the reading should be taken on the top of the meniscus (the rounded dome of the liquid column). The reading of a concave meniscus (Figure 2.7) should be taken on the bottom of the depression.

Figure 2.7 Reading a manometer (concave meniscus).

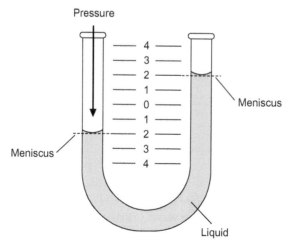

The most common type of manometer is the U-shaped tube. In this manometer, liquid or mercury fills two fluid chambers that are connected by the U-shaped tube so that the fluid is free to move between them. When additional pressure is applied to one chamber, the fluid in that chamber flows through the connecting tube into the other chamber, causing its level to rise until the head pressure created by the offset equals the pressure applied to the other chamber. Manometers are called *gravity balanced pressure devices* because of this characteristic.

U-shaped tubes are filled with a colored liquid. The density of the liquid is appropriate for the process being measured. The tubes are mounted against a scale proportional to the fluid used, and fluid is free to move between the two chambers. This type of manometer is a visual type. It is a reliable pressure measuring device used primarily as an indicating instrument. On a U-tube manometer, the meniscus is read on both sides of the tube, and the two readings are added together.

Pressure Gauges

Pressure gauges, like manometers, are used primarily as **local indicators**. They can be found throughout the plant attached to process piping and vessels. Pressure gauges provide technicians with a quick check of process conditions while they are walking through the unit. Critical pressures are usually monitored by means of a transmitter and control loop that sends a signal into the control room so that the pressure can be watched and/or controlled constantly.

Gauges often consist of a plastic or steel body, a metallic tube (called a Bourdon tube) (Figure 2.8A) that flexes (curls or uncurls) in response to pressure changes, a pointer that is connected to the linkage and the sensing element, and a scale marked in units of pressure. As the metallic tube flexes, the pointer moves up or down the scale to indicate the pressure. The scale markings determine what type of pressure the gauge measures. The Rosemount Smart Pressure Gauge (Figure 2.8B) features a robust design with industry-proven sensor technology to resist common traditional gauge failures. This pressure gauge replaces mechanical components and includes advanced wireless technology that delivers reliable field data communications.

> **Local indicators** instruments placed on equipment in the field only to be read in the field; may be used for comparison with transmitted instrumentation readings or used on noncritical processes to indicate process values.

Figure 2.8 A. Diagram of a pressure gauge (Bourdon tube). **B.** Photo of a "smart" pressure gauge.
CREDIT: B. © Emerson 2019.

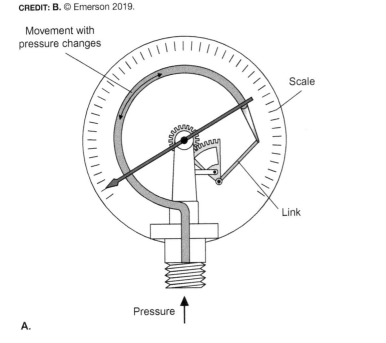

Process technicians are allowed to replace pressure gauges in certain situations. Every effort should be made to choose the appropriate gauge scale for the application. For example, if a pump normally operates at 150 PSIG, then the appropriate pressure gauge would have 150 in the middle of the scale. If a gauge were chosen that had a maximum pressure of 150, then the pump pressure would probably blow (overpressure) the gauge and render it inoperable.

For some applications, using a compound vacuum/pressure gauge (Figure 2.9) may be necessary when there is a need to measure pressure below one atmosphere, at atmospheric pressure, or above atmospheric pressure. The zero on a compound gauge is equivalent to 14.7 PSIA. From zero moving clockwise, the gauge is read in PSIG. From zero moving counterclockwise, the gauge is read in inches of mercury (in. Hg) vacuum.

Figure 2.9 Compound vacuum/pressure gauge.

CREDIT: Spaxiax/Shutterstock.

Figure 2.10 Differential pressure gauge.

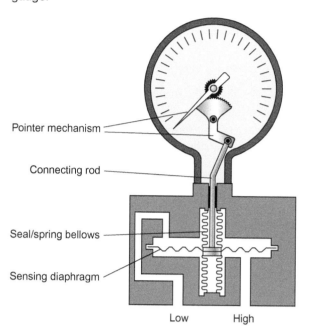

For other types of applications, a differential pressure reading may be desired. This type of reading may be obtained locally by a differential pressure gauge (Figure 2.10).

As with all instruments, there is a correct way to read gauges. First, look straight onto the faceplate making sure that your eyes are facing the pointer tip directly. Then, read the indication directly behind it. If you attempt to read it from the side, the angle of the view may induce an error, called the *parallax effect*.

Differential Pressure Cells

Differential pressure cell (D/P cell) an instrument that measures the difference between two related pressure points and produces a corresponding output signal that is proportional to the process being measured.

A **differential pressure (D/P) cell** is normally a simple mechanical type sensor found in transmitters. A more sophisticated transducer type may be found in electronic transmitters. Either type may be used to measure level, flow rate, and differential column pressure.

Differential pressure is defined as the difference in pressure between two related measurements. A differential pressure (D/P) transmitter measures the difference between two pressure points and produces a corresponding output signal.

Differential pressure transmitters (Figure 2.11) consist of a diaphragm capsule that can sense pressure on each side, a connecting rod, and the transmitter. The diaphragm, commonly filled with silicon but also made of steel, is connected to a rod which flexes in response to applied pressure from each side. The rod is attached to the transmitter mechanism which calculates the pressure differential (deflection) and converts that signal to an output signal for the control loop. With D/P measurements, the terms *high* and *low* are relative to each other and do not reflect the actual pressures sensed individually.

An example of using pressure to measure flow rate is as follow: If a D/P transmitter is connected across an orifice plate in a pipe that has a liquid flowing through it, the process pressure inside of the pipe may be 200 PSI, and for safety reasons this pressure could not be sent as a signal to the control loop and the control room. A signal representative of the process flow is calculated using the upstream and downstream pressures across the orifice plate which press

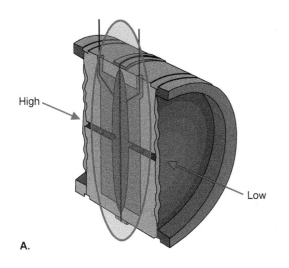

Figure 2.11 A. Diagram of a differential pressure (D/P) cell cross section (transducer highlighted). **B.** Photo of a differential pressure (D/P) cell.

CREDIT: B. © Emerson 2019.

against each side of a diaphragm capsule. The resulting differential pressure is converted in the transmitter to an output signal, or proportional signal, that represents the amount of flow rate.

As illustrated in Figure 2.11, the D/P cell itself may be a simple mechanical type sensor normally found in pneumatic transmitters (producing a signal range of 3–15 PSI output) or a more sophisticated transducer type found in electronic transmitters (producing a signal range of 4–20 mA output).

Strain Gauge Transducers

A transducer is a device that converts one form of energy into another form. A **strain gauge transducer** (Figure 2.12) consists of a group of wires that stretch, elongate, or strain when pressure is applied. The resulting strain converts one form of energy (process pressure) into an electronic signal representative of the process measurement. As current flows through the wires, resistance changes proportionally to the stress or strain applied. Strain is the act of changing the dimensions of a solid. When the dimensions of a strain gauge are changed, its resistance also changes. The greater the resistance, the greater the change in the process variable. This is the premise on which strain gauges operate. Strain gauges, commonly called load cells, measure tank levels using the material weight (pressure) applied to the strain gauge.

Strain gauges are used more frequently than other types due to their lower cost, resilience, and behavior dynamics. A process technician may never come in contact with an actual

Strain gauge transducer a pressure sensing and measuring device consisting of a group of wires that stretch when pressure is applied, creating resistance, thereby changing the process pressure into an electronic signal.

Figure 2.12 Strain gauge transducer.

CREDIT: Courtesy of Brazosport College.

Figure 2.13 Capacitance transducer.

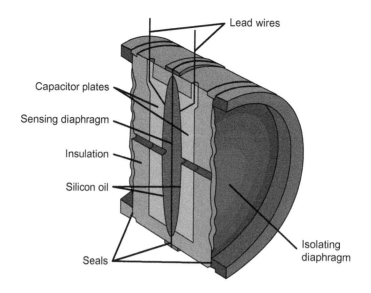

measurement transducer, but the concepts of how they operate and the technology involved in how they associate with other process instrumentation are important to understand.

Capacitance and Piezoelectric Transducers

Capacitance transducer an electrical device that contains a measurement diaphragm and capacitor plates; it converts an applied pressure to an electronic signal measurement that is proportional to the process being measured.

A **capacitance transducer** (Figure 2.13) is a device consisting of two parallel plates, usually fixed in position, that are separated by a dielectric medium such as air, gas, or silicon. It can be enclosed with protective diaphragms. They are second only to strain gauges in popularity as pressure measurement transducers. They are available in single or differential pressure designs.

When pressure is applied to the measurement diaphragm, the capacitor plates are caused to move closer together. Since capacitance is inversely proportional to the distance between the plates, as the plates get closer together, capacitance increases. That amount of capacitance is converted into a proportional signal representing the process variable (e.g., pressure).

Piezoelectric transducers are a type of electroacoustic transducer that converts the electrical charges produced by some forms of solid materials into energy. The word *piezoelectric* literally means electricity caused by pressure.

2.4 Pressure Conversions

Let us compare gauge pressure to absolute pressure (Figure 2.14). In the earlier definitions, a statement was made that the absolute pressure scale begins at true zero pressure (a total vacuum), whereas zero on the gauge pressure scale is equal to whatever atmospheric pressure is. The average pressure at sea level is 14.7 PSIA and 0 PSIG. Gauge pressure displays

Figure 2.14 Gauge and absolute pressure comparison.

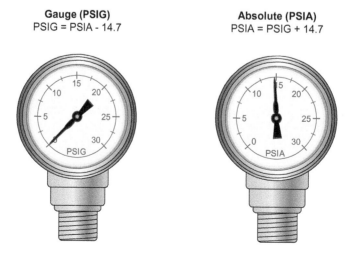

zero, which compensates for atmospheric pressure. For example, if the atmospheric pressure in Denver, Colorado, is 12.0 PSIA, the PSIG gauge would still read 0 PSIG, and the absolute pressure scale would read 12.0 PSIA.

So, why do we need two scales? The answer is that the gauge pressure scale was the only one that made sense at one point in time. How much pressure was inside a vessel compared to the outside pressure was important to know. In time, knowing and controlling pressure at an exact point, independent from atmospheric pressure variations, also became important. Scientists must use an absolute scale to be accurate.

Process technicians need to know how to move easily between the two scales. Here are the mathematical relationships:

$$PSIA = PSIG + 14.7 \text{ PSI}$$

$$PSIG = PSIA - 14.7 \text{ PSI}$$

Example pressure measurement conversions:

$$20 \text{ PSIG} + 14.7 \text{ PSIA} = 34.7 \text{ PSIA}$$

$$40 \text{ PSIA} - 14.7 \text{ PSIA} = 25.3 \text{ PSIG}$$

To accommodate low pressure measurements or small measurement spans, several scales have been developed. Two commonly used scales (Figure 2.15) are the inches water (in. H_2O) and inches mercury (in. Hg). Recall that manometers are highly accurate pressure measuring devices, there is a natural inclination to use their column heights as pressure units. These column inches are self-defining, and the pressure is equal to that exerted by one inch of their prospective liquid heights.

Here are the pressure equivalents compared to 1 PSI:

$$1 \text{ PSI} = 27.7'' \; H_2O = 2.04'' \text{ Hg}$$

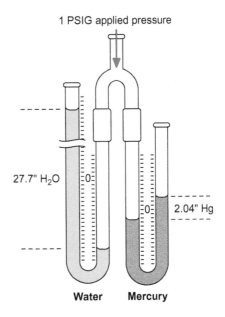

Figure 2.15 Inches water and inches mercury pressure.

Pressure Units

In a simpler world, there would be only one pressure unit, but this is no more reasonable than having one denomination of money. If the public had only one dollar bills to use for currency, people would have to pack a suitcase to purchase a car and would have to spend an entire dollar just to buy a stick of gum. This does not make any more sense than having only one pressure unit to fit all situations. Using PSI works for most pressure measurements, but smaller units of measurement are needed for vacuum applications. Consider the following example. A steam turbine's exhaust pressure is very often dropped to a vacuum to help its motive ability (speed). That vacuum is measured in inches of Hg (29.92 inches Hg = atmospheric pressure) which can easily be calibrated on a gauge. That same measurement in inches of H_2O would require a manometer with a height of almost 34 feet (about 10 meters). It would not be practical and could represent a safety hazard. Obviously, one pressure unit cannot possibly provide adequate resolution at both extremes. The process being measured will determine the appropriate unit of measurement used for calculation.

Pressure Conversion Calculations

Process technicians are responsible for learning to convert between the following pressure units: PSI, inches water column, and inches mercury. The following method takes a little longer to set up than simply multiplying with a factor. However, this method does allow the conversion of any pressure unit to another by simply knowing its value equivalent to 1 PSI. It is more efficient than trying to remember all of the factors for every conversion both to and from each unit. Process technicians need to memorize the general formula to always know how to perform these pressure conversions. Once memorized, this formula is a tool as reliable as a screwdriver or a wrench; it can be used over and over again.

This pressure conversion method is referred to as *the equivalent units method*. This method produces a *factor* each and every time a conversion problem is set up. The following values are quantitatively equal.

EQUIVALENT UNITS

$$1 \text{ PSI} = 27.7 \text{ in. water column} = 2.04 \text{ in. Hg}$$

Use these steps when converting pressure units:

1. Find or recall equivalent values for pressure units.
2. Place the value to be converted and the equivalent values into their proper place in the formula.
3. Perform calculation.
4. Check your answer to see that it makes sense.

GENERAL FORMULA FOR CONVERTING PRESSURE UNITS
Where:

$$1 \text{ PSI} = 27.7 \text{ in. } H_2O = 2.04 \text{ in. Hg}$$

$$\text{New units} = \frac{\text{(known units)}}{\text{(known equivalent units)}} \times \text{(new equivalent units)}$$

EXAMPLES

- Convert 10 in. Hg to in. H_2O.

$$\text{New units} = \frac{10 \text{ in. Hg} \times 27.7 \text{ in. } H_2O}{2.04 \text{ in. Hg}}$$

$$\text{New units} = 135.78 \text{ in. } H_2O$$

- Convert 3.6 PSI to in. Hg.

$$\text{New units} = \frac{3.6 \text{ PSI}}{1 \text{ PSI}} \times 2.04 \text{ in. Hg}$$

$$\text{New units} = 7.34 \text{ in. Hg}$$

Example pressure measurement conversion calculations:
Use the following formulas:

- PSIG to PSIA = PSIG + 14.7
- PSIA to PSIG = PSIA − 14.7
- PSIG to in. H$_2$O = PSIG × 27.7
- PSIG to in. of Hg = PSIG × 2.04

$$\text{inches of H}_2\text{O to PSIG} = \frac{\text{height of liquid}}{27.7}$$

$$\text{inches Hg of PSIG} = \frac{\text{height of liquid}}{2.04}$$

Note the following examples of conversion:

- 150 PSIG to PSIA (164.7 PSIA)
- 150 PSIA to PSIG (135.3 PSIG)
- 42 PSIG to in. H$_2$O (1163.40 in. H$_2$O)
- 55 PSIG to in. Hg (112.20 in. Hg)
- 300 in. of H$_2$O to PSIG (10.83 PSIG)
- 100 in. of Hg to PSIG (49.02 PSIG)

Summary

Pressure is a common process variable and is defined as the amount of force per unit of area. Mathematically, pressure is defined as Pressure = Force/Area or P = F/A. Mass applied to a unit of area depends on the density of the molecules within the mass and how heavy the molecules are. Gases and vapors apply pressure equally to all surfaces in a container while liquids and solids may vary according to the size or shape of the container.

When calculating pressure, factors that need to be considered include the following:

- Liquids and gases (considered to be fluid)
- Liquids and gases (different behaviors when force is applied)
- Total height of the liquid column in a given vessel
- Surface pressure
- Density of the material

Speed, mass, and density of molecules affect force. Temperature is the greatest factor affecting molecular speed. Increases in temperature are directly proportional to increases in speed. If the increased speed is confined, then pressure increases as well. The greater the collective amount of mass (either atomic or molecular weight) applied to a particular unit of area, the greater the pressure. The density of a substance determines the number of collisions occurring between the molecules. Where there are more molecules, there is more potential for a higher number of molecular collisions, creating more force. For gases, special pressurized containers must be used to contain the molecules, since gases fill up the entire space they are allowed to occupy. Specific gravity is the density of a substance relative to the density of water (expressed as 1.0). This makes the measurement of all other liquids or solids have a specific gravity either greater than or less than water.

Pressure measurements may be taken at unique points within processing operations. In some cases, knowing the difference between two points is important. This is called differential pressure, or delta P (D/P, ΔP) and is accomplished by forcing the process stream through a calibrated orifice plate. Sensors on either side of the orifice plate sense the pressure and send the reading to a differential pressure transmitter where the difference in the two pressures is calculated and the result is transmitted appropriately.

Typical measurement scales include pounds per square inch (PSI), pounds per square inch absolute (PSIA), pounds per square inch gauge (PSIG), atmospheres (atm) or bars, and inches of water (in. of H$_2$O) or mercury (in. of Hg). At sea level, 14.7 PSI is the standardized pressure measurement that is equivalent to 1 atmosphere, with zero on an absolute scale being equivalent to absolutely no pressure at all. Vacuum is most often measured in inches of mercury. Since the PSI scale originates at atmospheric pressure regardless of where you are, the ambient or surrounding pressure establishes zero on this scale with the 14.7 (sea level) atmospheric pressure already taken into account. This is the most common measuring scale used in modern processing facilities.

Pressure sensing and measuring instruments are commonly used to indicate absolute, gauge, and vacuum pressures in process facilities. The oldest type of pressure gauge still in operation is the manometer, and the most common of these is U-shaped. Manometer accuracy depends mainly on the reading being taken from the meniscus. Pressure gauges, like manometers, are primarily used as local indicators. Pressure gauges are generally made of steel bodies and have an internal metallic tube that flexes in response to pressure changes. This response is linked to a pointer by a mechanism that causes the pointer to move along a predetermined scale (PSIG, PSIA, vacuum, or compound). Other process situations require measuring pressure differential. In these cases, a differential pressure (D/P) cell and transmitter configuration may be used to sense the high and low side of a process stream. In this case, the output would be read remotely. Another type of pressure measuring device is a strain gauge transducer that converts one form of energy (applied pressure to a group of wires that stretch) into another form of energy (an electronic signal) when the device responds to strain. Similarly, a capacitance type transducer may be used as a pressure measurement transducer and is available in single or differential pressure designs. Capacitance is inversely proportional to the distance between the plates.

The two mathematical relationships between the two pressure measurement scales are:

$$\text{PSIA} = \text{PSIG} + 14.7 \text{ PSI}$$

$$\text{PSIG} = \text{PSIA} - 14.7 \text{ PSI}$$

Low pressure measurement scales use inches of H_2O or inches of Hg. When using manometers, their liquid column height can also be factored into the equation or scale.

Checking Your Knowledge

1. Pressure is described as _____ per unit area.
   a. flow
   b. pounds
   c. force
   d. inches

2. Pressure is increased when:
   a. the number of molecules per unit area is decreased.
   b. heavier molecules per unit area are introduced.
   c. molecules begin to move faster.
   d. the number of molecules are spread out over a larger unit area.

3. Atmospheric pressure at sea level is _____ PSIA.
   a. 0
   b. 2
   c. 14.7
   d. 29.92

4. Match the instruments listed below with their method of operation.

Instruments	Method of Operation
I. Manometer	a. Uses wires that stretch when pressure is applied
II. Capacitance transducer	b. Uses a measurement diaphragm to check pressure
III. D/P cell	c. Uses two sensors to measure a change in pressure
IV. Strain gauge	d. Uses a liquid filled tube to measure pressure

5. Convert 1 PSI into inches of H_2O.
   a. 0
   b. 2
   c. 14.7
   d. 27.7

6. Convert 14.7 PSI into inches of H_2O.
   a. 109.6
   b. 247.17
   c. 398.76
   d. 407.19

7. Convert 30 PSIG to PSIA.
   a. 44.7
   b. 15.3
   c. 60.2
   d. 59.4

NOTE: Answers to Checking Your Knowledge questions are in the Appendix.

Student Activities

1. Hold your hand on your chest while you take a very deep breath. What happened? Did you see or feel any expansion? Explain what happened when you did this.

2. Using an empty plastic gallon milk jug with a screw on top, fill it at least 1/4 full with very hot water, being careful not to burn yourself. Cap the jug quickly and let it stand for up to an hour. What happened?

3. Hold an empty plastic bottle upside down over a pan of boiling water. Ensure the bottle is positioned so as not to burn your hand by using tongs or a protective glove to hold the bottle in place for several minutes. The bottle will fill with steam. Remove bottle (still upside down) from boiling water area and immediately cap the bottle. As the bottle and the steam cool down, the bottle sides collapse. Explain the principles that cause this to occur.

4. Do you think atmospheric pressure is powerful enough to crush a can? Use the following materials and procedure to determine the answer:

Materials

- Hot plate
- Empty soft drink can with one tablespoon of water inside
- Tongs (to pick up the hot can)
- Shallow pan filled with cold water

Procedure

a. Turn on the hot plate to a medium temperature.

b. After ensuring a tablespoon of water has been inserted into the soft drink can, place it onto the hotplate.

c. Observe how the water temperature rises and fills the can with water vapor, allowing steam to escape from the top opening.

d. Using the tongs, quickly pick up the soda can, turn it upside down and place the opening into the shallow pan of cold water.

    NOTE: The can should immediately crush and a dull popping sound should occur. If this does NOT happen then start over with the procedure ensuring the tablespoon of water inside the can has had enough time to vaporize.

e. Explain what happened. Air pressure decreases as the water vapor condenses inside the can. The can crushed due to the air pressure inside the can becoming less than the atmospheric pressure outside the can.

5. Using a large mouth plastic bottle, insert an air-filled balloon taking care not to burst it. Cap the bottle. Make a small hole on the plastic bottle and force air into the bottle by using an air pump/injection needle apparatus (tire air pump). Describe what happens to the balloon.

6. Have students identify pressure sensing or measurement devices in the lab.

7. In a pilot plant or on a tabletop model, have students correctly read pressure gauges.

8. Use a vacuum pump with a pressure gauge hooked to a one gallon tin can with tubing. Start pump and note the pressure when the can implodes due to the pressure of the atmosphere.

9. Make a U-tube manometer from tubing and fill with colored water to measure very low pressures.

10. **Pressure Conversion Worksheet**

Instructions

Using the Conversion Chart, perform the calculations on a clean sheet of paper. Turn in your completed calculations as per your instructor's guidelines. Ensure problem numbers and corresonding answers are clearly written.

Conversion Chart	
Convert PSIG to PSIA	PSIG + 14.7
Convert PSIA to PSIG	PSIA − 14.7
Convert PSIG to inches of H_2O	PSIG = 27.7
Convert PSIG to inches of Hg	PSIG = 2.04
Convert inches of H_2O to PSIG	$\dfrac{\text{height of liquid}}{27.7}$
Convert inches of Hg to PSIG	$\dfrac{\text{height of liquid}}{2.04}$

Use this conversion chart to complete the following conversion exercises.

Pressure Conversions:

1. 3.00 PSI = _____ in. w.c.
2. 3.61 PSI = _____ in. w.c.
3. 9.00 PSI = _____ in. w.c.
4. 11.0 PSI = _____ in. w.c.
5. 15.0 PSI = _____ in. w.c.
6. 1.00 PSI = _____ in. Hg
7. 10.2 PSI = _____ in. Hg
8. 15.8 PSI = _____ in. Hg
9. 100 PSI = _____ in. Hg
10. 500 PSI = _____ in. Hg

11. 100 in. w.c. = _____ PSI
12. 74.7 in. w.c. = _____ PSI
13. 23.2 in. w.c. = _____ PSI
14. 50.0 in. w.c. = _____ PSI
15. 300 in. w.c. = _____ PSI
16. 74.7 in. w.c. = _____ in. Hg
17. 100 in. w.c. = _____ in. Hg

18. 145 in. w.c. = _____ in. Hg
19. 235 in. w.c. = _____ in. Hg
20. 5.00 in. w.c. = _____ in. Hg

21. 4.47 in. Hg = _____ PSI
22. 9.35 in. Hg = _____ PSI
23. 15.7 in. Hg = _____ PSI
24. 44.7 in. Hg = _____ PSI
25. 29.9 in. Hg = _____ PSI
26. 2.00 in. Hg = _____ in. w.c.
27. 16.0 in. Hg = _____ in. w.c.
28. 24.2 in. Hg = _____ in. w.c.
29. 29.9 in. Hg = _____ in. w.c.
30. 100 in. Hg = _____ in. w.c.

PSIA to PSIG Conversions at Standard Atmospheric Pressure:

31. 4.00 PSIA = _____ PSIG
32. 14.7 PSIA = _____ PSIG
33. 22.2 PSIA = _____ PSIG
34. 37.7 PSIA = _____ PSIG
35. 50.0 PSIA = _____ PSIG
36. 0.00 PSIG = _____ PSIA
37. 21.0 PSIG = _____ PSIA
38. 35.4 PSIG = _____ PSIA
39. 45.0 PSIG = _____ PSIA
40. 50.0 PSIG = _____ PSIA

11. Vacuum Pressure Measurement: Laboratory Procedure

Background Information

Pressure is defined as a force per unit area. A common unit of pressure is pounds per square inch, or simply PSI. At sea level, atmospheric pressure averages 14.7 PSI on the absolute pressure scale and zero (0) PSI on the gauge pressure scale. Atmospheric pressure is dynamic, meaning that it changes according to weather conditions. If a vessel is blocked in, the internal pressure drops below atmosphere and the pressure across its walls exceeds its rated capacity, then extensive damage to the vessel can occur in the form of crushing. Most vessels are not designed to withstand significant pressure applied from the outside inward.

Consider the following example. A storage tank is prepared for maintenance by removing all process materials from the vessel and then steaming out the vessel. All connections on the tank are closed. Then, cold wet rain drenches the tank for several hours during the evening time, causing the steam in the tank to condense. When the day shift process technicians make their first rounds on the following morning, they find the tank has been crushed on one side causing extensive damage.

Materials Needed

- vacuum source
- copper tubing
- vacuum gauge or a PSIG gauge that reads negative pressure or a PSIA gauge
- a vessel to vacuum source connecting apparatus
- a gallon tin can or plastic liter bottle

    NOTE: The materials needed for this apparatus can be found at most hardware stores.

Safety Requirements

Safety glasses are required in the lab due to the potential dangers associated with imploding or exploding vessels.

Procedure

a. Connect the container to be tested to a vacuum source. If the vacuum source does not have a pressure indicator mounted to it, then you will have to connect one to the apparatus with a tubing tee.

b. Make sure the gauge is indicating zero before starting the experiment.

c. Slowly apply vacuum to the plastic container.

d. Record the reading at the initial wall collapse and then again when it is completely collapsed.

e. Apply the same amount of positive pressure to the plastic bottle.

    SAFETY TIP: There is a danger of explosion

f. Record your observations.

g. Do this to several containers if time allows.

h. Complete the final report for this lab.

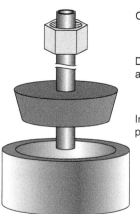

Connect this end to a vacuum pump

Drill a hole through the rubber stopper and insert the copper tubing

Insert the rubber stopper into the plastic bottle making a tight fit

Findings

Compile your observations and readings into a report that includes your recorded data, an explanation of the results of the experiment, and a conclusion. Include a paragraph explaining how a process technician could guard against the scenario provided in the background information.

Chapter 3
Process Variables, Elements, and Instruments: TEMPERATURE

 Objectives

After completing this chapter, you will be able to:

3.1 Explain temperature and the effects of heat energy on the movement of molecules. (NAPTA Process Variables . . . : Temperature 2*) p. 43

3.2 Calculate units of measure associated with temperature and temperature instruments.
differential (delta)
temperature scales
Fahrenheit
Celsius/Centigrade. (NAPTA Process Variables . . . : Temperature 1) p. 45

3.3 Illustrate common types, purpose, and operation of temperature sensing and measurement devices used in the process industries:
thermometer
bimetallic strip
resistance temperature detector (RTD)
thermocouple
thermistors

---

*North American Process Technology Alliance (NAPTA) developed curriculum to ensure that Process Technology courses will produce knowledgeable graduates to become entry-level employees in process technology. Objectives from that curriculum are named here in abbreviated form. For example, "(NAPTA Process Variables . . . : Temperature 2)" means that this chapter's objective relates to objective 2 of NAPTA's Process Variables, Elements, and Instruments content related to temperature measurement.

thermowell

temperature gauge. (NAPTA Process Variables . . . : Temperature 3, 4) p. 49

3.4 Using conversion formulas, complete Fahrenheit and Celsius conversions. (NAPTA Process Variables . . . : Temperature 5) p. 57

Key Terms

Bimetallic strip—two dissimilar strips of metal bonded together that expand and contract at different rates when exposed to temperature change, causing a rotating effect; used as the primary element in a temperature gauge or bimetallic thermometer, **p. 51.**

British thermal unit (BTU)—the amount of heat energy required to raise the temperature of 1 pound of water 1 degree Fahrenheit, **p. 44.**

Calorie—the amount of heat energy required to raise the temperature of 1 gram of water 1 degree Celsius (4.1868 joules), **p. 44.**

Celsius/centigrade—scale of measurement to determine temperature; freezing point of water is 0 degrees Celsius (C), and the boiling point of water is at 100 degrees C, **p. 47.**

Conduction—the transfer of heat through matter via vibrational motion; exchange media must be in direct contact, **p. 44.**

Convection—the transfer of heat through the circulation or movement of a liquid or a gas, **p. 44.**

Fahrenheit—scale of measurement to determine temperature; freezing point of water is 32 degrees Fahrenheit (F), and the boiling point of water is at 212 degrees F, **p. 47.**

Filled thermal system—a temperature sensing bulb filled with a liquid, vapor, or gas and connected by means of a capillary tube to a pressure measuring element, **p. 56.**

Glass stem thermometer—a temperature measuring device constructed of a glass bulb and a capillary (small) tube made of glass with a numbered scale; the bulb and the capillary tube contain a liquid such as mercury or colored alcohol that expands, allowing a reading to be made from the etched scale on the length of the tube; usually utilized in laboratory settings for calibration, **p. 49.**

Heat—added energy that causes an increase in the temperature of a material (sensible heat) or a phase change (latent heat); the transfer of kinetic energy to increase the temperature of a substance via the process of conduction, convection, or radiation, **p. 44.**

Infrared thermometer—a device that infers temperature using thermal radiation emitted by the substance being measured; also called *pyrometer*, **p. 50.**

Kelvin—scale of measurement that is sometimes called an *absolute scale* because 0 Kelvin (K) is the point at which no heat exists; freezing point of water is 273 K and boiling point of water is 373 K, **p. 47.**

Pyrometer—see infrared thermometer, **p. 50.**

Radiation—the transfer of heat energy through electromagnetic waves, **p. 44.**

Rankine—an absolute scale of measurement to determine temperature; freezing point of water is 492 degrees Rankine (R) and boiling point of water is 672 degrees R, **p. 47.**

Resistance temperature detector (RTD)—primary element that measures temperature changes in terms of electrical resistance, **p. 52.**

Temperature—the average kinetic energy of a material; the indication of heat energy available to flow between bodies of differing temperatures, **p. 43.**

Temperature differential (deltaT or ΔT)—difference between two related temperatures or between temperatures measured at two different points, **p. 46.**

Temperature gauge—an independent analog device with a sensing element such as a bimetallic strip, bourdon tube, or bellows that is linked to a pointer displaying the temperature on a calibrated face, **p. 56.**

Thermistor—a type of resistor used to measure temperature changes, relying on the change in its resistance with changing temperature, **p. 54.**

Thermocouple—primary element consisting of two wires of dissimilar metals connected at one end; when it is exposed to heat, it generates a voltage proportional to the change in temperature, **p. 53.**

Thermowell—a thick walled, typically stainless steel device shaped like a tube, which is inserted into a hole in piping or equipment; it is specifically prepared to house a temperature sensing and measuring element, **p. 55.**

3.1 Introduction

Industrial plants achieve the process of creating products by changing the temperature of various feedstocks in specialized process equipment. Understanding the nature of heat, heat exchange, and its effect on a process is critical to this success. This chapter describes temperature and heat energy as they relate to molecular movement. Boiling and freezing points on commonly used temperature measuring scales are discussed along with the various temperature measuring devices including their specific purpose and operation. Conversions between the various temperature measuring scales are also covered.

Temperature and Effects of Heat

Temperature is defined as the average kinetic energy of a material. Temperature is also the indication of heat energy available to flow between bodies of differing temperatures. The sun gives off heat energy (Figure 3.1) to its surrounding area and this heat radiates outward, flowing from extremely hot to what we feel here on earth on a daily basis.

Temperature the average kinetic energy of a material; the indication of heat energy available to flow between bodies of differing temperatures.

Figure 3.1 Temperature and heat.
CREDIT: VladisChern/Shutterstock.

As the temperature increases in a substance, its physical characteristics may also change. Water, with the increase of heat energy, becomes steam as the molecules increase in speed and overcome surface tension to vaporize (boiling point). Conversely, when taking heat away from water, the molecules slow down considerably and if enough heat is lost, the water turns to ice (freezing point).

This principle of physical characteristic change is used inside temperature measuring devices, such as the common thermometer, to determine the amount of temperature available in an external substance (e.g., performed by exposing the bulb of the thermometer to the process fluid). Inside a glass thermometer is a liquid that can be alcohol or mercury. When the liquid is heated, it immediately expands away from the bulb and into the connected capillary tube. The expansion of the liquid has been calibrated in the markings on the outside of the thermometer to indicate, or measure, the amount of heat energy (temperature) in the substance being measured. When the thermometer is taken out of the substance being measured, the measurement begins to move readily to the surrounding temperature. Temperature is determined by observing the markings etched on the glass and to where the liquid level has expanded.

Effects of Heat Energy on Molecular Movement

A temperature measurement is one that is taken at a *given point in time* with a specific measuring device that has been calibrated to a particular temperature scale. By comparison, heat is the amount of energy that flows *over time* between bodies of differing temperature. **Heat** always flows from a higher temperature to a lower temperature until temperatures are equal. Heat is also a measurable quantity.

Since heat is the energy that causes molecules to become more active, they move faster when heat is added and their temperature also increases. Because a temperature increase or decrease is dependent on the amount of heat energy moving from one substance to another, there is a direct relationship between the two. Units of heat measurement are based on this relationship. For example, a **British thermal unit (BTU)** is defined as the amount of heat required to raise the temperature of a pound of water 1 degree Fahrenheit. In the metric system, 1 **Calorie** is defined as the amount of heat required to raise the temperature of 1 gram of water 1 degree Celsius (4.1868 joules).

HEAT TRANSFER METHODS In process industries, heat transfer occurs in one of three distinct ways:

- Conduction
- Convection
- Radiation

Conduction is the transfer of heat through solid matter as it moves from one fixed molecule to the next. An example of this would be a common household iron. Heat generated by electrical energy moves through the soleplate of the iron by conduction. The soleplate does not change form or state, but the heat energy passes through the solid quite easily.

In process units, conduction occurs in a furnace in the solid walls of the furnace tubes. Hot flue gases, formed during the combustion of fuel gas in the burner area, flow upward in the furnace box and contact the metal furnace tubes carrying material. The heat energy is absorbed by the furnace tubes and is then transferred to the material passing through the tubes. The heat transfer process in the furnace tubes is conduction since the solid material does not change state but only passes the heat through itself.

Convection is the transfer of heat through the movement that occurs in a fluid when there is a difference in temperature from one region to another. The density differences between the hot and cold molecules cause this convection motion in fluids.

Using the same example of material moving through the tubes in a furnace, when the heat is transferred through the tubes (conduction) to the fluid contents of the tubes, the fluid transfers heat by convection and conduction throughout its volume.

Radiation is the transfer of heat by emitted radiant energy in the form of waves or particles. The sun is an example of a radiant energy source. The sun's heat energy travels through the vacuum of space.

Again using the furnace example, radiation occurs from the heat given off by the combustion process at the burners. Radiant heat flows from the burners to the outside walls of the piping and the furnace structure. The further the piping is from the burners, the less heat energy is transferred. This difference in the amount of heat energy is used advantageously as a measure for cost reduction and effectiveness in the use of burner fuel.

Heat added energy that causes an increase in the temperature of a material (sensible heat) or a phase change (latent heat); the transfer of kinetic energy to increase the temperature of a substance via the process of conduction, convection, or radiation.

British thermal unit (BTU) the amount of heat energy required to raise the temperature of 1 pound of water 1 degree Fahrenheit.

Calorie the amount of heat energy required to raise the temperature of 1 gram of water 1 degree Celsius (4.1868 joules).

Conduction the transfer of heat through matter via vibrational motion; exchange media must be in direct contact.

Convection the transfer of heat through the circulation or movement of a liquid or a gas.

Radiation the transfer of heat energy through electromagnetic waves.

HEAT TRANSFER: PHASE CHANGES Molecules within solids have little movement. The molecules have a force that tries to push them apart, and another force holding them together. When heat energy is applied, it increases the force pushing the molecules apart. As the temperature increases, the distance between the molecules in a solid also increases. The increased distance between molecules causes the volume within the solid to increase. The increased distance between molecules allows more room for the molecules to move. As the solid begins to melt, the molecules become active enough to lose much of their order and no longer have the strength between them to hold their shape against the force of gravity. When the substance is melting, adding heat energy does not raise the temperature. The temperature remains the same until all the substance melts.

When the solid has completely melted, the temperature of the liquid substance again increases. The increase in temperature increases the energy level of the molecules. They then begin to overcome the forces holding them together, and the liquid converts to a gas. At this point, all the thermal energy goes to increasing the energy of the molecules, causing no additional increase in temperature, and the liquid is boiling. Once the boiling liquid has completely changed phase to a gas and no liquid remains, then the continued addition of heat causes the temperature of the gas to rise. If temperature is increased further, the gas is called superheated. Superheated steam at high pressures is commonly used in process plants.

3.2 Temperature Measurement, Differential, and Scales

The concern for measuring temperature has been in the minds of humans for thousands of years. As early as 170 A.D., a Greek physician and writer by the name of Galen proposed a standard for temperature measurement. Galileo is credited with inventing the thermometer in 1592, but an accurate and repeatable mercury thermometer for fixed points on the low end of the temperature scale was not available until the early 1700s and was invented by Gabriel Fahrenheit (freezing point 32 degrees Fahrenheit and boiling point 212 degrees Fahrenheit). Then around 1742, Anders Celsius proposed that 0 should be used as the freezing point and 100 as the boiling point of water for his scale; therefore, the freezing point and boiling point of water became the benchmarks for both temperature scales.

In modern day industry, temperature measurement is extremely important to the operation and safety of plants due to its inherent relationship with pressure and volume. Remember the Ideal Gas Law in physics ($PV = nRT$ [where P is pressure, V is volume, n is the number of moles of gas, and R is the ideal gas constant]). Since R is a constant, if temperature is rising and volume is constant, then pressure must be increasing. Of course, liquids also expand when heated. Since process systems are contained in finite volume piping and vessels, process technicians have to pay close attention to the temperature, pressure, flow, and level indications that are displayed on their instruments. Each part of the process has specific operating parameters for these process variables (e.g., degrees Fahrenheit or Celsius for temperature).

Technicians work to ensure that process variables are maintained at specified values (both low and high temperatures) and do not exceed acceptable limits. If the temperature is rising in a (closed) gas process, then the pressure is rising somewhere else as well. If the temperature is rising and the pressure is not rising, there may be a path open to another volume. In the case of gas, the process is routed to a safe outlet such as a vent system, flare system, or the atmosphere if applicable. Technicians should be aware of any changes, but temperature readings should always stay at or close to desired values.

Temperature can be measured in the following ways:

- Single points
- Multiple points
- Multiple averaging points
- Temperature differential (delta T)

Temperature differential (deltaT or ΔT) difference between two related temperatures or between temperatures measured at two different points.

Single point measurement (Figure 3.2 top) is the most common. When the temperature of a fluid traveling through a line is needed, temperature can be measured. When the average temperature in a reactor zone is needed, a single temperature point may be installed at various points in the reactor. Single points can be observed independently or they can all be added together and an average calculated. When observing two different temperature points in a process, a **temperature differential (deltaT or ΔT)** measurement may be calculated. An example of a differential temperature measurement is the inlet versus the outlet temperature of a heat exchanger (see Figure 3.2 bottom).

Figure 3.2 Single point and differential (delta) temperature indicators.

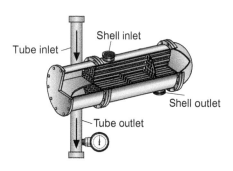

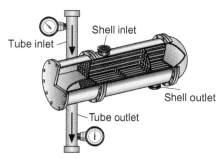

Temperature Measurement Scales

Each temperature measurement may be represented, or determined, on different scales (Figure 3.3). These scales include the following:

- Fahrenheit
- Celsius/centigrade
- Rankine
- Kelvin

Figure 3.3 Temperature measurement scale comparisons.

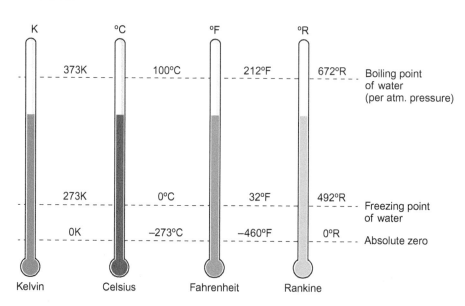

All of these four temperature scales are in use in the United States. The United States is the only community using Fahrenheit, while the rest of the world uses centigrade or its equivalent, Celsius. Kelvin is used worldwide. Absolute values of temperature (e.g., Rankine and Kelvin) must be used for Gas Law calculations.

FAHRENHEIT The **Fahrenheit** temperature scale is divided into 180 degrees between the freezing and boiling point of water. The freezing point of pure water is 32 degrees Fahrenheit and the boiling point of pure water is 212 degrees Fahrenheit at standard pressure. As a matter of interest, the temperature of 0 degrees Fahrenheit was the lowest attainable temperature under laboratory conditions in 1714, the year Fahrenheit introduced his thermometer.

CELSIUS/CENTIGRADE The **Celsius/centigrade** (C) scale is divided into 100 degrees between the freezing point and boiling point of pure water. On this scale, pure water freezes at 0 degrees C and boils at 100 degrees C at standard pressure.

KELVIN The **Kelvin** (K) scale is called an absolute scale because 0 K is the point at which theoretically, no heat exists. Theoretically there is no molecular motion and kinetic energy (KE = $\frac{1}{2}$ mv^2) is zero. The Kelvin scale uses the size of the Celsius degree unit. Water freezes at 273 K and boils at 373 K. Notice that neither the word "degree" nor a degree mark is used in front of the K. That is by standard practice.

RANKINE The **Rankine** scale (degrees R), an absolute scale, is the Fahrenheit equivalent of the Kelvin scale. It also has 0 as the point at which theoretically no heat exists. On this scale, the freezing point of pure water is 492 degrees R, and its boiling point is 672 degrees R.

Fahrenheit scale of measurement to determine temperature; freezing point of water is 32 degrees Fahrenheit (F), and the boiling point of water is at 212 degrees F.

Celsius/centigrade scale of measurement to determine temperature; freezing point of water is 0 degrees Celsius (C), and boiling point of water is at 100 degrees C.

Kelvin scale of measurement that is sometimes called an *absolute scale* because 0 Kelvin (K) is the point at which no heat exists; freezing point of water is 273 K, and boiling point of water is 373 K.

Rankine an absolute scale of measurement to determine temperature; freezing point of water is 492 degrees Rankine (R), and boiling point of water is 672 degrees R.

Boiling and Freezing Points of Water

All matter has a freezing point and a boiling point. Even though certain matter is thought of as eternally solid, that doesn't mean that the solid can't exist as a liquid or even as a gas. Rock is a solid, but rock exists in the earth's core as magma and flows from volcanoes as lava. Solid steel melts away at the tip of a cutting torch. In comparison, water exists in all three states: solid (ice), liquid (water), and gas (water vapor or steam).

The most important reason that water is used as the reference point for boiling and freezing is that pure water can be produced in almost any laboratory anywhere in the world. Historically, water was distilled to make it pure. This was easily accomplished hundreds of years ago. In addition, water has a relatively high thermal capacity that adds to its stability as a calibrating standard.

An interesting phenomenon occurs at the melting point of water and then again at its boiling point. At both of these points, as long as two states (ice and liquid or liquid and vapor) are present, a thermal equilibrium exists, stabilizing the temperature. The reason for this phenomenon is that as water or other materials change state, the latent (hidden) heat requirements have to be met before the sensible (measurable) heat can cause the temperature to change again. For example, if a thermometer is placed into a pot of water on a stovetop before turning the heat on, the thermometer should register ambient temperature. As the water heats up, its temperature changes steadily until it reaches the boiling point. At that point, the temperature becomes stable (doesn't change) until all the water boils off (vaporizes). The temperature reading would then begin to climb higher on the thermometer. In this example, where is all of the heat going during the state change of the water? The heat is going into the molecular structure of the substance (water) itself. This latent heat requirement to change a material's state makes the boiling and freezing points of a substance highly stable for calibration points.

A basic comparison of the temperature measuring scales is found in the boiling and freezing points (Table 3.1).

Table 3.1 Temperature Measurement Scale Comparisons at Atmospheric Pressure

Units	Absolute Zero	Melting/Freezing Point	Normal Boiling Point	Normal Room Temp	Normal Body Temp
°F	−460	32	212	77	98.6*
°C	−273	0	100	25	37
K	0	273	373	298	310
°R	0	492	672	569	591

* Average touch tolerance temperature for humans is 120 degrees Fahrenheit.

$$°R = °F + 460 \quad K = °C + 273 \quad (1.8)°C = °F - 32$$

Effects of Pressure and Contaminants on the Boiling Point of Water

At sea level, where pressure is 1 atm (14.7 PSIA or 760 mmHg), water boils at 212 degrees Fahrenheit. Water boils at a lower temperature at the top of a mountain where the atmospheric pressure is less. Figure 3.4 compares the boiling point of water at sea level to boiling points on two well-known mountain peaks.

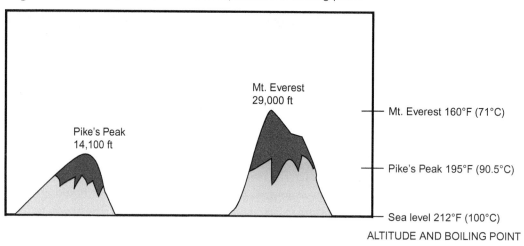

Figure 3.4 Atmospheric pressure comparisons of boiling point.

Water, fresh or salt, in lakes, rivers, oceans, etc., has other substances mixed in due to the solvent ability (solvency) of water. These substances affect the boiling and freezing points. They should be taken into consideration along with ambient temperatures when process operations depend on knowing the fluctuations in boiling and freezing points because solutes change the amount of energy the water needs to change states.

The boiling point of water is a function of pressure. Water at the top of Pike's Peak boils at 195 degrees Fahrenheit due to the lower atmospheric pressure (12 PSIA). Water in a boiler with a pressure of 150 PSIA boils at 358 degrees Fahrenheit.

3.3 Temperature Sensing and Measurement Instruments

While there are many places in a processing facility to measure temperature, there are a few common temperature sensing and measurement instruments. These devices include the following:

- Thermometer
- Bimetallic strip
- Resistance temperature device (RTD)
- Thermocouple
- Thermistor
- Temperature gauge

Thermometer

A thermometer is a self-contained instrument used to determine the temperature of a substance. Although the term thermometer includes all temperature sensing devices, the term usually implies a glass stem thermometer; however, two types of thermometers are discussed in this chapter: glass stem and infrared.

GLASS STEM A **glass stem thermometer** is constructed of a bulb and a capillary tube (Figure 3.5) with a numbered scale. The bulb and the capillary tube contain a liquid such as mercury or colored alcohol. The liquid in the sensing bulb responds to temperature changes by expanding or contracting. Since the bulb is connected to the capillary tubing, the additional liquid moves upward into the tube until the volume change has stopped. Volume expansion or contraction is directly proportional to the temperature increase or decrease. A temperature reading is taken directly from the etched calibration marks that flank the capillary tube.

To read a thermometer, view it so the top of the liquid column (meniscus) is at eye level (see Figure 3.5). Allowing a sufficient time period for the thermometer to reach a steady condition is also important. Getting an accurate reading requires process circulation to be adequate to represent a true picture of the process temperature.

Glass stem thermometer a temperature measuring device constructed of a glass bulb and a capillary (small) tube made of glass with a numbered scale; the bulb and the capillary tube contain a liquid such as mercury or colored alcohol that expands, allowing a reading to be made from the etched scale on the length of the tube; usually utilized in laboratory settings for calibration.

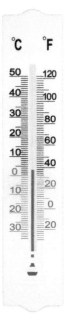

Figure 3.5 Celsius and Fahrenheit thermometer.

CREDIT: Tomas Ragina/Shutterstock.

Liquid-in-glass thermometers are usually found in indoor applications such as laboratories or on indoor process equipment. For measuring temperatures while gauging or sampling a storage tank, the liquid-in-glass thermometer may be placed in a holder that has a cup at the bottom to hold process fluids against the bulb until a reading can be taken. This apparatus, including the thermometer as lowered into the tank, helps ensure the temperature reading is accurate.

Mercury filled thermometers, when broken, require a stringent cleanup procedure as mandated by the EPA. Therefore, these types of thermometers are not generally found in process areas.

INFRARED An **infrared thermometer (or pyrometer)**, also called a laser thermometer, is used where a glass thermometer would not be practical. For example, an infrared device would be used for checking the surface temperature of furnace tubes where the equipment component is extremely hot. Since all objects having a temperature above absolute zero continuously emit energy by radiation in the form of electromagnetic waves or by conduction or convection, this energy can be measured and temperature determined. Infrared (IR) radiation means thermal energy, and it is the part of the electromagnetic spectrum that runs from high energy gamma and X-rays to long wavelength radio waves. IR wavelengths are usually 1 μm (micrometer) to 1 mm (Figure 3.6).

Infrared thermometer a device that infers temperature using thermal radiation emitted by the substance being measured; also called *pyrometer*.

Pyrometer infrared thermometer.

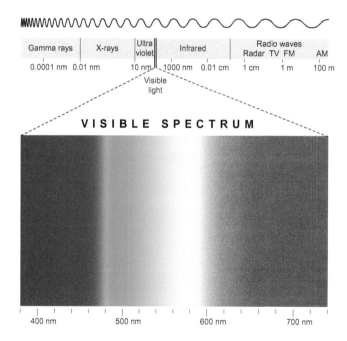

Figure 3.6 Electromagnetic spectra.

CREDIT: Peter Hermes Furian/Shutterstock.

When measuring the IR radiation, the hotter the object, the more the peak energy shifts to shorter wavelengths. The area under the radiation curve is the total energy emitted. IR radiation behaves similarly to light and can travel at or near the speed of light. IR radiation can also be reflected, refracted, absorbed, and emitted. Generally, the amount of IR energy an object radiates is more convenient to measure than temperature; then one can infer the temperature of the object. An IR thermometer can calculate the temperature of the object if the emissivity is known. *Emissivity* is a measure of the blackness of an object or how much it resembles a blackbody. (A *blackbody* is an ideal object that absorbs all radiation that hits it, so it is also a perfect emitter.)

A spot IR thermometer is noncontact and measures a spot on a target. The object being measured should be twice the size of the IR measurement beam or spot being applied to its surface. This device can be used to measure as low as −50 degrees Fahrenheit to as high as 6500 degrees Fahrenheit. This type of IR thermometer can be used to profile oven, furnace, and heater temperatures. Line scanners are also available that measure a process line by breaking it up into several spots. Both spot and line thermometers can be used on moving targets, while thermal imagers are generally used for stationary targets. Thermal imagers provide information on an entire target area and newer versions act much like night vision glasses by providing scanning capability of an outfall pond or pipe rack.

Another element or type of noncontact (human) temperature measuring device is the change-of-state sensor in which dots change color when temperature changes.

Silicon diode sensors are used to measure temperatures in the cryogenic range. Applications include the space industry, certain lab monitoring activities, or elements in the liquid state that normally occur as a gas. The cryogenic temperature range is defined as from −150 degrees Centigrade (−238 degrees Fahrenheit) to absolute zero (−273 degrees Centigrade or −460 degrees Fahrenheit), the temperature at which molecular motion comes as close as theoretically possible to ceasing completely.

Bimetallic Strip

A **bimetallic strip** consists of two dissimilar strips of metal that are bonded together. The operating principle of this type of temperature sensing and measuring device is based on choosing two dissimilar metals that have the attribute of easily expanding and contracting at different rates with the addition or subtraction of heat. The expression used to describe this effect is referred to as the *coefficient of thermal expansion*. The bimetallic element is often shaped into a spiral or helix for compactness.

Since all materials react to temperature changes at differing rates, the two metals in the bimetallic strip must have significantly different rates of expansion and contraction when exposed to temperature changes, which causes a bend or rotation to occur on the strip. When heated, the thermal expansion of the metal with the highest coefficient actually bends the bar in the direction of the other lower coefficient metal. If a short length of a compound bar (bimetallic strip) can bend enough to be measured, then a longer length of the same bonded metals could bend much more (Figure 3.7).

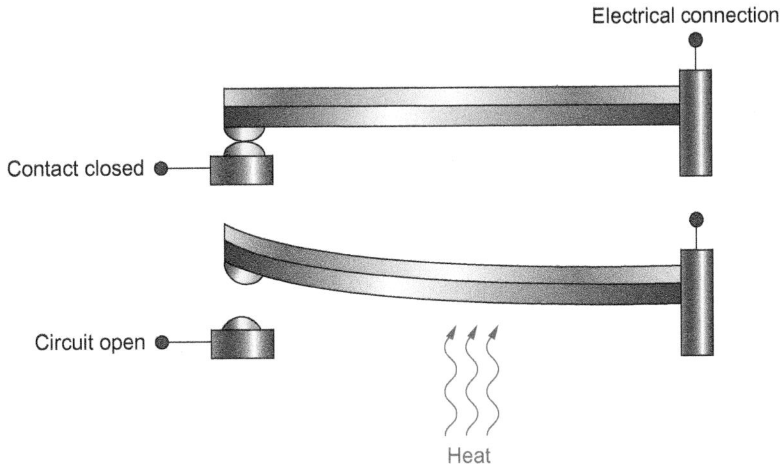

Bimetallic strip two dissimilar strips of metal bonded together that expand and contract at different rates when exposed to temperature change, causing a rotating effect; used as the primary element in a temperature gauge or bimetallic thermometer.

Figure 3.7 Bimetallic elements and effects of temperature change.

If a long bimetallic strip is coiled into a tight helix and placed within a metal tube where one end of the coil is attached to an indicator pen, then the expansion and contraction of the strip would move an indicator to represent how much temperature is being measured. The spiral is a flat shape much like an electric stove burner. This device is called a *bimetallic thermometer*. The bimetallic strip can be used in other applications such as temperature switches and thermal motor overloads where the strip's bending effect during temperature change causes a mechanical action in the form of a tripped switch at high temperature. It can also be used to shut off power to a motor by opening the electrical circuit at thermal overload.

Dial thermometers (Figure 3.8) are the most common thermometers in industrial use, and they are frequently of the bimetallic type with circular dials. Filled systems are also available. The dials come in a wide range of temperature scales and styles.

Figure 3.8 Dial type thermometer (bimetallic helix coil).

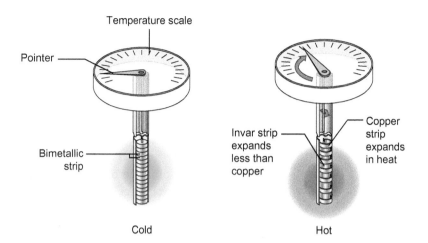

Thermometers, placed in thermowells and used with process equipment located in outdoor facilities, are generally bimetallic in nature due to their material structure and their compatibility with processes. They are reliable in regard to temperature measurement and are cost effective.

Resistance Temperature Detector (RTD)

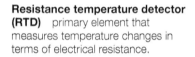

Resistance temperature detector (RTD) primary element that measures temperature changes in terms of electrical resistance.

A **resistance temperature detector (RTD)** is a primary element (calibrated resistor) that measures temperature changes in terms of electrical resistance (Figure 3.9). RTDs are mounted or inserted in thermowells like other thermometers, and they are second only to thermocouples as the most common electronic temperature sensing element in industry. RTDs are more linear than thermocouples and generally more accurate, but they lack the operating range of a thermocouple. The standard temperature range of an RTD is from −200 to 900 degrees Fahrenheit. Special ranges go as high as 1475 degrees Fahrenheit.

Figure 3.9 Resistance temperature detector (RTD) devices.

CREDIT: © Emerson 2019.

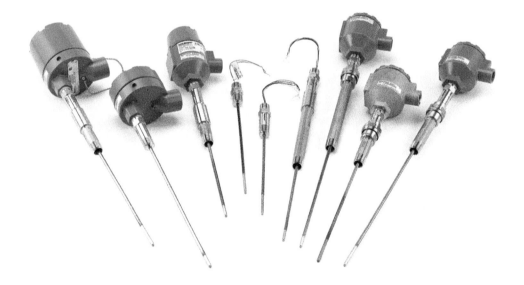

The fundamental operating principle of an RTD is that as the temperature of a conductor changes, so does its resistance. The most common RTD is no more than a fine platinum wire wound around a mandrel of nonconductive material. The mass of the wound wire is small and allows quick response to temperature changes.

All conductors respond to changes in temperature. For example, the resistance of slightly doped (where a minute amount of an impurity is added to a pure substance to alter its properties) industrial standard platinum changes by 0.00385 ohms/degrees C. What this means is that a platinum wire with a resistance of 1.0 ohm would change by 0.00385 ohms for every 1 degree Celsius change in temperature. If a platinum wire were long enough to have a resistance of 100 ohms, the wire would change by 0.385 ohms per degree C. This resistance-to-temperature relationship is known as its temperature coefficient of resistance. Other metals such as nickel, iron, and copper are also used as RTDs.

The most commonly used RTD in industry is the 100 ohm platinum type. This type is called a 100 ohm RTD because its resistance at 0 degrees C is 100 ohms. Since this RTD has a resistance change of 0.385 ohms per 1 degree C, then at the boiling point of water (100 degrees C) the resistance would be 138.5 ohms (100 ohms + 38.5 ohms).

$$\text{RTD resistance} = \text{base} + \text{delta °C}$$

In processing industries, many RTDs are connected to transmitters. The transmitter responds to a change in resistance from the RTD and produces a corresponding instrument signal. Since the transmitter cannot distinguish resistance produced by the RTD from the resistance added to it by the lead wires connecting them together, lead wire compensation must be provided. The transmitter normally does this compensation for us by providing input from a third or even fourth wire. These additional wires provide a solution to the lead wire problem by allowing the resistance of the lead wires to be measured and subtracted from the total sensed resistance. By eliminating the lead wire resistance, only the resistance of the RTD is available for processing.

Thermocouples

Thermocouples (Figure 3.10) are the most commonly used, and the simplest, electrical temperature sensing elements in industry today. A thermocouple may be placed in a thermowell and used to sense temperature in various locations such as furnace coil outlets, boiler tubes, steam lines, and on compressor suction and discharge lines. Thermocouples consist of two dissimilar metals joined together at one end such that when heated, they generate a small voltage that can be measured at the other junction. When the junction is exposed to temperature, a voltage (millivolts) is generated proportional to that temperature. This tiny generated voltage is reasonably linear to the temperature difference between the two junctions, making the thermocouple a good temperature sensing element.

Thermocouple primary element consisting of two wires of dissimilar metals connected at one end; when it is exposed to heat, it generates a voltage proportional to the change in temperature.

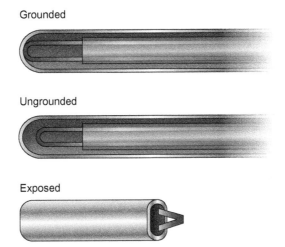

Figure 3.10 Thermocouple principle.

Thermocouple wires are color coded so that a technician can readily identify them according to their type. For example, a type J thermocouple has a white positive conductor and a red negative conductor. A type K thermocouple has a yellow positive conductor and a red negative conductor. The industry standard for thermocouple wire is to color the negative lead either red or a shade of red.

Thermocouples generate an electromotive force (EMF), or voltage, in accordance with the Seebeck effect. In 1821, Thomas J. Seebeck found that an electric current would flow in a circuit made of two dissimilar metals as long as the junctions of the metals were maintained at different temperatures.

The thermocouple has two junctions. The *measured junction* is connected to the process. The reference junction is connected to the transmitting and/or readout device. The temperature difference between the two junctions is proportional to the process temperature. If other wires are connected to the cold junction, these wires cause a false reading. In these cases, compensations must be made. This is called *cold junction compensation*.

Thermocouple junctions, wires, and welds are protected by the primary and secondary connection tubes of the thermowell.

Most thermocouples used today are prefabricated and sold or purchased in large quantities. Prefabricated thermocouples are butt-welded. Butt welding is popular because it provides a joint of minimum mass and yet provides good physical strength.

The specifics for providing a good method for cold compensation is beyond the scope of this text, but all the methods available must determine the temperature at the cold junction so that the temperature differences between the junctions can reflect an actual temperature measurement rather than a delta temperature. There may be other temperature measurement devices available that would be easier to use, but thermocouples are usually the best option since they have the widest temperature measuring ranges of all the temperature measuring elements. This fact alone makes selecting them more desirable.

Thermocouples usually fail at the hot junction, the measuring point where the positive and negative wires are welded together. The two dissimilar metals are likely to fail at the bonded joint. This failure mode produces an open circuit with the voltage going to zero. The result will be an immediate change in the controlled temperature reading which would cause the process equipment to operate incorrectly. Example: A furnace coil outlet temperature suddenly begins to read, as a rapid increase or decrease causes the fuel gas to close or open respectively. This could lead to off specification product or could also shut down the furnace. If the indicated, or controlled, reading suddenly changes, an instrumentation technician should be contacted to check the loop for failed components, including the thermocouple.

Thermistors

Thermistor a type of resistor used to measure temperature changes, relying on the change in its resistance with changing temperature.

A **thermistor** is a very small ceramic resistor with a high temperature coefficient of resistance. Thermistors are usually made of a sintered (compacted) mixture of metallic oxides. Thermistors are sensitive to very small changes in temperature. For a given change of temperature, the resistance of a thermistor changes approximately 10 times as much as the resistance of a platinum RTD. When mounted in small thermowells, thermistors respond very quickly because of their small thermal mass.

Thermistors are usually made in the form of a tiny bead that is encased in glass. Disc thermistors (Figure 3.11) are also available. Since thermistors are not interchangeable,

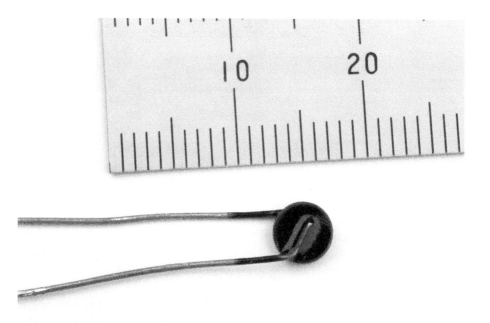

Figure 3.11 Disc thermistor.
CREDIT: David J. Green/Alamy Stock Photo.

temperature instruments must be calibrated to match the specific thermistor design. They are also noted for their long term drift due to aging. Accuracy and ambient temperature compensation are usually less than conventional temperature sensors.

A thermistor is a semiconductor that exhibits a large (usually negative) temperature coefficient of resistance. Thermistors are very sensitive and permit full scale operating ranges of less than 1 degree Fahrenheit. The upper operating temperature is determined by physical changes in the semiconductor material and is typically 200 to 750 degrees Fahrenheit (93 to 398 degrees Celsius).

Thermowell

Thermowells, while not actual temperature sensing and measuring devices, are protective devices that are used to isolate temperature sensing elements from the adverse conditions of the process. The design of the thermowell (Figure 3.12) also allows the actual sensing and measuring element to be removed from the process without having to shut down and isolate the equipment first. This is a very important advantage in continuous processing units.

Figure 3.12 Thermowell assembly.
CREDIT: © Emerson 2019.

A typical **thermowell** (Figures 3.13 A and B) is a thick walled (e.g., stainless steel) device shaped like a tube with threaded top. The device may be constructed of another type of material and may be attached by a flange or welded into place depending on the service where the thermowell is to be used.

Each thermowell, or protection well as it is called in some facilities, is constructed from a solid piece of bar stock that has been cut down on a lathe to its specified dimensions. A hole is then drilled from the top so that a temperature sensing element can be inserted. Where the thermowell is to be installed, a hole is drilled, and a Weldolet® taped with pipe threads is welded to the line or vessel. The thermowell is screwed into place, attached by a flange on a nozzle, or welded into place, depending on the service. The temperature sensing/measurement element is inserted into the thermowell and held in place by an attaching gland. A wiring terminal head may be screwed to the top of the thermowell to allow connection of the thermocouple wiring to the signal wiring so as to provide indication or control input.

Thermowells must be robust and corrosive resistant since they come directly into contact with the process. Good heat conducting characteristics through the proper selection of construction material (e.g., stainless steel) is required. If the process temperature is

Thermowell a thick walled, typically stainless steel device shaped like a tube, which is inserted into a hole in piping or equipment; it is specifically prepared to house a temperature sensing and measuring element.

Figure 3.13 **A.** Thermowell assembly (exploded view). **B.** Threaded thermowell sensor.

CREDIT: B. © Emerson 2019.

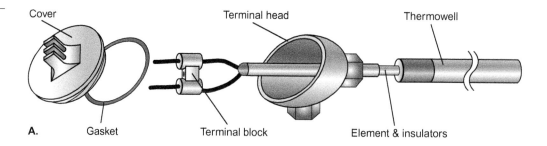

varying, there is a conductive time lag associated with the time taken for process heat to transfer through the thickness of the thermowell wall and to the actual element sensing the temperature. The advantages derived by having a thermowell usually outweigh any disadvantages or negative effects of the added response time. Several variations of the probe end of the thermowell have been designed to expedite the transfer of heat through the thermowell walls.

Temperature Gauge

Temperature gauge an independent analog device with a sensing element such as a bimetallic strip, bourdon tube, or bellows that is linked to a pointer displaying the temperature on a calibrated face.

A **temperature gauge** is generally described as an independent analog device with a sensing element such as a bimetallic strip, bourdon tube, or bellows linked to a pointer that subsequently displays changes in temperature on a calibrated face. Temperature gauges, like pressure gauges, are used in the processing area for local indication and can usually be seen from a distance. The gauge face and pointer provide an easy to read temperature indication from a distance. Temperature gauges may be used in an outside area as a secondary device for comparing temperature readings taken from thermocouples or RTDs.

Early industrial temperature gauges that may still be in use today were actually pressure gauges adapted to measure the pressure created by a *filled thermal system*. A **filled thermal system** (Figure 3.14) is a temperature sensing bulb filled with a liquid, vapor, or gas connected by means of a capillary tube to a pressure measuring element. This is where the term temperature gauge began and the faceplate was changed to reflect temperature units.

Filled thermal system a temperature sensing bulb filled with a liquid, vapor, or gas and connected by means of a capillary tube to a pressure measuring element.

The temperature bulb of a filled thermal system is usually inserted into a thermowell connected to the process. Filled thermal systems are closed. This means that changes in the process temperatures change the pressure in the volume of the fluid or gas within the temperature sensing bulb and directly affect the pressure measuring sensor in the gauge housing.

Temperature sensing devices such as bimetallic thermometers and any other temperature measuring system that can provide dial type, or analog, readouts are loosely referred to as temperature gauges. Common pocket type thermometers with a dial indicator are usually bimetallic. As was said, the measuring element in a bimetallic

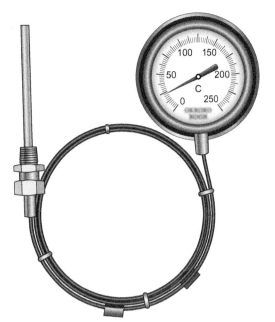

Figure 3.14 Filled thermal system.

CREDIT: © Emerson 2019.

thermometer is a helix-shaped bimetallic strip contained in a metallic sheathing (long slender tube capped at one end). As the temperature changes, the bimetallic coil twists in response. One end of the coil is attached to a shaft that is threaded down through the center of the bimetallic coil. The indicator pen is attached to the tip of the shaft and moves accordingly.

3.4 Temperature Conversion

Both the Fahrenheit and Celsius temperature scales are important since the Celsius scale is the preferred scale abroad and the Fahrenheit scale dominates the United States. The Fahrenheit scale has 180 units between the freezing point (32 degrees Fahrenheit) and the boiling point (212 degrees Fahrenheit) of water, whereas the Celsius scale has 100 units between freezing point and boiling point (0 degrees to 100 degrees Celsius). Both scales are linear, making them easier to convert from one scale to another, but since the zero value is different, another step is added.

When viewing the scales side by side and comparing the unit size, a difference can be seen. In actuality, the Celsius scale unit is 1.8 times larger than the Fahrenheit unit. The two scales also differ by 32 Fahrenheit units. This relationship is shown mathematically below.

$$(1.8)\,°C = °F - 32$$

The Kelvin scale is an absolute temperature scale that uses the Celsius degree unit. The freezing point of water is 273 K with the zero point on the scale equaling −273 degrees Celsius (see Figure 3.3). When writing Kelvin, the degree symbol (or the word) is not associated with the numerical value.

$$K = °C + 273\,K$$

The Rankine scale is similar to the Kelvin scale in that zero is absolute 0 in both scales. One degree Rankine is equal to one degree Fahrenheit. Conversion information for an example of degrees Kelvin to Rankine is on the following page.

$$R = (K) \times 1.8 \; or \; (K) \times 9/5$$

Conversion formulas are necessary in order to find the corresponding temperature on a different scale. In the following formulas, the scale to convert to is on the left side of the equation. This means that the known value is placed into the right side of the equation. To convert, place the known value into the equation and then use a calculator to do the math.

Converting Fahrenheit to Celsius

The following formula is used to convert Fahrenheit to Celsius:

$$°C = \frac{°F - 32}{1.8}$$

Example: Convert 68°F to °C.

$$°C = \frac{68°F - 32}{1.8}$$
$$°C = \frac{36}{1.8}$$
$$°C = 20$$

Converting Celsius to Fahrenheit

The following formula is used to convert Celsius to Fahrenheit:

$$°F = (1.8 \times °C) + 32$$

Example: Convert 40°C to °F.

$$°F = (1.8 \times 40°C) + 32$$
$$°F = (72) + 32$$
$$°F = 104$$

Converting Celsius to Kelvin

The following formula is used to convert Celsius to Kelvin:

$$K = °C + 273$$

Example: Convert 100°C to K.

$$K = 100°C + 273$$
$$K = 373$$

Converting Kelvin to Celsius

The following formula is used to convert Kelvin to Celsius:

$$°C = K - 273$$

Example: Convert 500 K to °C.

$$°C = 500 K - 273$$
$$°C = 227$$

Converting Kelvin to Rankine

The following formula is used to convert Kelvin to Rankine:

$$°R = (K) \times 9/5$$

Example: Convert 500 K to °R.

$$°R = (500) \times 9/5$$
$$°R = 900$$

Summary

In this chapter, temperature is described as the degree of hotness or coldness of a substance as indicated on a temperature measuring scale. The amount of heat (temperature) measured is the quantity of potential heat that could be given off from the substance. The giving off of heat equalizes with the surrounding substances until both are at the same temperature. When heat is added to a substance it may change state (solid → liquid → gas). The amount of heat required to change the state of a substance is called latent heat.

The effects of heat energy on molecular movement cause the molecules to become more active and to move faster. This increase in molecular movement also causes a corresponding temperature increase. Heat can be transferred from substance to substance by conduction, convection, and radiation.

Temperature measurements may be taken at single points, multiple averaging points, and/or places where differential (delta) temperatures are required. These measurements are read on temperature measurement scales with the most common ones being Fahrenheit (used almost exclusively in the United States), Celsius (used exclusively in other countries), and their corresponding absolute scales (Rankine and Kelvin, respectively). All four scales are based on the boiling and freezing points of pure water.

The boiling point for pure water is 100 degrees on a Celsius scale and 212 degrees on a Fahrenheit scale. The freezing point on a Celsius scale is 0 degrees, while the corresponding freezing point on a Fahrenheit scale is 32 degrees. When measuring temperatures at sea level, other things need to be considered when determining the appropriate temperature sensing device so that the appropriate scale may be applied, understood, and reported correctly for the environment.

The following are common types of temperature sensing, measurement, and supporting devices used in the process industries:

- *Thermowell*: a protective device usually constructed of stainless steel and used to isolate temperature sensing elements from adverse process conditions

- *Thermometer*: usually a glass stem or liquid-in-glass temperature measuring device where the liquid in the bulb has a much greater volume than the volume in the indicator and is inserted into the substance to be measured; the volume increases as the material in the bulb expands and moves upward in the indicator tube until it reaches a stable point where the value is read

- *Bimetallic strip*: two dissimilar metals bonded together so that when heat is applied one of the metals expands or contracts more readily than the other, causing a bend or rotation of the strip whereby the amount of movement is calibrated on a dial type readout of the measured temperature

- *Resistance temperature detector (RTD)*: a primary element connected to transmitters that measures temperature changes in terms of electrical resistance of a conducting material such as platinum

- *Thermocouple*: the most commonly used temperature sensing element in industry; it consists of two dissimilar metals joined together at one end so when they are heated, they generate a small voltage proportional to the temperature change that can be measured at the other junction

- *Thermistor*: a small ceramic resistor with a high temperature coefficient of resistance that is sensitive to very small changes in temperature

- *Temperature gauge*: an independent analog device with a sensing element such as a bimetallic strip, bourdon tube, or bellows linked to a pointer that displays temperature on a calibrated face; a filled thermal system may also be used as a temperature gauge in a closed system where internal pressure does not affect the volume of fluid or gas within the temperature sensing bulb.

The process technician must understand and be able to apply basic formulas for converting between the different temperature scales.

Checking Your Knowledge

1. Temperature is defined as the _____ kinetic energy of a material.
   a. total
   b. average
   c. maximum
   d. minimum

2. _____ is the manner in which heat moves through solid matter.
   a. Conduction
   b. Convection
   c. Radiation
   d. All of the above

3. Which temperature scale(s) contain(s) 100 degrees between the freezing point of water and the boiling point of water? (Select all that apply.)
   a. Centigrade/Celsius
   b. Fahrenheit
   c. Rankine
   d. Kelvin

4. Where would water boil at a lower temperature?
   a. In a sealed container
   b. At sea level
   c. At the top of a mountain
   d. Water always boils at the same temperature.

5. Which of the following is the most commonly used, and simplest, electrical temperature sensing element?
   a. Thermistor
   b. Thermowell
   c. Infrared
   d. Thermocouple

6. Calculate and select the conversion of 212 degrees F into degrees C.
   a. 0
   b. 32
   c. 100
   d. 373

7. Calculate and select the conversion of 60°C to Fahrenheit.
   a. 108
   b. 140
   c. 330
   d. 410

NOTE: Answers to Checking Your Knowledge questions are in the Appendix.

Student Activities

1. Assign students to determine the temperature sensing element/device found in at least four appliances, vehicles, or other types of equipment used on an everyday basis. Share findings in class discussion.

2. Using a dial readout (with a pointer) thermometer commonly used in outdoor gardens, perform the same activities as the liquid-in-glass thermometer. Compare your results. In a lab or simulator, look at a cutaway of the dial readout device and see how it was made.

3. Cut a circle out of common copy paper and then starting at an angle, cut a spiral around and around until the center is reached. Punch a small hole in the center. Tie a knot in the end of a piece of string and thread the other end through the hole in the spiral. Light a small candle and hold the spiral by the string over the candle taking care NOT to put the paper too close to the flame. Observe what happens to the spiral. Is this heat transfer radiation, convection, or conduction? How would this experiment apply to a plane flying over a refining facility?

4. Identify temperature sensing and measurement devices in the lab.

5. Have students use a multimeter and hook up leads to a thermocouple. Heat the thermocouple with a heat gun (the type used on carpet because it gets much hotter). Have students read the mV output on the meter. After the mV output is reached, have students use conversion tables to get from millivolts to temperature for the type of thermocouple used (J or A).

6. Use sling psychrometer to measure wet and dry bulb temperature to determine dew points and percentage of relative humidity.

7. **TEMPERATURE CONVERSION WORKSHEET**

   ### Instructions
   - Using the temperature conversion chart, perform the calculations on a clean sheet of paper.
   - Turn in your completed calculations as per your instructor's guidelines.

- Ensure problem numbers and corresponding answers are clearly written.
- Use the following conversion chart to complete the following conversion exercises:

Temperature Conversion	
What Is to Be Converted:	Formula to Be Used:
Fahrenheit to Celsius	°C = $\frac{°F - 32}{1.8}$
Celsius to Fahrenheit	°F = (1.8 × °C) + 32
Celsius to Kelvin	K = °C + 273
Kelvin to Celsius	°C = K − 273
Kelvin to Rankine	°R = K × 9/5

- Perform the following temperature conversions on a sheet of paper and then write an e-mail listing the problem number and your answer. Send it to your instructor.

Temperature Conversion Problems

1. 0°F = ___ °C
2. 32°F = ___ °C
3. 68°F = ___ °C
4. 100°F = ___ °C
5. 212°F = ___ °C
6. 0°C = ___ °F
7. 25°C = ___ °F
8. 37°C = ___ °F
9. 50°C = ___ °F
10. 200°C = ___ °F
11. 0°C = ___ K
12. −50°C = ___ K
13. 100°C = ___ K
14. 400 K = ___ °C
15. 50 K = ___ °C
15. 50 K = ___ °R

8. TEMPERATURE MEASUREMENT: Laboratory Procedure

Background Information

The student must be capable of connecting a digital multimeter to a thermocouple and then reading the millivolt scale appropriately.

A thermocouple is one of the most common temperature sensing elements in the process industry. Thermocouples are nothing more than two dissimilar wires joined (usually fused) at one end and connected to a measuring device at the other. A thermocouple operates according to the Seebeck Effect, which states that a voltage is produced when there is a temperature difference between two junctions of dissimilar metals in a circuit. The measuring junction is also called the *hot* junction and the reference junction is called the *cold* junction. Since the voltage (actually millivolt) output from the thermocouple is based upon the difference in temperature between the two junctions, the thermocouple actually measures differential temperature. If a thermocouple is to provide an actual temperature on a scale such as the Fahrenheit or Celsius, the reference junction temperature must be known. In fact, the conversion tables that are used to convert the raw mV signal generated from the thermocouple into a temperature is referenced to the freezing point of water (also known as icepoint). Modern electronic transmitters have the ability to compensate for the temperature difference between the real reference junction temperature and true ice point, whereas the original thermocouple work was based upon dipping the *cold* junction into ice water. In this lab, there is no luxury of ice point compensation so ice water is used instead.

Another useful bit of information is that thermocouple wires are color coded so that a technician can readily identify them according to their type. For example, a type J thermocouple has a white positive conductor and a red negative conductor. A type K thermocouple has a yellow positive conductor and a red negative conductor. The industry standard for thermocouple wire is to color the negative lead either red or a shade of red.

Materials Needed

- type J thermocouple or type J extension wire
- ice
- 250 mL beaker (minimum size)
- hair dryer or other regulated heat source
- digital multimeter with a millivolt selectable position or scale
- millivolt-to-temperature table for type J thermocouples
- glass thermometer

Safety Requirements

Students should always wear safety glasses in the lab.

Simplified Procedure

a. Identify the sensor or instrument.
b. Describe the function of the sensor or instrument.
c. Calculate an expected result of the sensor or instrument.
d. Conduct an experiment to verify the calculated results.

Detailed Procedure

To perform this lab, the following steps should be performed:

a. Connect a multimeter to the lead wires of a type J thermocouple as illustrated.

Record the mV reading: _____ mV

NOTE: Connect the ampmeter and the positive and negative thermocouple leads (see Additional Information below) according to their proper polarity.

b. Place the thermocouple into a well stirred beaker of mostly ice in water.

Record the mV reading: _____ mV

c. Determine the ice water temperature according to the conversion table.

Temperature derived from the conversion table: _____ °C

d. Compare this temperature to a glass thermometer reading ambient temperature.

Temp: _____ °C

e. Using a heat source such as a hair dryer, heat the thermocouple again. Wait until the reading has had time to stabilize, then record the mV reading.

Record the mV reading: _____ mV

f. Determine the temperature of the hot air generated by the hair dryer.

Hot air temperature: _____ °C

Additional Information

If manufactured thermocouples are not available, twisting the two conductors of thermocouple lead wire together works.

Findings

Write an observational conclusion to this lab. In this conclusion, be sure to explain what happens to a thermocouple when the hot and cold junctions are at equal temperatures and when they are not. Include an explanation about ice point and how it affects the raw millivolt reading taken by the digital multimeter. Discuss your findings with the class or your instructor according to given directions.

Chapter 4
Process Variables, Elements, and Instruments: LEVEL

 ## Objectives

After completing this chapter, you will be able to:

4.1 Classify terms associated with level and level instruments:

- level measurement
- innage
- ullage (outage)
- direct and indirect level measurement
- interface level
- meniscus
- liquid (hydrostatic) head pressure. (NAPTA Process Variables . . . : Level 1, 3*) p. 65

4.2 Predict the relationship between temperature and level measurement as it relates to the density and volume of liquid. (NAPTA Process Variables . . . : Level 3, 4) p. 71

4.3 Identify the most common types, purposes, and operation of level-sensing/measuring devices used in the process industries:

- gauge or sight glass
- float gauge
- tape gauge
- differential pressure (D/P) cell

---

*North American Process Technology Alliance (NAPTA) developed curriculum to ensure that Process Technology courses will produce knowledgeable graduates to become entry-level employees in process technology. Objectives from that curriculum are named here in abbreviated form. For example, "(NAPTA Process Variables . . . : Level 1, 3)" means that this chapter's objective relates to objectives 1 and 3 of NAPTA's Process Variables, Elements, and Instruments content related to level measurement.

bubbler

displacer

ultrasonic device

radar electromagnetic device

nuclear device

loaded cell. (NAPTA Process Variables . . . : Level 2) p. 73

Key Terms

Bubbler—a level measurement system used in open or vented containers that measures head pressure and that then converts the measurement to a level reading; this allows the measurement of head pressure in a liquid without the pressure sensor coming in contact with the process fluid, **p. 78.**

Continuous level measurement—monitoring of all level points in the tank from the 0 percent level (bottom of the measuring device) to 100 percent full; used in processes with control loop instrumentation to provide ongoing measurement, **p. 67.**

Decanting—a process for the separation of mixtures of immiscible liquids or of a liquid and a solid mixture such as a suspension, **p. 69.**

Depth—the distance from the surface of a liquid down to a zero reference point, **p. 66.**

Direct level measurement—measures the process variable directly in terms of itself (e.g., using a sight glass or dipstick), **p. 67.**

Displacer—a sealed cylindrically shaped tube that uses the principle of buoyancy to displace an amount of fluid equal to the container level, **p. 80.**

Flat-glass level gauge—there are three types of flat-glass level gauges: reflex, transparent, and welded pad; the heavy, thick-bodied glass promotes safety in high temperature and high-pressure hydrocarbon or steam service, **p. 74.**

Float gauge—a local measuring device that uses cables, pulleys, levers or any other mechanism to convey the position of a float to a liquid level, **p. 75.**

Height—a distance from a zero reference point to the surface, **p. 65.**

Inches of water column—a pressure measurement scale that can be used for liquid level in low pressure containers, generally open or vented to the atmosphere; accuracy is dependent on the liquid's specific gravity; the measurement can be converted to pounds per square inch (PSI), **p. 73.**

Indirect level measurement—measures another process variable (e.g., head pressure or weight) in order to determine level, **p. 68.**

Innage—the measurement from the bottom of a tank to the surface of the product, **p. 66.**

Interface level—the surface at which two immiscible fluids meet; the interface can be between two liquids, a liquid and a slurry, or a liquid and a foam, **p. 68.**

Level measurement—the act of establishing the height of a liquid surface in reference to a zero point, **p. 65.**

Liquid (hydrostatic) head pressure—the pressure exerted by the height of a column of liquid; the most common indirect level measurement, **p. 73.**

Load cell—a level measuring device consisting of a transducer that measures force or weight (called a strain gauge), which is bonded to a robust support column called a force beam; generally built into the supporting structure of a vessel, **p. 83.**

Meniscus—the curved upper surface of a column of liquid having either a convex or concave shape, **p. 69.**

Nuclear device—uses a tightly controlled gamma radiation source with a detector to inversely infer a tank's level; used when other technologies are unsuccessful;

located on the outside of a tank and impervious to the effects of adverse process conditions; common gamma radiation sources are the radioisotopes cobalt 60 and cesium 137, **p. 82.**

Percent level measurement—a measurement of level based upon percentage, with 0 percent level being the lowest measuring point and 100 percent level being the maximum measuring point, **p. 66.**

Point level measurement—level is measured at one distinct point in a tank, **p. 67.**

Pounds per square inch (PSI)—a pressure measurement scale often used in liquid level measurement, using the liquid's hydrostatic head pressure; also expressed as inches of water column, **p. 73.**

Reflex level gauge—a local measuring device that has a single flat-glass panel capable of refracting light off a prism like backside, creating a silvery contrast above the liquid level that allows it to be seen from a distance, **p. 74.**

Tape—a narrow strip of calibrated ribbon (steel) used to measure length, **p. 76.**

Tape gauge—a local level-measuring device consisting of a metal tape that has one end attached to an indicator and the other end attached to a float residing on top of the liquid level, **p. 76.**

Transparent level gauge—a local measuring device that has two transparent flat-glass panels, front and back. These flat-glass panels, in conjunction with the metal sides, form a vertical chamber where the process level can be seen and measured, **p. 75.**

Tubular type sight glass—a local measuring device that has a reflex or clear glass tube open into a vessel on top and bottom, **p. 74.**

Ullage (outage)—the measurement from the surface of the product to a reference point at the top of the tank, **p. 66.**

Ultrasonic and radar device—an accurate distance measuring instrument usually inserted into the top of a vessel; emits a pulse of energy that is reflected off the surface of the material back to the receiver; may be either a single unit (transponder) or two separate units (transmitter and receiver), **p. 81.**

Welded pad gauge—a local measuring device generally made of flat glass and integrally mounted to the vessel by either a welded or flanged connection with threaded pipe; extremely rugged to prevent vessel drainout, **p. 75.**

4.1 Introduction to Level and Terms

Level is the height or depth of a process material within a piece of process equipment. This chapter defines the process variable identified as level, as well as the most common terms associated with level measurement. Many types of process equipment have storage or running capacity amounts that require monitoring and controlling in order to ensure the process runs smoothly and meets production specifications. The most common types of level-measuring devices are discussed as to their respective purpose and operation. This is followed by a discussion of relationships among the variables of level, temperature, density, and volume, which brings everything together to show the effects of each variable on the others.

What Is Level?

Gravity causes the surface of an undisturbed liquid to be a flat level plane. In the process industry, the word *level* describes this plane in terms of its height above a reference point. Accordingly, **level measurement** can be defined as the act of establishing the height of a liquid surface in reference to a zero point.

Level measurements are usually concerned with quantifying the position of the surface of the liquid as compared to a reference point. The most common way to express level is in liquid height. **Height** is the distance (Figure 4.1) from a zero reference point

Level measurement the act of establishing the height of a liquid surface in reference to a zero point.

Height a distance from a zero reference point to the surface.

Depth the distance from the surface of a liquid down to a zero reference point.

in a container to the surface of the material. **Depth** is a distance measurement starting at the surface (Figure 4.1) of a liquid and then extends downward. Depth measurements are uncommon in industry, but the process technician should be able to differentiate between the depth of a liquid and the more common height measurement. In the United States, the actual height of a liquid in a tank is expressed in feet and/or inches (e.g., 10 feet 3 inches). Level measurements may also be expressed in feet with decimal subunits (e.g., 10.25 feet).

Figure 4.1 Height and depth.

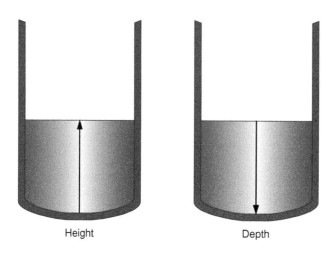

Percent level measurement a measurement of level based upon percentage with 0 percent level being the lowest measuring point and 100 percent level being the maximum measuring point.

The most common way to indicate the level of a tank is by **percent level measurement**. This measurement of level is based upon percentage with a zero percent level indicating empty, a 50 percent level indicating half full, and a 100 percent level indicating that the tank is full.

Innage and Ullage (Outage)

Process technicians should always be cognizant of tank levels. Tanks can be overfilled or underfilled. Tanks are often charged (filled) with several different materials and should be monitored accordingly. Depending upon the specific needs of a process, one tank may require a larger vapor space than another. Concerns such as these require knowledge of two terms that are used in and around industry that express this very concept. They are innage and ullage. **Innage** is the measurement from the bottom of the tank to the surface of the product and **ullage**, also known as **outage**, is the measurement from the surface of the product to the top of the vessel (Figure 4.2).

Innage the measurement from the bottom of a tank to the surface of the product.

Ullage (outage) the measurement from the surface of the product to a reference point at the top of the tank.

Figure 4.2 Innage/ullage (outage) measurement.

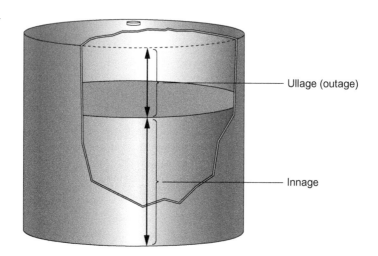

Level Measurement Categories and Methods

Level measurements are divided into two distinct categories: point level measurement and continuous level measurement (Figure 4.3).

- **Point level measurement:** Where level is measured at one distinct point in a tank. The most common reason for providing point measurements is to establish high and low level alarm points.

Point level measurement level is measured at one distinct point in a tank.

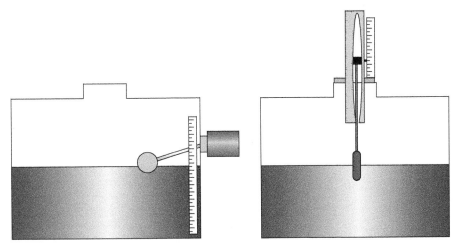

Figure 4.3 Point level and continuous level measurements.

- **Continuous level measurement:** Used to monitor and control all level points in the tank from 0 percent at the bottom of the measuring device to 100 percent full.

Continuous level measurement monitoring of all level points in the tank from the zero percent level (bottom of the measuring device) to 100 percent full; used in processes with control loop instrumentation to provide ongoing measurement.

The measurement terms *direct* and *indirect* are differentiated by how they actually measure a variable such as level. A **direct level measurement** (Figure 4.4) determines the process variable in terms of itself.

Direct level measurement measures the process variable directly in terms of itself (e.g., using a sight glass or dipstick).

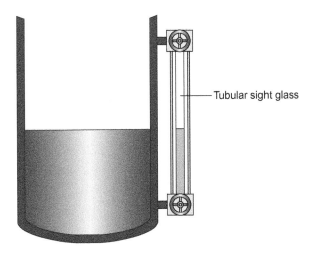

Figure 4.4 Direct level measurement.

The following are examples of how the direct level method may be applied to measure level:

- A float inside a storage tank tracks the surface level of the liquid contents by floating on top of the surface (a continuous level measurement).
- A sight glass on the side of a vessel (see Figure 4.4) that allows technicians to see the actual level as long as it is in the range (top to bottom) shown by the sight glass because liquid seeks its own level between two joined compartments (i.e., tank and sight glass).
- A dipstick, when inserted into a tank vertically until the end of the dipstick rests on the bottom (a point level measurement), is wetted from the bottom end of the dipstick to the top of the liquid surface. When the dipstick is raised, the actual measurement can be read by observing the place where the liquid stopped wetting the stick.

Indirect level measurement measures another process variable (e.g., head pressure or weight) in order to determine level.

Indirect level measurements are characterized by how they measure one property of the contained material such as head pressure or weight to infer another measurement, in this case the level. As shown in Figure 4.5, the pressure gauge at the bottom of the vessel indicates the total amount of head pressure. The greater the amount of pressure observed, the higher the level is in the tank. Load cells and displacers are examples of instruments from which indirect measurements are translated to levels. These devices are described in detail in the section Level Sensing and Measurement Instruments.

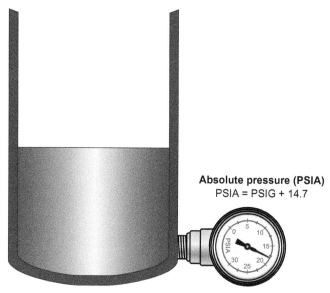

Figure 4.5 Indirect level measurement.

The term *datum*, or *datum line*, is used to describe the zero reference point for level measurement. For a dipstick, this is the bottom of the stick where the measurement starts at zero. For a sight glass, the zero reference point is the lowest point at which the scale can be read. For a transmitter, the zero reference point, or datum line, is the calibrated point at which the zero point is established.

Interface Level

Interface level the surface at which two immiscible fluids meet; the interface can be between two liquids, a liquid and a slurry, or a liquid and a foam.

The common boundary, or surface, between two immiscible (i.e., incapable of mixing) liquids is called an **interface level**. This interface (Figure 4.6) can be between two liquids, a liquid and a slurry, or a liquid and a foam. If the oil and water are combined in a clear jar

Figure 4.6 Interface in a vessel.

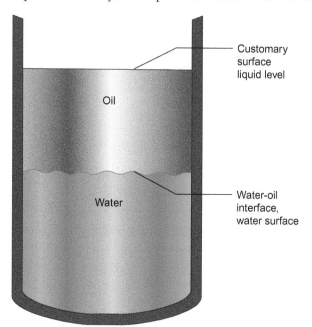

and then shaken, the two liquids almost immediately separate. The water (heavier liquid) naturally flows to the bottom of the jar and the oil moves to the top. Once the two liquids have settled, there are two surfaces in the jar. One surface is the water surface at the water and oil interface, and the other is the customary surface, the air and oil interface.

In industry, if separating immiscible liquids becomes necessary, the mixture would be transferred into a separating tank (Figure 4.7) where a process called **decanting** would take place. One example of equipment for the decanting operation is a separator where lighter (less dense) liquid spills over a weir (which is like a dam or holding plate) into a separate compartment for removal, while heavier (denser) liquid is removed in front of the weir through a pipe connected to the bottom of the tank. Keeping track of the liquids' interface ensures that the outflows, both top and bottom, can remain free of the other material.

Decanting a process for the separation of mixtures of immiscible liquids or of a liquid and a solid mixture such as a suspension.

Figure 4.7 Separating tank.

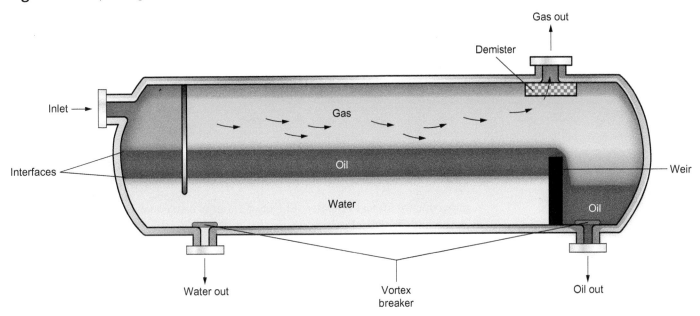

Meniscus

When taking direct measurements, understanding the concept of a meniscus is important. A **meniscus** (Figure 4.8) is the curved upper surface of a column of liquid. A meniscus can have a concave shape (swaying downward and read at the lowest portion of the curve) or a convex shape (bulging upward and read at the upper point of the curve). In either case, the

Meniscus the curved upper surface of a column of liquid having either a convex or concave shape.

Figure 4.8 Reading a meniscus: concave reading level *(left)* and convex reading level *(right)*.

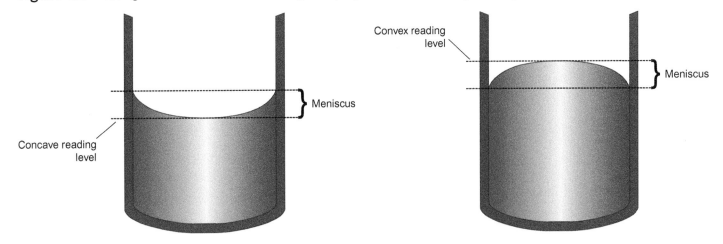

reading should be taken by lining up the high or low point of the meniscus curve with the visual indicator. The most common place to observe a meniscus is in a glass tube manometer or in a tubular sight glass. For many years, manometers were associated with storage tanks. Manometers are still in use today; however their numbers have diminished significantly.

Calculating Head Pressure

Head pressure can be calculated if the height of the liquid and its density are known. The density of a material may be converted to reflect its specific gravity. Remember that the specific gravity (SG) of a liquid is the ratio of its density to the density of water (standard reference). This makes calculating the head pressure of any liquid easy. As noted, one inch of water exerts a one-inch water column (w.c.) pressure. So, to calculate head pressure, take the height of a liquid in inches and multiply it by its specific gravity. Head pressure here is calculated in standard (imperial) units, not in metric units.

Formula for calculating head pressure:
Where:

$$P = \text{hydrostatic head pressure (inches w.c.)}$$
$$h = \text{level in inches}$$
$$SG = \text{specific gravity}$$
$$w.c. = \text{water column}$$
$$P = \frac{(1 \text{ in. w.c.})(h)(SG)}{\text{in.}}$$

Examples

- A tank contains 10 feet of water; 10 feet of water is also equal to 120 inches of water, expressed mathematically as:

$$10 \text{ ft} \times \frac{12 \text{ in.}}{1 \text{ ft}} = 120 \text{ in.}$$

How much head pressure is exerted by the full height of the liquid?

$$P = \frac{(1 \text{ in. w.c.})(120 \text{ in.})(1.0)}{\text{in.}} = 120 \text{ in. w.c.}$$

Note: This example incorporates the 1 inch water column pressure per 1 inch height factor. This constant can now be assumed for all future examples and word problems.

- An open top tank holds a liquid with a specific gravity of 1.350. The liquid has a vertical height of 230 in. How much head pressure (in inches water column) does this liquid exert?

$$P = (230 \text{ in.})(1.350)$$
$$P = 310.5 \text{ in. w.c.}$$

Calculating the Height of a Liquid

Using this same simple approach to solve for head pressure, solve for liquid height instead. To do this, the total head pressure and the density of the liquid must be known. Liquid height is equal to head pressure divided by its density ($h = P/SG$).

Formula for calculating height of a liquid:

$$P = \text{hydrostatic head pressure (inches w.c.)}$$
$$h = \text{level in inches w.c.}$$
$$SG = \text{specific gravity}$$
$$w.c. = \text{water column}$$

$$h = \frac{P}{SG}$$

$$h = \frac{(\text{head pressure})}{(\text{specific gravity})}$$

Examples

- A pressure gauge attached to the bottom of a tank (vented to atmosphere) is indicating 200 in. w.c. The specific gravity of the liquid in the tank is 1.25. What is the height of the liquid?

$$h = \frac{(200 \text{ in. w.c.})}{(1.25)}$$

$$h = 160 \text{ in.} = 13 \text{ ft } 4 \text{ in.}$$

- A pressure gauge attached to the bottom of a tank (vented to atmosphere) is indicating 450 in. w.c. The specific gravity of the liquid in the tank is 1.725. What is the height of the liquid?

$$h = \frac{450 \text{ in. w.c.}}{1.725}$$

$$h = 260.9 \text{ in. w.c.}$$

4.2 Level, Temperature, Density, and Volume

Level measurement was defined earlier as the act of establishing the height of a liquid surface in reference to a zero point. Recall also that when the mercury in a sealed glass bulb thermometer is heated, it expands up into the capillary tube. When the water in the radiator of a car heats up (Figure 4.9), the water volume increases as indicated by the HOT versus COLD line marks on the cooling system overflow reservoir. Fluids characteristically act this way.

Figure 4.9 Liquid level—cold versus hot.

Temperature (Figure 4.10) affects the density of all matter. The physical manifestation of this phenomenon is that a fixed amount of matter experiencing an increase in temperature requires more room because its volume increases proportionally. If its volume has increased and it still has the same mass, then its density must have changed to compensate for the new relationship. Recall that density is defined as mass per unit of volume. Again, if the fixed amount of mass occupies a larger volume, then its density has to decrease.

There are several ramifications to this physical phenomenon. For example, a tank is filled to its capacity and then heated, its contents may expand to a problematic level. Under normal

Figure 4.10 Liquid level before and after heating comparisons.

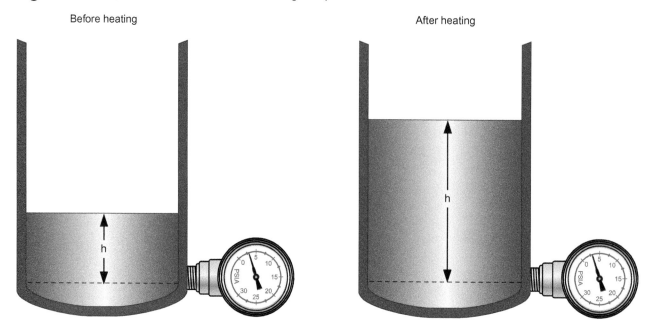

conditions, standard operating procedures would generally take this into account; however, a new process technician may not completely understand the ramifications of this action.

This overfill could also cause an instrumentation problem. Level measuring instruments, such as differential pressure instruments, rely on certain characteristics of the process remaining constant. Recall that head pressure is equal to the height of the liquid times its density. If the density of the liquid in the tank decreases, which means it occupies more volume, the liquid has to respond by expanding upward in the tank, increasing its height. The head pressure remains the same because head pressure is a function of the height of the liquid times its new density.

Density and Hydrostatic Head Pressure

Density is calculated as mass per unit volume. The density of a material may be converted to reflect its specific gravity. The specific gravity of a liquid is the ratio of the density of a liquid to the density of water for liquid measures.

Formula for calculating liquid (hydrostatic) head pressure:

$$P = \text{hydrostatic head pressure (inches w.c.)}$$
$$h = \text{level in inches}$$
$$SG = \text{specific gravity}$$
$$\text{w.c.} = \text{water column}$$
$$P = \frac{(1 \text{ in. w.c.})(h)(SG)}{\text{in.}}$$

Examples

- A tank contains 10 feet of water. 10 feet of water is also equal to 120 in. of water (10 feet × 12 in./foot = 120 in.). How much head pressure is exerted by the full height of the liquid?

$$P = \frac{(1 \text{ in. w.c.})(120 \text{ in.})(1.0)}{\text{in.}}$$

$$P = 120 \text{ in. w.c.}$$

This example incorporates the 1 inch water column pressure per 1 inch height factor. This is a constant that equates to "1" which is the specific gravity of water and is the standard reference for future examples and word problems.

- An open top tank holds a liquid with a specific gravity of 1.350. The liquid has a vertical height of 230 in. How much head pressure (in inches water column) does this liquid exert?

$$P = (h)(SG)$$
$$P = (230 \text{ in.})(1.350)$$
$$P = 310.5 \text{ in. w.c.}$$

In the process industries, most liquid level measurements are typically provided through indirect methods. The most common indirect level measurement method is **liquid (hydrostatic) head pressure**. Head pressure is the pressure exerted by a column height of liquid. Liquid head pressure (Figure 4.11) is usually expressed in **inches of water column** (in. w.c.) or in pounds per square inch (PSI).

Liquid (hydrostatic) head pressure the pressure exerted by the height of a column of liquid; the most common indirect level measurement.

Inches of water column a pressure measurement scale that can be used for liquid level in low pressure containers, generally open or vented to the atmosphere; accuracy is dependent on the liquid's specific gravity; the measurement can be converted to pounds per square inch (PSI).

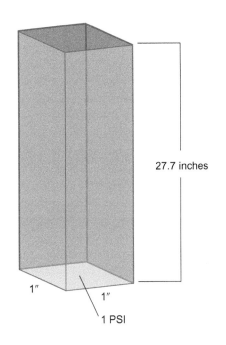

Figure 4.11 Liquid head pressure.

One inch of water column pressure (1 in. w.c.) is the amount of pressure exerted by the height of 1 in. of water. So, 2 in. of water exerts 2 in. water column pressure and 100 in. of water exerts 100 in. of water column pressure. As you can see, this pressure unit is self-defining and very easy to use when calculating head pressure.

As stated above, liquid head pressure is also expressed in the form of **pounds per square inch (PSI)**. If the height of the liquid and its density are known, one can calculate the equivalent head pressure in PSI. One foot of water exerts 0.433 PSI of head pressure. Hence, 10 feet of water exerts 4.33 PSI.

Pounds per square inch (PSI) a pressure measurement scale often used in liquid level measurement using the liquid's hydrostatic head pressure; also expressed as inches of water column.

4.3 Level Sensing and Measurement Instruments

The most common types of level sensing and measuring devices used in the process industry include the following:

- Gauge or sight glass
- Float gauge

- Tape gauge
- Differential pressure (D/P) cell
- Bubbler
- Displacer
- Ultrasonic device
- Nuclear device
- Load cell
- Radar electromagnetic device

Gauge or Sight Glass

A common sense law of physics states that a liquid seeks its own level in joined containers. Gauge glasses operate just like U-tube manometers. If they have an equal pressure applied to both sides, their levels are the same. Since gravity applies equally to both sides, if the height of the liquid is greater on one side, the height naturally adjusts by moving liquid between the two chambers. For the most part, gauge glasses are very accurate. However, phenomena exist that can cause a slight offset in column height between the two chambers. One such phenomenon is capillary action, and the other is an extreme temperature difference that causes the density of the material contained in the gauge glass (which is on the *outside* of the vessel) to be heavier or lighter.

Gauge glasses are among the oldest direct level measurement devices still in use today. The fundamental purpose of a gauge glass is to indicate level. Gauge glasses serve as a *window* into the process. This can be very helpful when troubleshooting the system. There are two basic types: the tubular glass type and the flat glass type. Also, as described below, there are several variations of the flat glass type.

The externally mounted tubular, reflex, and transparent gauge glass types should have block valves between their top and bottom connections, and the tank. In fact, most of these block valves are equipped with an internal ball check valve. The ball check valve seats off (i.e., closes) if there is an unusually large amount of flow through rate. These valves are added as a safety feature so that if the glass breaks, only a small amount of fluid is able to get out.

Tubular type sight glass a local measuring device that has a reflex or clear glass tube open into a vessel on top and bottom.

TUBULAR TYPE SIGHT GLASS The **tubular type sight glass** (see Figure 4.4) is the oldest type of gauge glass still in use today and serves as a window into the process. The tubular type sight glass has a reflex or clear glass tube and is open into a vessel on the top and the bottom. They are relatively inexpensive but fragile, which is why they are used less frequently. Tubular type sight glasses operate just like U-tube manometers; the sight glass level matches the liquid in the vessel. If they have equal pressure applied to both sides, their levels will be the same.

Flat-glass level gauge there are three types of flat-glass level gauges: reflex, transparent, and welded pad; the heavy, thick-bodied glass promotes safety in high temperature and high-pressure hydrocarbon or steam service.

Reflex level gauge a local measuring device that has a single flat-glass panel capable of refracting light off a prism like backside creating a silvery contrast above the liquid level that allows it to be seen from a distance.

FLAT GLASS LEVEL GAUGES **Flat glass level gauges** (Figure 4.12) are designed with heavy bodies and thick glass to promote safety in high temperature and high-pressure hydrocarbon service. Special valves are commonly used at the top and bottom of these gauges so the gauge glass can be cleaned and/or removed while material is still in the tank.

There are three types of flat-glass level gauges: reflex, transparent, and welded pad.

Reflex level gauges have a single flat-glass panel capable of refracting light off a prism like backside creating a silvery contrast above the liquid level. The level in a reflex gauge glass can be seen from a distance, which is helpful when making rounds in a darkened process area. Reflex gauges are not capable of distinguishing between immiscible liquids.

Figure 4.12 Flat-glass level gauge. **A.** Flat-glass level gauge structure. **B.** Gauge mounted to vessel.
CREDIT: B. Pakkalin Chitratorn/Shutterstock.

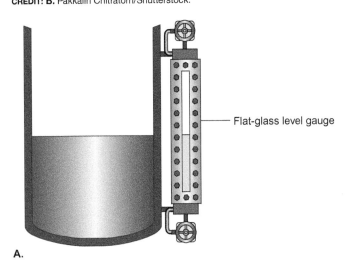

A. B.

Transparent level gauges have two transparent flat-glass panels, front and back, that, in conjunction with the metal sides, form a vertical chamber. The process level can be seen and measured using this vertical chamber. Transparent level gauges are not as easy to read as the reflex level gauge even though one can see right through them. However, transparent gauges are required to measure an interface level.

Welded pad gauges are generally made of flat glass, but they are integrally mounted to the vessel (Figure 4.12B). They can be welded, flanged, or connected with threaded pipe. This type must be extremely rugged because unlike the others, if the weld is compromised, all the material in the tank can drain out.

Floats and Float Gauges

A *float* is a level sensing element that tracks the surface of a liquid by floating on it. The float must be buoyant enough to float on the surface and heavy enough to provide the force necessary to actuate the level tracking mechanism.

A **float gauge** is an instrument that uses cables, pulleys, levers, or any other mechanism to convey the position of a float to a liquid level. Instruments that are called float gauges are usually point level devices, which can indicate, measure, or track a level. Most point level floats are used to actuate high/low level alarms and shutdown circuits.

Float level gauges operate on two simple principles: buoyancy and mechanical action. The float senses the surface of the liquid, and then a mechanism or transducer equates its position into a level measurement.

Floats can also directly actuate a control valve, similar to a simple toilet bowl float and valve mechanism. If you connect the industrial float to one end of a metal rod that pivots on a fulcrum point, then the other end can actually open and close a control valve.

Magnetic level indicators (Figure 4.13) consist of a chamber, a magnet equipped float which rises and lowers with the fluid level, and an indicator mounted to the chamber. The indicator houses a column of small flags, which indicate the level of the fluid in the chamber, based on the position of the float. As the fluid level rises and lowers, the float rises and lowers as well, and the flags are tripped from one orientation to the other; typically the red side indicates the liquid level and the silver side indicates the vapor space. As the float rises and falls with the process level, tripping the flags, it also stimulates any attached transmitters and switches, providing a signal back to the control system.

Transparent level gauge a local measuring device that has two transparent flat-glass panels, front and back. These flat-glass panels, in conjunction with the metal sides, form a vertical chamber where the process level can be seen and measured.

Welded pad gauge a local measuring device generally made of flat glass and integrally mounted to the vessel by either a welded or flanged connection with threaded pipe; extremely rugged to prevent vessel drainout.

Float gauge a local measuring device that uses cables, pulleys, levers or any other mechanism to convey the position of a float to a liquid level.

Figure 4.13 Magnetic level indicator.

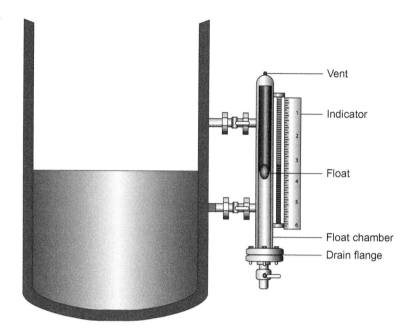

In a continuous level measuring instrument, the float and the following device are also equipped with a magnet so that they stay magnetically coupled with each other. As the float rises and falls with the surface of the liquid, the magnetic follower moves with it. The following mechanism is isolated from the process fluid by a nonmagnetic housing. A cable to a readout device is connected to the magnetic follower that can then provide a level indication.

Although there are applications where a direct float-operated continuous measuring device is found, most of them fall under the category of tape gauges and are discussed below.

Tapes and Tape Gauges

A **tape** is usually defined as a narrow calibrated strip or ribbon (steel) used to measure length. A **tape gauge** is a level-measuring device consisting of a metal tape that has one end attached to an indicator and the other end attached to a float.

Metal tape gauges are generally used to measure continuous level changes in a tank. Specifically, metal tapes are used because they do not stretch under normal conditions. A taut tape provides an absolute distance measurement between the surface of the level and the indicator mechanism. Consider this: A storage tank with a diameter of 50 feet has 1,224 gallons of material per inch of level. If the material in this tank were to sell for several dollars per gallon, the accuracy of this level gauge would be critical.

Float and gauge board devices (Figure 4.14) are simple in construction and function. This gauge is comprised of a float that rides on the surface of the liquid, a connecting metal tape (or cable) and an indicator board. Notice that the indicator board is numerically inverted because the indicator is attached to the float by means of a specific length of tape. As the level in the tank decreases, the float rides down with the level, pulling the indicator upward on the board. Conversely, as the level increases, the indicator slides downward toward the 100 percent mark on the indicator board.

As the level increases, the more sophisticated tape gauges rely on a spring assisted reeling action to take up the slack in the tape instead of a counterweight. The indicators in these mechanisms can possess a digital readout instead of an indicator board. In either case, the tape gauge works the same. As the level changes, the length of the tape is carefully measured, providing an accurate level measurement.

Tape gauges can also measure solids in a vessel. Instead of the float constantly sitting on the surface, a plumb bob or sounder can be lowered from a drum and sensor mechanism located on the top of the tank to the surface of the solid. The drum is connected to a

Tape a narrow strip of calibrated ribbon (steel) used to measure length.

Tape gauge a local level-measuring device consisting of a metal tape that has one end attached to an indicator and the other end attached to a float residing on top of the liquid level.

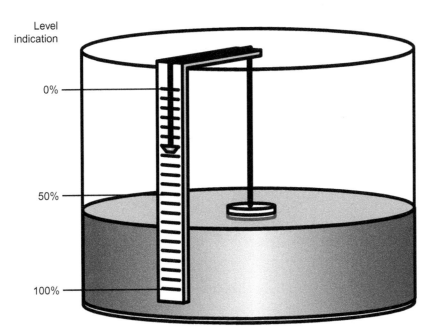

Figure 4.14 Tape and tape gauge.

reversible servo motor (a motor specially designed for high dynamics) and torque sensor. When the sensing bob hits the surface, the torque sensor notes the spot and the reversible motor rewinds the cable and sensing bob back to the top of the vessel.

Differential Pressure (D/P) Cells

In a previous chapter, a differential pressure cell (Figure 4.15) was described as a special type of pressure sensor that simultaneously measures two different pressure points in reference to each other and then produces a corresponding output signal (e.g., for tank level). One of the two inputs to the D/P cell is considered as the high side pressure input and the other as the low-side pressure input.

Figure 4.15 Differential pressure (D/P) cells.

CREDIT: © Emerson 2019.

When a D/P transmitter is used to measure level (Figure 4.16), the high pressure side of the D/P cell is connected to the bottom of the tank and the low pressure side is connected to the top of the tank. The high side measures total pressure that is a combined head pressure of the liquid plus any pressure applied to its surface. The low side measures only vapor

Figure 4.16 Differential pressure (D/P) cell (tank).

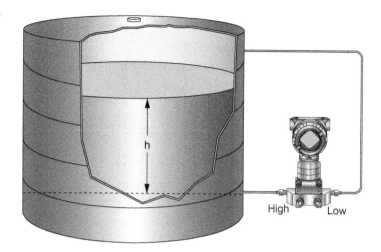

space pressure. Since a D/P cell measures the difference between two pressures, any pressure applied to both sides of the cell cancels out. This effectively eliminates the vapor space pressure from the measurement. With the vapor space pressure eliminated, the only pressure remaining is the head pressure exerted by the liquid.

The low pressure leg may be filled with a reference fluid to compensate for tank vapor with properties that may allow it to condense and fill the leg, altering its ability to read correctly. See Figure 4.17, a D/P cell level transmitter.

Figure 4.17 A D/P cell level transmitter. **A.** Dry leg. **B.** Wet leg. **C.** Atmospheric leg.

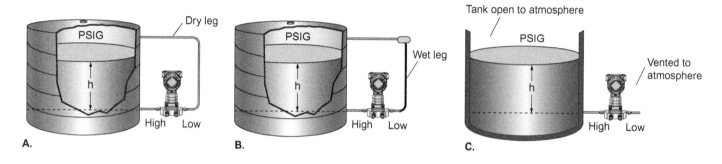

Note that a vessel indicates empty if the liquid level is below the lower tap and below the measuring range of the level instrument. The zero range suppression is required to indicate the correct value. See an example of this in Figure 4.18.

Head pressure is the most common variable used to determine level in industry. Recall that head pressure is the pressure exerted by a column height of liquid. If all tanks were open to atmosphere, then measuring head pressure could be done with a simple pressure-sensing instrument. However, in reality, almost every tank in industry is a closed environment where a certain amount of pressure is trapped above the liquid in the vapor space. As the temperature (or level) changes, the vapor space pressure also changes. The D/P cell (transmitter) works best in this type of situation because of its ability to measure the head pressure produced by the liquid in reference to the varying pressure in the vapor space.

Bubblers

A **bubbler** is a special kind of head pressure measuring method used on nonflammable processes. It allows measurement of the head pressure of a liquid without the pressure sensor coming in contact with the process fluid. Corrosive properties or other potentially

Bubbler a level measurement system used in open or vented containers that measures head pressure and that then converts the measurement to a level reading; this allows the measurement of head pressure in a liquid without the pressure sensor coming in contact with the process fluid.

Figure 4.18 A D/P cell level transmitter—zero range suppression.

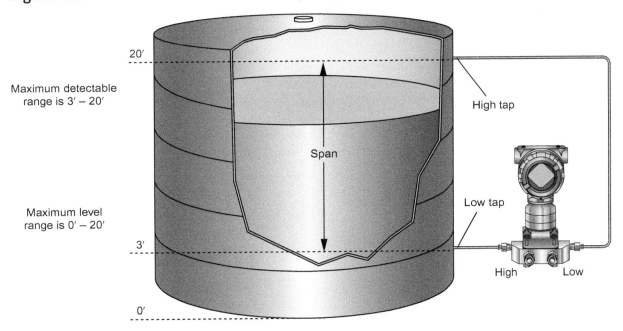

problematic physical properties of the process fluid can be lessened or eliminated by using a bubbler system instead of a directly connected sensing device. A bubbler consists of a purging gas source, pressure and flow regulating device(s), a pressure-sensing device, and an open-ended tube, called a dip tube (Figure 4.19).

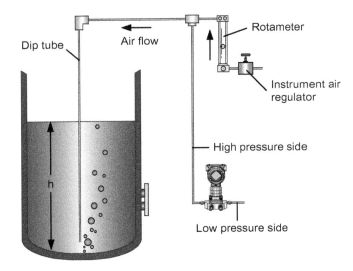

Figure 4.19 Bubbler.

The dip tube is usually extended downward from the top of the tank to about an inch or two (2.5–5 cm) above the bottom. Inert gases, such as nitrogen (N_2), are supplied to this otherwise closed tube system so that it flows very slowly down the tube and escapes out of the open end of the tube into the fluid, creating bubbles. The pressure of the gas in the tube must equal the total pressure exerted by the liquid at the end of the tube if it is to bubble out. Also, the diameter of the dip tube must be large enough so that the pressure throughout the tube is the same. A D/P transmitter or other pressure-sensing device is connected to the dip tube providing a measurement of the back pressure. The transmitter is calibrated so that its output represents the measured pressure as a level measurement.

Displacer a sealed cylindrically shaped tube that uses the principle of buoyancy to displace an amount of fluid equal to the container level.

Displacer and Transmitter

Displacers are among the oldest level sensing devices in industry. A **displacer** is a sealed cylindrically shaped tube that uses the principle of buoyancy (Archimedes' Principle). *Buoyancy* is defined as an upward force on a submerged body that is equal to the weight of the displaced fluid. The operating principle of a displacer (Figure 4.20) level device is based upon Archimedes' principle of buoyancy that states that an object immersed in a fluid is buoyed by a force equal to the weight of the displaced fluid. A partially or completely covered displacer exerts a buoyant force equal to the weight of the displaced fluid. As the liquid rises over the length of the displacer, its effective weight changes due to the volume of the fluid displaced. This resultant weight change is directly proportional to the change in level.

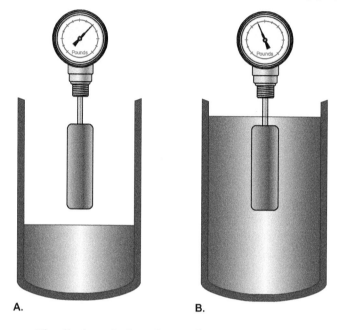

Figure 4.20 Displacer principle of operation. **A.** Gauge in pounds in air. **B.** Gauge in pounds in liquid.

The displacer is the primary element. It is connected to a transmitter that produces the proportional signal representing the level. The displacer is hung from an extension arm connected to a device called a torque tube (Figure 4.21), which acts in the same manner as a weight measuring spring. Another mechanism or transducer converts the change in rotation of the torque tube (in response to the weight change) into a level indication or signal output. Displacers are part of a complete liquid level measuring system.

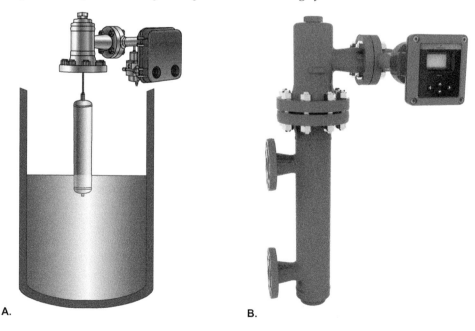

Figure 4.21 Displacer torque tube. **A.** Illustration. **B.** Photo.

CREDIT: B. © Emerson 2019.

Displacers are used to measure level indirectly by measuring the change in the buoyancy experienced by the displacer as the liquid rises over its length. They are used in refineries and chemical plants throughout the world to provide point and/or continuous level measurement. Displacers can provide an interface level measurement as long as there is a sufficient difference in densities between the two liquids.

The cylinder shape body of the displacer provides a consistent buoyancy change per incremental change in level. Also, note that the displacer must weigh more than the fluid it displaces so that it always remains in a vertical position. The basic premise of a displacer is that it displaces the fluid and does not float on it. Another important factor to remember is that displacers do not measure the true bottom level or the true topmost level, but rather the low or high level as determined by the displacer placement within the vessel according to the process design requirement.

Distance Measuring Devices

Ultrasonic and *radar* electromagnetic distance measuring devices (Figure 4.22) are instruments that are usually inserted into the top of a vessel, although ultrasonic types can be located on the bottom. **Ultrasonic and radar devices** are very accurate. Both instruments emit a pulse of energy that is reflected off of the surface of the material back to the receiver. They can be two separate units, a transmitter and a receiver, or they may be a single unit, called a transponder.

Ultrasonic and radar device
an accurate distance measuring instrument usually inserted into the top of a vessel; emits a pulse of energy that is reflected off the surface of the material back to the receiver; may be either a single unit (transponder) or two separate units (transmitter and receiver).

A.

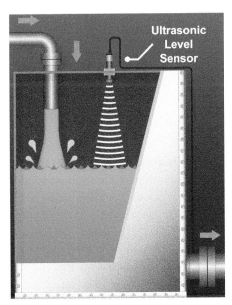

B.

Figure 4.22 Ultrasonic or radar device. **A.** Ultrasound device, **B.** Illustration.

CREDIT: **A.** © Emerson 2019. **B.** Rumruay/Shutterstock.

Ultrasonic and radar devices make excellent noncontact level measuring instruments. Since neither of these devices come in direct contact with the process material, they can be used when many other level sensing types cannot. Some process fluids demand a noncontact sensing device because of their chemical or physical properties. These fluids may be highly corrosive, which would reduce the life of the sensor, or they may simply have a tendency toward caking (i.e., depositing solids onto the sensor). In either case, the effectiveness of the sensor is diminished.

The word *ultrasonic* describes sound waves with frequencies higher than those detectable by the human ear. Radar devices use pulses or electromagnetic radio waves. In both cases, the instrument calculates the time it takes for the pulse to reach the surface of the material, bounce off, and return back to the receiver. If the density of the vapors in the space above the liquid (or solid) is known, then a simple distance calculation can be provided, resulting in an inferred level measurement.

Nuclear Device

A nuclear level instrument (Figure 4.23) is a device that uses a tightly controlled gamma radiation source with a detector to infer a level in a tank. The two most commonly used gamma radiation sources are the radioisotopes cobalt 60 and cesium 137.

CAUTION: Nuclear devices are VERY dangerous and must be handled properly at all times.

Figure 4.23 Nuclear level reading device.

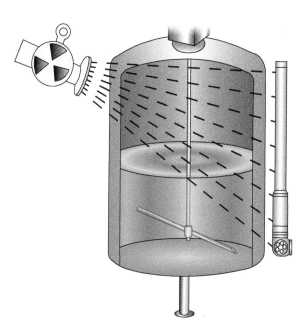

Nuclear device uses a tightly controlled gamma radiation source with a detector to inversely infer a tank's level; used when other technologies are unsuccessful; located on the outside of a tank and impervious to the effects of adverse process conditions; common gamma radiation sources are the radioisotopes cobalt 60 and cesium 137.

Nuclear devices can be used for point and continuous level measurements. When other technologies are unsuccessful, nuclear devices may be the only choice. Nuclear devices are located on the outside of the tank within a sealed cabinet making it impossible for adverse conditions within the tank to affect them and to protect personnel working near the tank.

The gamma rays emitted from a nuclear source penetrate the vessel walls and the process material inside and are detected on the other side of the vessel. The level measurement can be continuous level or point measurement. Level is measured by how many gamma rays get absorbed

Some of the properties of gamma radiation are similar to visible light. However, unlike visible light, the short wavelength and higher energy gamma rays can penetrate the vessel walls and the process material inside. Imagine standing in a dark room with a bright light shining through a translucent plastic tank containing a liquid. Your eyes would probably be able to see a diminished, yet detectable amount of light from the other side of the tank. This is similar to how a gamma source and detector operates. The radiation from the source that is received by the detector is inversely proportional to the change in level.

The typical installation of a nuclear level system is to place the source (a point source or a strip source) on one side of the vessel and the strip detector on the other side. The strip detector can discern the point where the gamma radiation is diminished (shadowed) by the liquid in the tank and then produces an output corresponding to the level.

Nuclear level detectors are only used in processes for which conventional level measurement does not work well due to the nature of the process. These systems are very expensive and require licensing from the Nuclear Regulatory Commission (NRC).

Load Cells

A **load cell**, an indirect level measuring device, is a transducer used to measure force or weight. A typical load cell (Figure 4.24) consists of a strain gauge bonded to a robust support column called a force beam. This force beam is usually a short piece of solid steel capable of supporting the entire weight of a vessel and its contents. As weight is applied to the load cell, a readout and/or transmitting device provides a corresponding output measurement. As weight is applied to the load cell, the internal strain gauge measures the deformity of the force beam. This produces a change in resistance that can be measured by an electronic circuit. Recall that a strain gauge under strain has a higher resistance than when it is not under a strain. This change in resistance can be correlated to a weight change and therefore, a level change (see Figure 4.24).

The most common process variable inferred by a load cell is level. If the vessel is weighed when it is empty and then again when it is full, a relationship between its weight and level can be established. If the tank is cylindrically shaped and standing upright, then a linear relationship between weight and level exists. For example, when 20 percent of the total tank weight minus the tare weight is measured, then the level in the tank would also be 20 percent. Load cells are capable of measuring either tension (hanging weight) or compression (supporting weight). Hoppers (special vessels that enhance the movement of solids) are filled with solids and usually hang from a tension (pulling or stretching) type load cell, while tanks containing liquids usually sit on a compression type load cell.

Electronic load cells are the most common providers of direct weight measurement in a plant. Weighing tank trucks and rail cars transporting materials into and out of a plant are two of the many direct weight measuring applications found in the process industry. Weight can also be used to infer the flow rate. A solid material moving across a weigh point (load cell) as it speeds down a conveyor belt can provide a weight per unit of time measurement.

Load cell a level measuring device consisting of a transducer that measures force or weight (called a strain gauge), which is bonded to a robust support column called a force beam; generally built into the supporting structure of a vessel.

Figure 4.24 Load cell. **A.** Illustration. **B.** Photo of load cell.

CREDIT: B. Courtesy of Brazosport College.

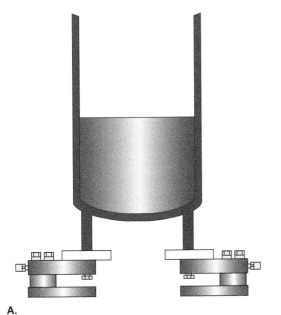

A.

B.

Summary

The process industry defines level as the act of establishing the height, a flat level plane, of an undisturbed liquid surface in reference to a zero point. The measurement of the liquid height may be made in several methods to include measuring what is inside (the innage) a container (e.g., storage tank) or referencing the space above the level (the ullage/outage) in the container. Height and depth may also be described as a percentage, such as 50 percent full. Liquid level measurements may be taken at only one point (point level) in a vessel or at all level points (continuous level) from 0 to 100 percent. Measurements may also be directly taken by measuring the level against itself, such as a float on the surface of the liquid level or indirectly taken through the use of one or more properties (head pressure or weight) of the liquid.

Liquids may separate into layers (stratify) due to their density. An example is a waste tank containing both oil and water. Water is generally heavier and it falls out to the bottom while oil is lighter and floats on top. The point at which the two materials join is called the interface. Both liquids combine for a total level in the container, but each liquid has its own percentage level. Process technicians must ensure that the water draw takes only the water portion while sending the oil portion to reclamation. It is important the interface level reads correctly and is maintained within operational parameters.

When observing a liquid level in a jar or other small container such as a glass tube manometer, understanding how the liquid surface tends to have either a convex or concave appearance is very important when taking a reading. Level readings should be taken on the highest or lowest part of the surface.

In the process industries, most liquid level measurements are typically provided through indirect methods. The most common indirect level measurement method is head pressure. Head pressure is the pressure exerted by a column height of liquid. Liquid head pressure is usually expressed in inches water column (in. w.c.) or in pounds per square inch (PSI). To determine head pressure, the density must also be known. Density may be converted and expressed as specific gravity, which is a ratio of the density of the liquid to the density of water.

There are many common level sensing and measurement instruments. One of the most common direct level measuring devices is a level gauge commonly known in the field as a sight glass. This instrument has a glass-covered portion allowing the process technician to read the level by observation. Most often the technician is only interested in knowing that the level is not too high or low. These level gauges come in three basic types: reflex, transparent, and welded pad.

Floats and float gauges are point level devices usually used to actuate high or low level alarms and shutdown circuits. These level gauges operate on the simple principles of buoyancy and mechanical action. If the float gauge is used in a continuous level measuring application, the gauge has a magnet.

Simple calibrated metal tapes are used to manually determine level. These tapes may also be connected to an indicator on one end and a float on the other. This apparatus measures continuous level changes in the associated tank or vessel. Accuracy on this type of level gauge is very important. When used to measure solids in a vessel, the float is replaced with a plumb bob and sounder mechanism.

When a D/P cell is installed to measure level, the high pressure side is connected to the bottom of the tank and the low pressure side is connected to the top of the tank. The high side measures total pressure (combined head pressure plus surface pressure) while the low side measures only the vapor space pressure above the liquid level. The vapor pressure measurement is cancelled out from both measurements leaving only the pressure exerted by the head pressure of the liquid itself.

Bubblers are special head pressure measuring devices where the pressure sensor does not come into contact with the process fluid. This level measuring system prevents, lessens, or eliminates problems from corrosive or other problematic physical properties of the substance being measured. A dip tube is positioned inside the vessel to about one or two inches (2.5 to 5 cm) above the bottom surface. A D/P transmitter or other pressure-sensing device is connected to the end of the tube to measure back pressure. This device is generally used on low pressure tanks that are vented to the atmosphere.

Displacers measure buoyancy by using a sealed cylindrically shaped tube (equal to the weight of the fluid it displaces) submerged into the liquid. Displacers measure the level indirectly by measuring the change in buoyancy experienced by the displacer as the liquid rises over the length of the displacer. The resultant (apparent) weight change is directly proportional to the change in level.

Ultrasonic and radar devices are distance measuring instruments usually inserted into the top of a vessel. They emit a pulse of energy that is reflected off the surface of the material back to the receiver. Since this device type is noncontact, it can be used where many other types cannot be used due to the chemical or physical properties of the substance. Ultrasonic devices use a sound wave while radar devices use pulses or electromagnetic radio waves.

Nuclear level measurement devices use a tightly controlled gamma radiation source with a detector to infer tank levels. The most common gamma radiation sources are the cobalt 60 and cesium 137 radioisotopes. These devices can be used in either point or continuous level measurements. Typical installations have a strip on the inside and outside where detection is determined by the shadowing of the light when comparing the liquid covering the strip to where it is not covering the strip. These devices, due to their nature, are only used on processes that cannot be conventionally measured with other instrument systems. They require registration and certification by the Nuclear Regulatory Commission.

Load cells are indirect level measuring devices which utilize a transducer to measure force or weight. A strain gauge, contained within the load cell, connected to a force beam measures the resistance to strain that correlates to a weight change and therefore a level change.

There is a great correlation among level, temperature, density, and volume. When one changes, the others also change. Temperature affects the density of all matter causing an increase or decrease in volume. Volume increases caused by density fluctuations do not increase head pressure since the total amount of head pressure is based on the specific gravity of the substance being measured. Knowing these characteristics helps to understand why the levels may change in process vessels.

Checking Your Knowledge

1. Level can be defined as the _____ of a material's surface distance from a zero reference.
   a. location
   b. width
   c. point
   d. height

2. Match the terms below with their proper definitions or descriptions:

Term	Definition
I. Ullage	a. The measurement of another property of a contained material to determine level
II. Innage	b. The measurement from the bottom of a tank to the surface of a material
III. Indirect level measurement	c. The measurement from the surface of a material to the top of a tank
IV. Direct level measurement	d. The measurement of the point where two immiscible materials meet
V. Interface level	e. The measurement of the process variable

3. The hydrostatic head pressure exerted by 36 in. of water is equal to _____.
   a. 36 in. w.c. ÷ 1.25
   b. 36 in. w.c.
   c. 36 in. w.c. ÷ 1.350
   d. 12 in. w.c.

4. If the mass remains constant and the density of a liquid in a tank decreases, the level of the liquid in the tank _____.
   a. increases
   b. decreases
   c. remains constant
   d. does nothing because density and level are not related

5. When the volume of a liquid in a tank decreases, the level of the liquid in the tank _____.
   a. increases
   b. decreases
   c. remains constant
   d. does nothing because density and level are not related

6. Which of the following is the oldest type of gauge glass still in use today?
   a. Flat glass level gauges
   b. Reflect level gauges
   c. Transparent level gauges
   d. Tubular type sight glass

7. What method is used to measure nonflammable processes?
   a. Bubbler
   b. Displacer
   c. Transmitter
   d. D/P cell

NOTE: Answers to Checking Your Knowledge questions are in the Appendix.

Student Activities

1. Take a large container and fill it with water. Determine a point on the container to record the highest water level. Pour out differing amounts of water (¾ full, ½ full, ¼ full) for each set of level measurements. Using a metal calibrated measuring tape, extend the tape into the water until it reaches the bottom. Ensure the tape is as straight vertically as possible and read the level of the water. Also, place the end of the tape at surface level and read the measurement at the top of the tape where the highest liquid level could be obtained. Which measurement would be considered innage and which one ullage (outage)? Do the two measurements add up to the total measurement (top to bottom of the container) that was taken earlier?

2. Using the conversion chart, perform the calculations on a clean sheet of paper. Turn in your completed calculations as per your instructor's guidelines. Ensure problem numbers and corresponding answers are clearly written. *Use this conversion information to complete the following conversion exercises.*
Level Conversion Formula

$$P = \frac{(1 \text{ in. w.c.})(h)(SG)}{\text{in.}}$$

$$P = h \times SG$$

$$h = \frac{P}{SG}$$

Perform the following level conversions on a sheet of paper, and then write an email listing the problem number and your answer. Send it to your instructor.

Head Pressure Problems

a. An open tank contains a liquid with a specific gravity of 1.735. If the height of the liquid is 10 feet, how much head pressure, in inches w.c., will it exert?

b. An open top tank has a liquid with a specific gravity of 0.95. If the height of the liquid is 200 in., how much head pressure, in inches w.c., will it exert?

c. An open top tank is filled with a liquid that has a specific gravity of 1.735. The liquid exerts a head pressure of 450 in. w.c. What is the level in inches?

d. An open top tank is filled with a liquid that has a specific gravity of 0.873. The liquid exerts a head pressure of 200 in. w.c. What is the level in inches?

e. A sealed tank has a liquid with a specific gravity of 1.00 in it. The pressure in its vapor space is 1 PSI. If the height of the liquid is 10.0 feet, how much head pressure, in inches w.c., will it exert?

3. Identify pressure-sensing and measurement devices in the lab.

4. Take apart a level instrument to locate the displacer float. Hook up air to see how the transmitted signal varies with displacer position.

5. Use cutaways of gauge glass valve(s) to examine the internals (especially ball check).

Chapter 5
Process Variables, Elements, and Instruments: FLOW

 Objectives

After completing this chapter, you will be able to:

5.1 Define terms associated with flow and flow measurement:
fluids (gases and liquids)
metered displacement
laminar flow
turbulent flow
weight/mass measurement.
direct and indirect flow measurement
positive displacement flow measurement
percent flow rate
volumetric flow units
mass flow units. (NAPTA Process Variables . . . : Flow 1*) p. 89

5.2 Identify the most common types, purpose, and operation of flow sensing and measuring devices used in the process industries:
orifice plate
Venturi tube
flow nozzle
pitot tube

---

*North American Process Technology Alliance (NAPTA) developed curriculum to ensure that Process Technology courses will produce knowledgeable graduates to become entry-level employees in process technology. Objectives from that curriculum are named here in abbreviated form. For example, "(NAPTA Process Variables . . . : Flow 1)" means that this chapter's objective relates to objective 1 of NAPTA's Process Variables, Elements, and Instruments content related to flow measurement.

multiport pitot tubes

rotameter

magmeter (electromagnetic meter)

turbine meter

mass flow meter

differential (D/P) transmitters

vortex meter

ultrasonic meter. (NAPTA Process Variables . . . : Flow 2, 3) p. 92

5.3 Explain the difference among total volume flow, flow rate, mass flow, and volumetric flow. (NAPTA Process Variables . . . : Flow 4, 5) p. 101

Key Terms

Differential pressure (D/P, Δp)—the difference between two related pressure measurements; usually used in measurements of process variables (pressure, temperature, level, and flow), **p. 91.**

Electromagnetic flow meters (magmeters)—magnetic flow meters designed to determine volumetric flow of electrically conductive liquids, slurries, and corrosive and/or abrasive materials, **p. 96.**

Flow measurement—flow rate (an instantaneous flow measurement) and total flow (a summation of instantaneous flow rates over a time interval or an accumulation of counts provided by a positive displacement device); measured in volume or mass units without respect to time, **p. 91.**

Flow nozzle—a device similar to a Venturi tube but with an extended tapered inlet commonly installed in a short piece of pipe called a spool piece, **p. 94.**

Flow rate—the specific amount of fluid moving past a given point per unit of time, usually expressed in volume units per unit of time, such as gallons per minute (GPM) or cubic feet per minute (CFM), **p. 89.**

Fluid—substances, usually liquids or vapor, that can be made to flow, **p. 89.**

Indirect flow measurement—measuring one variable of a process to infer another, **p. 91.**

Laminar flow—a condition in which fluid flow is smooth and unbroken; viewed as a series of laminations or thin cylinders of fluid slipping past one another inside a tube, **p. 89.**

Mass flow meter—a meter that eliminates the need to compensate for typical process variations such as temperature, pressure, density, and even viscosity; the most common type of true mass flow meter is the Coriolis, **p. 98.**

Mass flow units—a unit of measure for the weight being passed through a certain location per unit of time; usually expressed in pounds per unit of time, as in pounds per minute, **p. 91.**

Multiport pitot tube—a multiport tube having four impact points spaced across the pipe and facing the flow, with another tube sensing static pressure; measures the average pressure produced by the four impact points; common brand name Annubar®, **p. 95.**

Orifice flanges—flanges with holes drilled in them through to the pipe to allow pressures upstream and downstream of an orifice plate to be measured, **p. 93.**

Orifice plate—a piece of ⅛ in. to ½ in. thick metal with a calibrated hole drilled (or cut) through it; types of orifice plates include the concentric plate, eccentric bore, and segmental plate, **p. 92.**

Percentage flow rate—a common way to indicate a flowing process, with 100 percent equating to an actual quantity such as specified gallons per minute (GPM), **p. 91.**

Pitot tube—an L-shaped tube that is inserted into a pipe with its open end facing the flow and another tube sensing static pressure, **p. 94.**

Positive displacement flow measurement—measurement of flow in absolute volumes where the flowing material is admitted into a chamber of known volume and then transferred to a discharge point; a counter registers the number of times the chamber fills and discharges, **p. 91.**

Positive displacement meters—piston, oval gear, nutating disk, and rotary vane types of positive displacement flow meters, **p. 91.**

Reynolds number—a mathematical computation describing the quality of the flow of fluids numerically, **p. 90.**

Rotameter—a direct read variable area (tapered) flow tube in which fluid enters through the bottom, then flows upward, lifting a free floating indicator plummet (float); the position of the float against the calibrated marks on the glass tube indicates flow rate, **p. 95.**

Turbine meter—a flow tube containing a free spinning turbine (fan) wheel where the revolutions per minute (rpm) are proportional to flow, **p. 97.**

Turbulent flow—a condition in which the fluid flow pattern is disturbed, so there is considerable mixing, **p. 90.**

Venturi tube—a primary element used in pipelines to create a differential pressure (D/P) such that when converted to flow units (e.g., GPM), the flow in the line is measured, **p. 93.**

Volumetric flow units—a unit of measure for the volume being passed through a certain location in a pipe or channel per unit of time; usually measured in gallons per minute and cubic feet per minute, **p. 91.**

5.1 Introduction to Flow

Flow is a fluid in motion. Gases, like liquids, flow from one point to another depending on pressure and temperature. When **fluids** flow, they mix easily because of the continual movement between molecules. Material always flows from a high pressure area to a low pressure area. The rate that a fluid flows is a measure of how much fluid is flowing or moving through a pipe or channel within a given period of time. The term *flow* is often used interchangeably with **flow rate** in industry.

Molecules that are flowing continually change how they move among themselves. This is a basic characteristic of fluid flow whether it is a liquid or gas. Fluid movement may be either laminar or turbulent depending on the path taken. When the path is smooth and without obstruction, the flow is called laminar since the molecules line up in a smooth flowing pattern. This type of movement is characteristic of long pipelines where the molecules have a chance to settle and move in an orderly fashion. **Laminar flow** (Figure 5.1) is also called streamlined flow. The velocity of the flowing fluid changes smoothly and equally from the

Fluid substances, usually liquids or vapor, that can be made to flow.

Flow rate a quantity of fluid that moves past a specific point within a given amount of time, usually expressed in volume units per unit of time, such as gallons per minute (GPM) or cubic feet per minute (CFM).

Laminar flow a condition in which fluid flow is smooth and unbroken; viewed as a series of laminations or thin cylinders of fluid slipping past one another inside a tube.

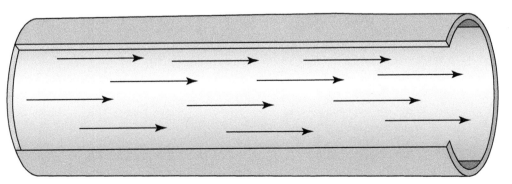

Figure 5.1 Laminar flow.

Laminar flow

Turbulent flow a condition in which the fluid flow pattern is disturbed, so there is considerable mixing.

pipe wall inward to the center of the pipe where the velocity is at its highest. Imagine the annular rings of a tree defining the velocity zones as they increase toward the center.

Piping bends, corrosion, valves, or any other obstruction to the flow inside a line causes turbulence or **turbulent flow** (Figure 5.2). If a stream of water passes over rock formations, the water looks like it is boiling and may even cause bubbles to form as air is entrained into the water from the turbulence. If an airplane goes through a section of turbulent air flowing by the plane, then the plane moves violently in response to the turbulence. At first this may seem undesirable, but in most cases in process industries, the opposite is true. A turbulent flow is consistent and therefore manageable. A turbulent flow mixes fluid and prevents solids from settling out, thus supplying more consistent feedstock to the process.

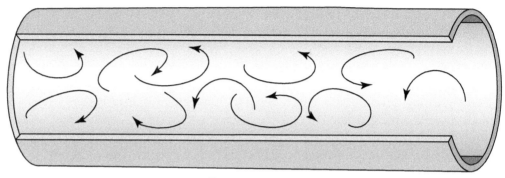

Figure 5.2 Turbulent flow.

Turbulent flow

Reynolds Number

Reynolds number a mathematical computation describing the quality of the flow of fluids numerically.

The **Reynolds number** can be used to identify whether a flow is laminar or turbulent. The Reynolds number is a mathematical computation known as the fluid velocity profile (Figure 5.3) that describes flowing fluids numerically. As mentioned, flowing fluids fall into one of two categories (laminar and turbulent), although there is a transitional zone where characteristics of both overlap. To calculate the Reynolds number, the following factors must be known:

- Velocity
- Density
- Viscosity of the fluid
- Inside diameter of the pipe

The following equation describes how the Reynolds number is obtained:

$$\text{Reynolds number} = \frac{(\text{velocity of fluid})(\text{inside diameter of pipe})(\text{density of fluid})}{(\text{absolute viscosity of fluid})}$$

Figure 5.3 Fluid velocity profile.

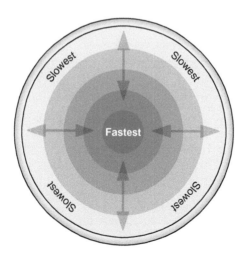

In this equation, an increase in fluid velocity, the inner diameter of the pipe, or the fluid density, makes the Reynolds number larger. An increase in absolute viscosity makes the Reynolds number smaller.

When a flowing process material has a Reynolds number of over 4,000, the flow is turbulent. To be fully turbulent, the Reynolds number should be greater than 10,000. When the Reynolds number is less than 2,000, the flow is laminar. The region above 2,000 and below 4,000 is said to be transient. This is the range that has characteristics of both turbulent and laminar types.

Direct and Indirect Flow Measurement

Total flow is a direct form of flow measurement as determined by totalizers. Totalizers display the flow rate from either an analog or pulse output flowmeter, as well as the accumulated total and grand total volume over time. Flow indicators such as flappers and paddlewheels seem to be direct, but they indicate the result of an impinging fluid force, so they are categorized as indirect flow measurement devices.

Flow rate is measured indirectly. **Indirect flow measurement**, similar to indirect level measurement, measures one variable of a process to determine another. For example, **differential pressure (D/P)** can be measured, then flow rate can be inferred. Although there are devices known as direct read flow meters (such as rotameters, weirs, and so on), these are still indirect methods when examined closely.

Flow measurement for liquids or gases may be made in exact or absolute quantities, percentages, and volumetric equivalents. Solids use these and also mass (weight) units.

Positive Displacement Flow Measurement

Positive displacement flow measurement is used within industries where the measurements of absolute volumes are required. Flow meters are used to determine exact quantities of material going into a tank (e.g., additives) and to measure the transfer of products between adjacent plants.

Positive displacement meters (metered displacement) operate by admitting a flowing material into a chamber with a known volume capacity and then transferring (discharging) all the contents to another point (e.g., gasoline blending tank). A counter registers the number of times the chamber(s) fills and discharges. A total amount of metered material can be read directly from the counter. Positive displacement meter types include piston, oval gear, nutating disk, and rotary vane.

Percentage Flow Rate

Percentage flow rate is a common way to indicate a flowing process. A 100 percent flow rate is equated to an actual quantity such as a specified number of GPM or liters per minute (L/min). If 100 percent flow rate is equal to 200 GPM (757 L/min), then a 50 percent flow rate equals 100 GPM (379 L/min), and so on.

Volumetric Flow Units

Volumetric flow units are instantaneous flow rate measurements for how much volume passes through a certain location in a pipe or channel per unit of time. Volumetric flow rates are measured in units such as gallons per minute (GPM) or cubic feet per minute (CFM).

Mass Flow Units

Mass flow rate is another instantaneous measure of how much actual mass passes a certain location per unit of time. Since mass in a constant gravity field has a definite weight, then **mass flow units** can be expressed in pounds per unit of time (e.g., lb/min). Mass flow rate is similar to volumetric flow measurements, with the exception of quantifying the total mass occupying the measured volume. This is because the mass (density) of any given volume of

---

Indirect flow measurement measuring one variable of a process to infer another.

Differential pressure (D/P, Δp) the difference between two related pressures; usually used in measurements of process variables (pressure, temperature, level, and flow).

Flow measurement flow rate (an instantaneous flow measurement) and total flow (a summation of instantaneous flow rates over a time interval or an accumulation of counts provided by a positive displacement device); measured in volume or mass units without respect to time.

Positive displacement flow measurement measurement of flow in absolute volumes where the flowing material is admitted into a chamber of known volume and then transferred to a discharge point; a counter registers the number of times the chamber fills and discharges.

Positive displacement meters piston, oval gear, nutating disk, and rotary vane types of positive displacement flow meters.

Percentage flow rate a way to indicate a flowing process, with 100 percent equating to an actual quantity such as specified gallons per minute (GPM).

Volumetric flow units a unit of measure for the volume being passed through a certain location in a pipe or channel per unit of time; usually measured in gallons per minute and cubic feet per minute.

Mass flow units a unit of measure for the weight being passed through a certain location per unit of time; usually expressed in pounds per unit of time, as in pounds per minute.

liquid depends on its temperature. For example, a gallon of water weighs 8.345 lb (3.785 kg) at its most dense temperature, 39.2 degrees Fahrenheit (4 degrees Celsius) and weighs 7.998 lb (3.627 kg) at 212 degrees Fahrenheit (100 degrees Celsius). The increase in temperature causes the water molecules to gain energy and move more rapidly, which results in water molecules that are farther apart and an increase in water volume. When water is heated, it expands, or increases in volume. When water increases in volume, it becomes less dense.

5.2 Types of Flow Measurement Devices

A primary flow element is a device that creates a measurable variable that is proportionally equal to flow rate. The most common primary flow elements measure differential pressure (D/P). This differential pressure (D/P) can be equated to a flow rate.

The following are the most common types of primary flow sensing and measuring devices used in the process industry:

- Orifice plates
- Venturi tubes
- Flow nozzles
- Pitot tubes
- Multiport pitot tubes (Annubar® tubes)
- Rotameters
- Electromagnetic meters (magmeters)
- Turbine meters
- Mass flow meters.

Orifice Plates

Orifice plate a piece of ⅛ in. to ½ in. thick metal with a calibrated hole drilled (or cut) through it; types of orifice plates include the concentric plate, eccentric bore, and segmental plate.

An **orifice plate** (Figure 5.4) is a piece of ⅛ in. to ½ in.-thick metal with a precise hole, or bore, drilled (or cut) through it. These holes are referred to as *metering holes*. This device is the simplest and most practical way to create a pressure drop in a flowing process.

There are several types of orifice plates. The most common type is the *concentric* plate, which has the bore hole drilled directly in the center. In contrast, in an *eccentric* plate, the bore is off center. Another type of plate is called a *segmental* plate; it has a half-moon shaped opening cut into it. Both the concentric and the segmental plates are designed to allow solids to pass freely through them rather than to accumulate in front of them. Lastly, there are quadrant edged and conical plates. These differ from the others because of their upstream face. The quadrant edged plate has a smooth, rounded inlet; the conical plate has a beveled or cone-shaped inlet. The quadrant and conical plates are used with more streamlined and viscous flows.

Figure 5.4 Orifice plates.
CREDIT: © Emerson 2019.

All these plates have a flat upstream face and most have a beveled downstream side. Manufacturers stamp or stencil the word INLET on the upstream side. Other information such as line and bore size is usually imprinted there as well.

An orifice plate often has a small hole drilled next to the metering hole called a *weep hole*. This small hole allows vapor or liquids that are likely to be stopped by the plate to pass through, keeping trapped materials on the upstream side of the plate from reducing the accuracy of the meter.

Orifice plates are generally mounted in a special set of flanges called **orifice flanges**. These flanges have pressure measuring taps drilled through the flange into the pipe, allowing the pressure upstream and downstream of the orifice plate to be read by a differential pressure (D/P) transmitter.

Orifice flanges flanges with holes drilled in them through to the pipe to allow pressures upstream and downstream of an orifice plate to be measured.

WHAT IS THE EFFECT OF CHANGE IN THE DIFFERENTIAL PRESSURE ACROSS AN ORIFICE PLATE ON THE FLOW OF A MATERIAL? A flow rate change causes a change in differential pressure; reduced flow results in decreased differential pressure.

RELATIONSHIP: FLOW AND DIFFERENTIAL PRESSURE In the 1700s, a scientist by the name of Daniel Bernoulli described many principles of fluid flow. One of them was how the total energy of a material flowing through a pipe would react to a change in pipe diameter. His principle, known as the Bernoulli principle, states that as the speed of a moving fluid increases, the pressure within the fluid decreases. So, within a horizontal flow of fluid, points of higher fluid speed will have less pressure than points of slower fluid speed. Through experimentation, scientists have found that the flow rate of a fluid is proportional to the square root of the differential pressure drop across a restriction such as an orifice plate. This concept is widely used in industry today.

Venturi Tubes

A **Venturi tube** (Figure 5.5) is a primary element used in piping to create a differential pressure (D/P). When converted to flow units (such as GPM), the flow in the line can be measured. Venturi tubes cost more than orifice plates, but their day-to-day operating cost can be significantly less because their design creates less permanent pressure loss. Pressure loss in any restrictive primary device is expected. However, in some larger applications, permanent pressure loss is a significant factor in determining which primary device should be used.

Venturi tube a primary element used in pipelines to create a differential pressure (D/P) such that when converted to flow units (e.g., GPM), the flow in the line is measured.

The shape of a Venturi tube is characterized by its smooth, cone shaped inlet and outlet components. As the fluid speeds up in the smaller-throat-diameter section, the pressure is reduced. According to Bernoulli's principle, if a fluid's static pressure decreases, its velocity increases. A tap at the inlet measures the high pressure, and a tap at the outlet measures the low pressure. A D/P transmitter can be connected to these taps to measure/compare the difference in pressure.

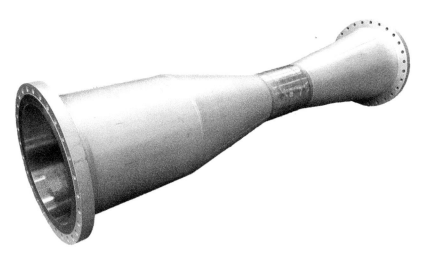

Figure 5.5 Venturi tube.
CREDIT: © Emerson 2019.

Flow Nozzles

Flow nozzle a device similar to a Venturi tube but with an extended tapered inlet, commonly installed in a short piece of pipe called a spool piece.

A **flow nozzle** (Figure 5.6) is similar to a Venturi tube in that a flow nozzle has an extended tapered inlet. The differences between a Venturi tube and a flow nozzle are that a flow nozzle can be inserted into a pipe at a flange connection, and the flow nozzle does not have a recovery (outlet) cone. A flow nozzle may also be installed in a short piece of pipe called a spool piece.

Figure 5.6 Flow nozzle.
CREDIT: © Emerson 2019.

Flow nozzles perform the same function as all other restriction devices; that is, they provide a differential pressure (D/P) that can be used to infer a flow rate. Because of their design, they are capable of allowing more flow to pass through them per unit of pressure drop than an orifice plate. They are also a good choice for use with slurries because their sloped inlets allow solids to pass freely through them.

The operation of a flow nozzle is the same as a Venturi tube on the front side and like an orifice plate on the downstream side. The smooth conical shape of the inlet is the primary reason why the flow nozzle is able to handle more than twice the flow of an orifice plate to create the same pressure drop. The cost of a flow nozzle is more than an orifice plate and less than a Venturi tube.

Pitot Tubes

Pitot tube an L-shaped tube that is inserted into a pipe with its open end facing the flow and another tube sensing static pressure.

The **pitot tube** (Figure 5.7) is a primary element used in pipelines to create a differential pressure (D/P) which, when converted to flow units (such as GPM), measures the flow in the line. A pitot tube is shaped like an "L" and is inserted into a pipe with the open end facing directly into the flow and another tube measuring or sensing static pressure in the same vicinity. By comparing the impinging pressure produced by the flowing fluid to the static line pressure, flow rate can be inferred.

Figure 5.7 Pitot tube.

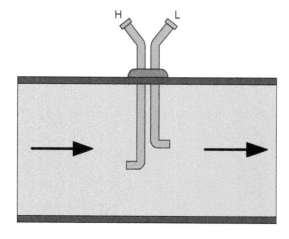

A disadvantage of the pitot tube is that it detects the velocity of a flowing fluid at only one point in the pipe. If the Reynolds number of the flowing material is high enough, then any point away from the wall represents the average flow rate. However, if the Reynolds number is too low (suggesting a laminar flow characteristic), then a pitot tube cannot provide an accurate average velocity measurement.

Multiport Pitot Tubes

A **multiport pitot tube** (Figure 5.8) is a tube with four impact points spaced across the pipe facing the flow and another tube sensing static pressure. The multiport tube has a smaller tube inside it that is designed to measure the average pressure measurement produced by four impact points across a pipe. Annubar® is one common trade name for a multiport pitot tube.

multiport pitot tube a multiport tube having four impact points spaced across the pipe and facing the flow, with another tube sensing static pressure; measures the average pressure produced by the four impact points.

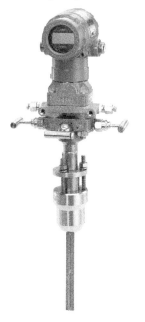

Figure 5.8 Multiport pitot tube.

CREDIT: © Emerson 2019.

The multiport pitot tube measures velocity like the pitot tube, but it has multiple impact points rather than one. The multiport tube takes an average of all the impact points to produce a single measurement. Like the pitot tube, a static pressure measurement is compared to the average impact pressure, producing a differential pressure (D/P) that can be measured and converted into a flow rate.

Rotameters

A **rotameter** (Figure 5.9) is a direct read variable area (tapered) flow tube in which the fluid enters through the bottom, then flows upward, lifting a free floating indicator plummet (also called a float). The position of the *float* can be referenced to the calibrated marks on the glass tube to indicate flow rate.

Rotameter a direct read variable area (tapered) flow tube in which fluid enters through the bottom, then flows upward, lifting a free floating indicator plummet (float); the position of the float against the calibrated marks on the glass tube indicates flow rate.

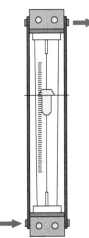

Figure 5.9 Rotameter.

Rotameters are designed to provide a flow indication. They can be made of glass, plastic, or metal. Glass and plastic types give the process technician an opportunity to see the process fluid. This is helpful when troubleshooting for process problems. The disadvantage of using glass and plastic is that the materials are fragile. Glass is especially prone to breakage. The armored, or metal, tube type eliminates this problem, but it hides the fluid from inspection. Generally speaking, rotameters are used for indication, but the metal tube types can be equipped with a magnetic follower both to indicate a signal and to transmit a signal.

The operating principle of the rotameter is based on gravity and impinging flow pressure. Physically, the smallest diameter of the tube is on the bottom, with the diameter increasing steadily as it goes up. The *plummet* (Figure 5.10), or *float*, requires that a minimal amount of impinging force be applied to the float for it to rise (float) in the flowing fluid. As the fluid velocity increases, it pushes the plummet upward to a new position in the tube. This new position is directly related to the impinging force applied to the plummet. At any point in time, the velocity at the smaller end of the tube is the highest, requiring less flow to float the plummet. As the diameter of the tube increases, however, the velocity slows down. This requires more fluid flow through the tube to push the plummet higher. The tube is marked in flow units providing a flow indication.

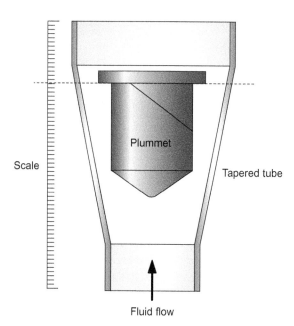

Figure 5.10 Rotameter plummet.

The shape and weight of the plummet, along with the size and shape of the flow tube, provide an accurate flow rate. Each rotameter is sized to indicate a specific type of liquid or gas. If the rotameter is improperly sized or replaced with a different tube and/or plummet, accuracy suffers.

Magmeters (Electromagnetic Flow Meters)

A magnetic flow meter is designed to determine volumetric flow of electrically conductive liquids, slurries, and corrosive and/or abrasive materials. **Electromagnetic flow meters (magmeters)** are used to measure flowing liquids having at least some ability to conduct electricity (Figure 5.11). Generally, this means *aqueous* (water based) materials.

Magmeters are important for several reasons. One is that they are designed with a completely obstruction free flow path. Another is that they are noninvasive, meaning that there are not any sensing parts protruding into the process. The noninvasive characteristic is particularly important to the food and drug industry where residual materials deposited in and around sensors cannot be tolerated.

The magmeter operates according to Faraday's law of induction, which states that an electrical potential is produced when a conductor moves at a right angle through a magnetic field.

Electromagnetic flow meters (magmeters) magnetic flow meters designed to determine volumetric flow of electrically conductive liquids, slurries, and corrosive and/or abrasive materials.

Process Variables, Elements, and Instruments: FLOW

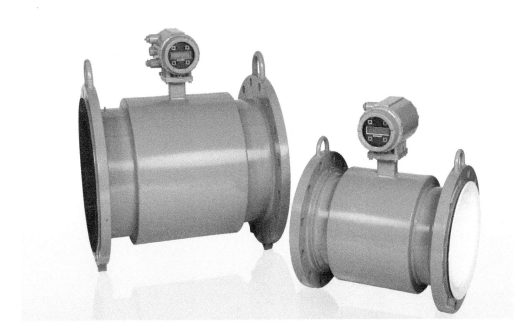

Figure 5.11 Magmeter (electromagnetic meter).

CREDIT: © Emerson 2019.

The magmeter produces a magnetic field that penetrates the nonferrous (nonmagnetic) flow tube. The liquid flowing through the tube is the conductor that Faraday describes as moving at a right angle to the magnetic field. This action creates an electrical potential, which is sensed by two electrodes that are positioned across from each other in the flow tube wall. The faster the flow, the greater the electrical potential generated.

Turbine Meters

A **turbine meter** (Figure 5.12) is a flow tube containing a free spinning turbine (fan) wheel where the revolutions per minute (rpm) are proportional to flow rate. The faster the flow, the faster the turbine spins. The output from the meter can be fed directly into an instrument that can indicate either flow rate or total flow.

The operation of the turbine meter is simple. As the process material flows through the meter, it deflects off the turbine blades, causing them to rotate about an axis like a toy

Turbine meter a flow tube containing a free spinning turbine (fan) in which where the revolutions per minute (rpm) are proportional to flow rate; the faster the flow, the faster the turbine spins; the output is fed directly into an instrument to indicate either flow rate or total flow.

Figure 5.12 Turbine meter. **A.** Illustration. **B.** Photo.

CREDIT: **B.** © Emerson 2019.

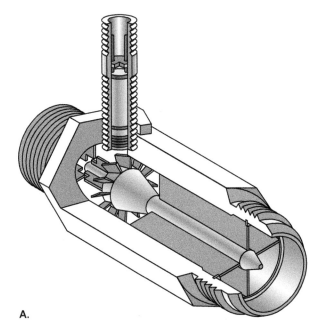

A.

B.

pinwheel. With an induction type sensor, one of the turbine blades is magnetic; when the blade rotates past an induction pickup coil located on the outside of the tube, the break in the magnetic field produces a pulse. Each pulse represents a specific volume of material. Turbine meters must be calibrated occasionally so that an adjusted K factor can be established. A K factor is the number of pulses generated per gallon of flowing material. Properties like viscosity and specific gravity can affect flow meter calibration.

Mass Flow Meters

Although *volumetric flow meters* can be combined with a *density sensor* (which may be a pressure and temperature measurement device) to infer mass flow, there are some meters that read mass flow directly. The most common type of true **mass flow meter** (Figure 5.13) is the Coriolis meter. True mass flow meters eliminate the need to compensate for typical process variations such as temperature, pressure, density, and even viscosity.

Mass flow meter a meter that eliminates the need to compensate for typical process variations such as temperature, pressure, density, and even viscosity; the most common type of true mass flow meter is the Coriolis.

Figure 5.13 Mass flow meter.
CREDIT: © Emerson 2019.

In mass flow meters, a drive coil causes the tube to vibrate. Two velocity sensors detect the natural frequency of the filled tube to establish a baseline measurement. As material starts flowing through the tube, the vertical motion produced by the drive mechanism causes the fluid entering the tube to push against the tube wall in one direction and push in the opposite direction leaving the tube. These resultant forces cause the tube to twist, producing a displacement and velocity difference that can be measured by the two sensors. Mass flow is determined by the difference between the two velocity sensor signals. Most mass flow meters read flow, temperature, and density. The mass flow meter is generally not affected by changes in fluid temperature, pressure, or density.

Differential Pressure (D/P) Transmitters

Differential pressure (D/P) transmitters (Figure 5.14; see also Figure 5.16 later in the chapter) are the single most commonly used flow measuring devices in the processing industry. Most fluid flow rates are measured by differential pressure (D/P). Process technicians need to know how to read and interpret the output signals from all flow transmitters while making their rounds. The knowledge of D/P versus flow rate is particularly helpful when troubleshooting a process problem.

The D/P transmitter responds to the pressure created by the primary device (orifice plate, Venturi, flow tube, etc.). The pressure signal is then converted into a standard output signal that can represent either differential pressure (D/P) or flow rate. When a differential pressure (D/P) cell responds to the high and low pressure measurements received from the

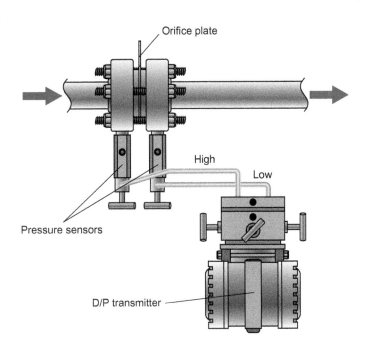

Figure 5.14 Differential pressure (D/P) transmitter.

primary element, a corresponding output signal is produced that is linear and proportional to the differential pressure (D/P). The electronic circuits in the transmitter are capable of producing two different types of output: output that is linear with respect to differential pressure (D/P) or output that is linear with respect to flow rate. The standard analog differential pressure (D/P) transmitter output is linear with respect to differential pressure (D/P).

An important concept inherent in all differential pressure (D/P) flow measurements is the square root relationship that exists between the measured differential pressure (D/P) and the inferred flow rate. If a flow rate changes by a number, then the pressure drop changes to that number squared. For example, if the flow rate doubles, the pressure drop is 2 squared, or 4. If the flow rate triples, the pressure drop is 3 squared, or 9. This is usually converted into a percentage. For example, if 25 percent of the calibrated differential pressure (D/P) measurement is sensed across the orifice plate, then 50 percent of the flow rate is moving through the differential pressure (D/P) cell.

To change percent D/P into percent flow rate, perform the following steps:

Step	Calculation
1. Insert the percent D/P into the equation:	Percent Flow Rate = $\dfrac{\sqrt{\% \, D/P}}{100} \times 100$
2. Change percent into a decimal form:	Percent Flow Rate = $\dfrac{\sqrt{25\%}}{100} \times 100$
3. Convert the percent to a decimal:	Percent Flow Rate = $\sqrt{.25} \times 100$
4. Take the square root:	Percent Flow Rate = 0.5×100
5. Make the decimal a percent again and record your answer:	Percent Flow Rate = 50%

Or,

Step	Calculation
1. Use the percent D/P equation:	Percent Flow Rate = $\dfrac{\% \, D/P}{100} \times 100$
2. Insert the percent D/P into the equation:	$\dfrac{25\%}{100} \times 100$
3. Change percent into a decimal form:	0.25×100
4. Take the square root:	0.50×100
5. Make the decimal a percent again and record your answer:	50%

Miscellaneous Flow Meters

Other types of flow meters that may be found in some process facilities include the following:

- Vortex
- Target
- Integral
- Ultrasonic

VORTEX FLOW METER A vortex flow meter (Figure 5.15) measures most process fluids and causes only minor pressure loss. Vortex flow metering does rely on the formation of vortices in the fluid flow and is not suitable for high viscosity fluids or for most low flow applications. A vortex flow meter uses vortex shedding to measure flow. Vortex shedding occurs when a *bluff body* (a flat fronted bar tapered toward the back) is placed in the flow path. This object causes the fluid to separate and form small eddies, called vortices, that are shed on the sides of the bluff body. These alternating vortices cause a fluctuation in pressure from side to side. A sensor inside or at the top of the bluff body registers these pressure fluctuations and sends a signal that is then converted by a transmitter to a standard signal that is proportional to the volumetric flow rate.

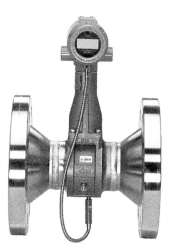

Figure 5.15 Vortex flow meter.

CREDIT: © Emerson 2019.

TARGET FLOW METER A target flow meter has a circular disc placed in the center of the flow path. Flow through the annular space between the target disc and the inside diameter of the meter causes a force on the disc that is proportional to the velocity head pressure. This force causes mechanical motion that is transferred to the top of the instrument through a connecting arm. The force is measured by an element that responds to the amount of motion produced by the arm.

INTEGRAL ORIFICE FLOW METER An integral orifice flow meter is a modified differential pressure (D/P) transmitter used to measure small fluid flows. The fluid flows into the body of the transmitter through a calibrated orifice that creates a differential pressure (D/P).

Some types of integral orifice flow meters are permanently installed in a meter run (Figure 5.16; see also Figure 5.14), which is installed as a whole in a process line of appropriate size. The integral orifice plate is built into the meter; it is welded in place and precalibrated. Pressure sensor taps are positioned vertically on either side of the integral orifice. The differential pressure (D/P) measurement is sent to the transmitter that converts the signal to either a pneumatic or electronic signal proportional to the differential pressure (D/P).

Figure 5.16 Integral orifice flow meter–meter run.

CREDIT: © Emerson 2019.

ULTRASONIC FLOW METER An ultrasonic flow meter measures the velocity of a fluid by measuring the amount of time that elapses between two pulses of an emitted beam of ultrasound. Ultrasonic flow meters can be inline (intrusive, wetted) or clamp-on (nonintrusive) designs. The meters are designed to measure differences in temperature, suspended particulates, viscosity, and density that can affect readings. Ultrasonic flow meters can also measure the frequency shift of the Doppler effect, reflecting the frequency of sound, light, or other waves as the fluid moves toward (or away from) the signal.

Ultrasonic effects can also be used in an open channel, measuring a specific level in the flume or weir. This is done by emitting a pulse of sound from the front of the sensor to the surface of the open channel flow stream and measuring the time for the echo to return, like sonar. An advantage of ultrasonic flow meters is their lack of moving parts, which translates to inexpensive use and low maintenance costs.

5.3 Flow Rate, Total Volume Flow, and Volumetric Flow

There are two kinds of flow measurement: flow rate and total flow. Flow rate is an instantaneous flow measurement. Total flow is the sum of instantaneous flow rates over a time interval or an accumulation of counts provided by a positive displacement device. In either case, total flow is measured in volume or mass units over a specified time interval (e.g., gallons per day, liter per minute).

The units associated with each of these types of flow include the following:

- Volumetric flow: gallons per minute (GPM) or liters per minute (L/min)
- Mass flow: pounds per minute (kilograms per minute)
- Total flow: gallons, pounds, cubic feet (liters, kilograms, cubic meters) over a specified time interval.

Liquid actively flowing into a drum is metered in volumetric flow units (e.g., GPM), while the total flow (the accumulated instantaneous flow value) is measured in gallons (or liters).

Summary

Flow is either a gas or a liquid in fluid motion rather than static, or standing still. Flowing fluids mix easily since the molecules are also moving amongst themselves. How fast the flowing fluid moves is the flow rate and this flow rate is a measure of how much volume of the fluid is moving over a given time period, such as a minute or an hour. Fluid movement is either laminar or turbulent, depending on the characteristics of the path in which the fluid moves. The Reynolds number is a mathematical computation describing how laminar or turbulent a fluid movement is.

There are several different types of direct and indirect flow measurement devices that include positive displacement, percentage, volumetric, and mass flow units. The rate of flow in a fluid is always measured indirectly in each of these types of devices, even when the name of the instrument sounds like it could be of the direct measurement type.

Metering the flow for specific measurements usually involves a positive displacement method, such as when using a piston, oval gear, nutating disk, and/or rotary vane type meter. A common method of measuring flow rate is by percentages, such as 50 or 100 percent gallons per minute (GPM). Mass flow measures the weight (lb or kg/min) of the material flowing through a particular measuring point. The mass weight depends on the temperature of the fluid being measured; the greater the temperature, the lower the weight. Density, temperature, viscosity, and pressure (head) are other factors that affect flow.

Many types of flow sensing and measuring devices are used in the process industry. They fall into a few categories based on their method of operation and how they are used in a process. The more common of these devices measure differential pressure (D/P). Bernoulli's principle, which states if a fluid's static pressure decreases its velocity increases, is used to determine if differential pressure (D/P) in a flowing fluid equates to a flow rate.

The first of these types of differential pressure (D/P) devices is the orifice plate. There are several types of orifice plates but they operate on the principle of restricting the flow of a fluid and allowing it to pass through a calibrated hole with a specific inlet and outlet side. Similar to orifice plates in operation, a Venturi tube costs more but is used in installations when pressure loss across the apparatus is unacceptable. The Venturi tube is cone-shaped on either end with a smaller throat in the middle. Another device, a flow nozzle, is also similar to the orifice plate but it has an extended tapered inlet. A flow nozzle can be inserted into a pipe at a flange since the nozzle does not have a recovery outlet cone like a Venturi tube.

Two other differential pressure (D/P) flow measuring devices are the L-shaped pitot tube and the multiport pitot (Annubar®) tube. The open end of a pitot tube is inserted into the flow and another tube measures static pressure in the same vicinity. The difference between pitot and Annubar® is that the pitot measures only one place in the flow while the multiport Annubar® tube has several ports and takes an average of all the differential pressure (D/P) points. The latter is more accurate.

Rotameters, which can be made of glass, plastic, or metal, allow process technicians the ability to see the movement of the process fluid since it pushes the calibrated plummet farther upward as pressure increases in the fluid movement. This basic type is for an indication of flow only, but a metal tube may be equipped with a magnet follower so that the flow rate can be both indicated locally and transmitted via a signal to a remote location such as a control room.

Magmeters measure the volumetric flow of electrically conductive liquids and slurries whether they are corrosive and/or abrasive materials. This type of flow measuring device is noninvasive to the process or the fluid flow path. Magmeters produce a magnetic field that penetrates a nonferrous flow tube and goes into the flowing liquid at right angles, creating electrical potential that is sensed by electrodes positioned across from each other in the flow tube wall.

Turbine meters consist of a flow tube with a fan type wheel. The number of times the fan spins (rpm) is proportional to flow rate. Output information from the turbine meter is fed directly to another instrument that indicates either flow rate or total flow (percentage). Mass flow meters, such as Coriolis, eliminate the need to compensate for process variations such as temperature, pressure, density, and even viscosity. Two sensors measure the forces causing the tube to twist, producing a displacement and velocity difference. Mass flow is the difference between the two signals.

Differential pressure (D/P) cells and transmitters are secondary flow meter devices that respond to the signals created by the primary devices described previously. The signals are converted to an output signal that may represent either the differential pressure (D/P) or flow rate. Flow rate is proportional to the square root of the differential pressure (D/P).

When comparing flow measurements, there are two kinds: flow rate and total flow. Flow rate is an instantaneous flow measurement, and total flow is a summation of instantaneous flow rates over a time interval. Flow rates may also be an accumulation of counts provided by a positive displacement device. Total flow is measured in volume or mass units without respect to time.

Checking Your Knowledge

1. To calculate the Reynolds number, which of the following variables is/are needed to perform the calculation? *Select all that apply.*
   a. Inside diameter of the pipe
   b. Viscosity of the fluid
   c. Density of the fluid
   d. Velocity of the fluid
   e. Laminar flow number

2. An increase in the absolute viscosity makes the Reynolds number ____.
   a. larger
   b. smaller
   c. unchanged
   d. irrelevant

3. The temperature of the flowing fluid must be considered when measuring ____ flow rates.
   a. positive displacement
   b. percent
   c. volumetric
   d. mass
   e. volumetric and mass

4. Which type of orifice plate has a half-moon shaped opening cut in it?
   a. Concentric
   b. Eccentric
   c. Segmental
   d. Pivotal

5. Match the terms below with their proper definitions or descriptions:

Term	Definition
I. Orifice plate	a. Measures flow using a free floating indicator plummet
II. Flow nozzle	b. Measures flow using an L-shaped tube and another tube that compares the impinging pressure with static pressure
III. Venturi tube	c. Measures flow using a metal disc containing a drilled opening.
IV. Pitot tube	d. Measures flow using a tapered inlet device inserted into a flange connection (or spool piece)
V. Rotameters	e. Measures flow using a cone-shaped device with inlet and outlet components

6. Calculate the percent flow rate flow rate for a D/P of 28%.
   a. 25
   b. .28
   c. .30
   d. 52.9

7. Match the terms below with their proper definitions or descriptions:

Term	Definition
I. Mass flow meter	a. Measures flow using a flow tube containing a spinning fan
II. Magmeter	b. Measures flow by sensing electrical potential
III. Turbine meter	c. Measures flow using a U tube and frequency vibrations
IV. D/P transmitter	d. Measures flow by responding to the signal created by the primary device and converting it into an output signal

8. If the flow through an orifice doubles, the differential pressure (D/P) ____.
   a. doubles
   b. triples
   c. quadruples
   d. stays the same

9. Special flanges to mount orifice plates are called ____ flanges.
   a. pipe
   b. orifice
   c. concentric
   d. Annubar®

10. Which unit is associated with mass flow?
    a. gallons per minute
    b. liters per minute
    c. pounds per minute
    d. cubic feet per minute

11. Properly identify each instrument by choosing the appropriate image for each flow measuring device.
    I. Orifice plate
    II. Flow nozzle
    III. Venturi tube
    IV. Pitot tube
    V. Multiport pitot tube (Annubar®)

a.   b.   c.   d.   e.

12. Properly identify each instrument by choosing the appropriate image for each metering device.
    I. Electromagnetic flow meter (magmeter)
    II. D/P transmitter
    III. Turbine meter
    IV. Rotameter
    V. Mass flow meter

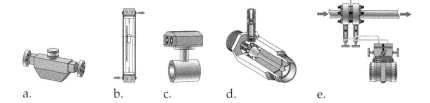

a.   b.   c.   d.   e.

NOTE: Answers to Checking Your Knowledge questions are in the Appendix.

Student Activities

1. Adjust water flow on a garden hose by opening the valve wide. When the flow is at maximum, place your thumb over varying portions of the nozzle opening. What happens when you do this? Does pressure or velocity increase? What thumb position makes the water project out the furthest?

2. Identify flow sensing and measurement devices in the lab.

3. Use a trainer with supply and process tanks. In small groups, start up the trainer and control flow and level loops to circulate the system.

4. Hook up a rotameter to a water connection and vary the rate to observe the inventory changes in a large container. Compare instantaneous rate to volume.

5. Fill a 1 gallon container with water from a faucet. Time how long it takes to fill the container. Calculate the rate in gallons per minute. Weigh the filled container. Calculate the rate in pounds per minute.

6. Perform the calculations on a clean sheet of paper. Turn in your completed calculations as per your instructor's guidelines. Ensure problem numbers and corresponding answers are clearly written. Convert percent differential pressure (D/P) into percent flow rate.

% Differential Pressure (D/P)	% Flow Rate
50% D/P	_____ % flow rate
75% D/P	_____ % flow rate
80% D/P	_____ % flow rate
90% D/P	_____ % flow rate
60% D/P	_____ % flow rate
32% D/P	_____ % flow rate
66% D/P	_____ % flow rate
45% D/P	_____ % flow rate
83% D/P	_____ % flow rate
92% D/P	_____ % flow rate

7. **FLOW MEASUREMENT: Laboratory Procedure**

Background Information

The term *flow* describes the movement of matter from one place to another. Flow can be classified as either instantaneous flow rate or total flow. Instantaneous flow rate is time dependent whereas total flow is not. The rate at which a material is flowing through a pipe will probably be measured as an instantaneous flow such as gallons per minute whereas the water meter at your home measures total flow (gallons).

Under normal automatic process control, a process technician should expect that the control instruments both measure and then stop a flowing material that is filling a tank. There are, in fact, metering instruments with preset shutoff capabilities that can mechanically accomplish this task, as well as more sophisticated control schemes that are designed to do the same. In either case, the process technician should know how long it takes for the tank to fill so that he or she can be available to monitor the shutoff. For example, if the tank is to be charged with 1,000 gallons of a material and that material is flowing into the tank at 80 gallons per minute, then it will take 12 minutes and 30 seconds to charge. The process technician should check to make sure that the charge is completed accurately.

Materials Needed

- Low flow rate rotameter with a throttling valve
- Containment vessel with graduated volume marks on it (the marked units should match the rotameter units if possible).
- Water supply and wet sink or drain apparatus
- Timekeeping device that indicates minutes and seconds
- Fittings and tubing for connecting the apparatus

Safety Requirements
- Safety glasses are required in the lab.

Procedure
a. Set up the flow rate apparatus as illustrated below.

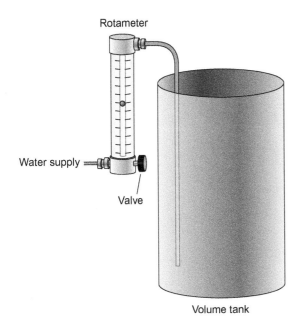

b. Place marks at 25 percent, 50 percent, 75 percent, and 100 percent volume on the vessel wall; 100 percent should equal a known volume.

c. Connect a water supply to the input of the rotameter and a piece of tubing to its outlet side.

d. With the open end of the tube placed in a basin or drain, fill the supply line, rotameter, and discharge tubing with water and adjust the flow rate to some point about midrange. Let it stabilize at that rate.

e. At the same time, place the discharge tubing from the rotameter into the vessel and start the timing device.

f. Fill the vessel until it reaches the 25 percent, 50 percent, 75 percent, and 100 percent volume marks. Record the time at each interval. Stop the timing device and remove the tubing from the rotameter.

Percent Level	Elapsed Time
25	
50	
75	
100	

g. Calculate the average flow rate.

$$\text{Flow Rate} = \frac{\text{Volume}}{\text{Time}}$$

Where: Volume units are the same as the rotameter units.
Example:

$$\text{Flow Rate} = \frac{2.0 \text{ gallon}}{6.5 \text{ minutes}}$$

$$\text{Flow Rate} = 0.3077 \text{ GPM}$$

h. Compare the calculated flow rate to the indicated flow rate on the rotameter.

i. Complete the final report for this lab.

Findings
Compile your observations and readings into a report that includes a sketch of the equipment, the recorded data, an explanation of the results of the experiment, and a conclusion. Include a paragraph explaining the difference between instantaneous flow versus total (accumulated) flow. Reference your class notes if necessary.

Chapter 6
Process Variables, Elements, and Instruments: ANALYTICS

Objectives

After completing this chapter, you will be able to:

6.1 Define terms associated with analytical instruments:
pH (acid or base) and ORP
conductivity
optical measurements
chromatography
combustion (O_2 and CO)
total organic carbon analyzer. (NAPTA Process Variables . . . : Analytics 1, 2*) p. 109

6.2 Identify the most common types and purpose of analytical devices used in the process industries and their purposes:
optical analyzer (turbidity analyzer or meter and/or opacity analyzer or meter)
color analyzers
conductivity meter
pH and ORP meters
oxygen analyzer

*North American Process Technology Alliance (NAPTA) developed curriculum to ensure that Process Technology courses will produce knowledgeable graduates to become entry-level employees in process technology. Objectives from that curriculum are named here in abbreviated form. For example, "(NAPTA Process Variables–Analytics 1, 2)" means that this chapter's objective relates to objectives 1 and 2 of NAPTA's Process Variables, Elements, and Instruments content related to analytics.

Process Variables, Elements, and Instruments: ANALYTICS 107

Lower Explosive Limit (LEL)—Upper Explosive Limit (UEL)

gas or liquid chromatograph and mass spectrometer

spectrophotometer (UV/VIS and/or infrared [IR]). (NAPTA Process Variables . . . : Analytics 2, 3, 5) p. 111

6.3 Identify common types of miscellaneous measuring devices:

vibration meter

proximity sensor

accelerometer

load cells

amp meter

decibel meter. (NAPTA Miscellaneous Measuring Devices 1, 2) p. 121

6.4 Discuss how analytical data affects the role of the process technician and how the process technician affects the operation of analytical instrumentation. (NAPTA Process Variables . . . : Analytics 4) p. 124

Key Terms

Accelerometer—a vibration measuring device (e.g., piezoelectric type), **p. 122.**

Acceptable limits—the operating range within which a piece of rotating equipment can operate without causing excessive wear to the bearings or other types of catastrophic failure, **p. 122.**

Analyzer measurements—qualitative and quantitative, **p. 111.**

Chromatogram—a graphic record of the separated components, **p. 111.**

Chromatography—a process where molecular components of a liquid or gas are separated and identified by means of a tube, called a column, **p. 111.**

Color/optical analyzers—photometers that operate in the visible light spectrum (400–800 mμ); two types include the visual color analyzer and the photometer, or spectral color analyzer, **p. 114.**

Combustion—the rapid consumption of fuel resulting in its conversion to heat, light, and gases, **p. 111.**

Conductivity—a measurement of the ability of a material to conduct an electrical current; the measure of the ability of a liquid (or any solution) to conduct electricity, **p. 110.**

Conductivity analyzer—a device that measures all ions in an aqueous solution, **p. 114.**

Conductivity meter—a device that measures the conductivity of a process sample by comparing it to a known value or standard cell, **p. 114.**

Electrolyte—a nonmetallic substance that is an ionic conductor; the greater a substance conducts an electrical charge, the more ionic the substance becomes, **p. 114.**

Gas chromatograph (GC) analyzer—an analyzer that provides the necessary means to accomplish the chromatographic separation and analysis, **p. 117.**

Gas chromatography—an analytical method that can provide a molecular separation of one or more individual components in a sample, **p. 117.**

Inline analyzer—an analyzer (either continuous or contiguous) installed directly in a process line that may be either a sample system or of the newer probe type, **p. 109.**

Mass flow rate—amount of mass passing through a plane per unit of time, **p. 123.**

Mass spectrometer (MS)—a device capable of separating a gaseous stream into a spectrum according to mass and charge, **p. 118.**

Opacity analyzers—optical analyzers used to determine how much particulate matter is in a gas sample, **p. 116.**

Optical measurement—a measurement that uses reflection, refraction, or absorption properties of light to measure the chemical or physical properties of a sample, **p. 111.**

Optical meters—any or all of the following: turbidity meters, opacity meters, and color meters, **p. 114.**

ORP analyzer—an electrochemical analyzer that measures a small electromotive force (EMF) across a hydrogen-ion-sensitive glass bulb; an ORP (oxidation reduction potential) meter also measures a small voltage (EMF) across its electrode and has the capability of detecting all oxidizing and reducing ions in the solution, **p. 113.**

Overspeed—a dangerous condition that can occur in a turbine or other type of equipment that moves too fast, **p. 123.**

Oxidation reduction potential (ORP)—a measure of redox potential created by the ratio of reducing agents to oxidizing agents present in the sample, **p. 113.**

pH—a measurement of the hydrogen ion concentration in a solution that can react and indicate whether a substance is an acid or a base, **p. 110.**

pH meter—an electrochemical instrument that measures the acidity and alkalinity of a solution; an analyzer that measures hydrogen concentration in a solution, **p. 112.**

Photometers—color analyzers that operate in the visible light spectrum (400–800 μm), **p. 114.**

Photometry system—an automated system used to determine the color of a sample, **p. 115.**

Process analyzers—unattended analytical instruments capable of continuously monitoring a process stream, **p. 109.**

Properties—characteristics of a substance, such as pH, ORP, conductivity, optical measurement, composition, and/or combustion capability, **p. 110.**

Qualitative—a measurement of the properties of a substance, **p. 110.**

Quantitative—a measurement of the amount of properties within a substance, **p. 110.**

Rectilinear speed—linear speed expressed in distance per unit of time (e.g., feet per second), **p. 123.**

Representative sample—a sample that contains portions of the process stream combined together over time or distance, **p. 111.**

Rotational speed—number of revolutions per unit of time (e.g., revolutions per minute, or RPMS), **p. 123.**

Sample system—the various components of a sampling apparatus that obtains, transports, and returns the sample with the analyzer itself conditioning and analyzing the sample, **p. 109.**

Spectrometer analyzer—an analyzer used to detect chemical components in a process sample by measuring variations in transmittance (or absorption) of a spectrum of light passed through the sample; may use visible light (VIS), ultraviolet light (UV), or infrared light (IR and NIR) as a source and detection measurement, **p. 119.**

Speed—the distance traveled per unit of time irrespective of direction (e.g., feet per second), **p. 123.**

Speed monitor—a device that measures speed; comprises a speed sensor and a readout/receiving device, **p. 123.**

Total carbon analyzers—analyzers used to determine how much carbon is in a sample; used to detect carbon based contaminants in steam condensate and wastewater, **p. 119.**

Turbidity analyzer—an optical analyzer used to determine the cloudiness of a liquid, **p. 115.**

Turbidity and opacity meters—optical meters; measure the transmittance or absorption of light passing through a sample, **p. 115.**

Velocity—the distance traveled over time or change in position over time, **p. 123.**

Vibration—the periodic motion of an object, **p. 121.**

Vibration meter—a device used to measure displacement, velocity, or acceleration due to vibration; consists of a pickup device, an electronic amplification circuit, and an output meter, **p. 121.**

Vibration sensors or monitors—device used to sense the effects of vibration and send a signal to a meter or monitor, or to shut down a device if operating limits are exceeded, **p. 122.**

Visual color analyzers—compare a physical sample with a standard by shining visible light through a sample; the color comparison is made by the human eye/brain, **p. 115.**

6.1 Introduction to Analytical Instruments and Terms

There are many types of instruments that analyze various streams within a process facility. Analytical instruments are designed to monitor the chemical and physical properties of the process stream (Figure 6.1). **Process analyzers** (meters) are laboratory instruments specifically designed to withstand the rigorous environmental conditions presented by the processing area. A process analyzer can be defined as an unattended analytical instrument that is capable of continuously monitoring a process stream.

A properly designed **inline analyzer** (continuous and contiguous) system is capable of doing the following steps for each sample taken:

1. Obtaining a representative sample from the process stream
2. Transporting the sample to the analyzer in a way that maintains the physical and chemical integrity of the sample
3. Conditioning the sample so that it can be introduced into the analyzer
4. Analyzing the sample
5. Returning the sample to the process or discarding the sample in another suitable manner

Of the five important design components listed above, a **sample system** needs to do three of them: Steps 1, 2, and 5. A sample system accounts for most of the problems encountered in an analyzer loop (see Figure 6.1) because of the job it must do. The sample system

Process analyzers unattended analytical instruments capable of continuously monitoring a process stream.

Inline analyzer an analyzer (either continuous or contiguous) installed directly in a process line that may be either a sample system or of the newer probe type.

Sample system the various components of a sampling apparatus that obtains, transports, and returns the sample with the analyzer itself conditioning and analyzing the sample.

Figure 6.1 Analytical instruments (typical online mounted unit, can be "indoor" or "outdoor").

CREDIT: © Emerson 2019.

can get plugged, contaminated, bent, broken, or compromised in any number of other ways. Many new inline, or probe type, analyzers eliminate the need for the problematic and expensive sample system. However, most analyzers still require a well designed sample system.

Analyzers are capable of measuring both **qualitative** and **quantitative** properties of a process. A qualitative analysis is a determination of the composition of a substance. For example, determining whether methane, ethane, propane, and butane are present in the sample is a qualitative test. A quantitative analysis is a determination of the amount or proportions of a substance. A typical quantitative measurement would be reported in terms of percentage or in parts per million (ppm) units.

There are hundreds of different kinds of analytical instruments designed to analyze many different chemical and physical properties, but the **properties** that are of greatest interest within the process industry where analytical instruments are used to measure or monitor include the following:

- pH
- Oxidation reduction potential (ORP)
- Conductivity
- Optical measurement
- Chromatography (concentration of components in process streams)
- Combustion (concentration of components in flue gas)

pH

pH is a measurement of the hydrogen ion concentration [H$^+$] of a solution that indicates how acidic or basic a substance is. Solutions having a pH below 7.0 are acidic. Solutions with a pH above 7.0 are basic. A pH of 7.0 is neutral—neither acidic nor basic.

$$pH = -\log_{10}(H^+)$$

This means pH is the negative base 10 logarithm of the hydrogen ion concentration of a solution. To calculate it, take the log of the hydrogen ion concentration and reverse the sign to get the answer.

Oxidation Reduction Potential (ORP)

ORP, also called redox, is a measurement of potential created by the ratio of reducing agents to oxidizing agents that are present in the sample. Where pH measurement is specific to hydrogen ion concentration, ORP measurement can be thought of as being specific to free electron concentration. This concept is described in more depth later in this chapter under the section Common Types of Analyzers.

Conductivity

Conductivity is a measurement of the ability of a material to conduct an electrical current. Industry uses conductivity as a means of determining the level of dissolved solids in a liquid, which in turn tells them when they need to blow down or treat a process fluid. Otherwise, damage to equipment or processes may occur.

$$\sigma = \frac{1}{\rho}$$

Where:

σ is the conductivity of the material in siemens per meter, $S \cdot m^{-1}$; ρ is the resistivity of the material in ohm meters, $\Omega \cdot m$

Conductivity is the inverse of resistivity and is measured in siemens/cm^2 and micromhos.

Optical Measurement

An **optical measurement** uses reflection, refraction, or absorption properties of light (electromagnetic waves) to measure the chemical or physical properties of a sample. The extent to which light is bounced back, bent, or absorbed provides information about the sample.

Optical measurement a measurement that uses reflection, refraction, or absorption properties of light to measure the chemical or physical properties of a sample.

Chromatography

Chromatography is the process in which molecular components of a liquid or gas are separated and identified by means of a tube, called a column. A special packing material, located in the column, selectively retains molecules of components of the **representative sample** as the sample propagates through the column when a carrier gas pushes the sample. The sample components move through the packing material according to component affinity for the packing and individual vapor pressures. A detector located downstream of the column is then able to sense these separated components as they emerge from the column. It produces a graphic record of the separated components, which is called a **chromatogram.**

Chromatography a process where molecular components of a liquid or gas are separated and identified by means of a tube, called a column.

Representative sample a sample that contains portions of the process stream combined together over time or distance.

Chromatogram a graphic record of the separated components.

Combustion

Combustion is the rapid oxidation of a substance resulting in its conversion to heat, light, and gases. During complete combustion, hydrocarbon fuels such as natural gas (CH_4) convert entirely to carbon dioxide (CO_2), water, and heat. In a boiler or furnace, monitoring of excess oxygen (O_2) or the remaining carbon monoxide (CO) in the flue gases can be used to enhance the combustion process. It does so by providing input into a control system which in turn limits the amount of excess air entering the firebox (the combustion compartment of the boiler or furnace). Oxygen is the only component in air that affects the combustion reaction, but it accounts for only about one-fifth (20.9 percent) of its total volume. The rest (primarily nitrogen at 78.1 percent) occupies space and steals heat energy from the system by carrying it out the flue. Being able to measure the concentration of components in the flue gas allows technicians to optimize the combustion process and comply with environmental permits.

Combustion the rapid consumption of fuel resulting in its conversion to heat, light, and gases.

6.2 Analytical Sensing or Measuring Instruments

Purpose of Analytical Devices

Analyzers are used within industry for reasons such as the following:

- Environmental monitoring and reporting
- Mechanical integrity of fixed equipment
- Economics
- Product quality assurance

ENVIRONMENTAL MONITORING AND REPORTING Analyzers, commonly referred to as *continuous environmental monitoring systems* (CEMS), may be used to report emissions directly to the Environmental Protection Agency (EPA). Ensuring the safety of both the environment and its citizens within the surrounding community is critical, since dangerous levels of contaminants emitted into the atmosphere could pose an emergency or life threatening condition. Also, EPA guidelines are becoming increasingly confining. To be sure that emissions coming from each unit fall within a given spectrum, **analyzer measurements** must be accurate and reliable to prevent potentially debilitating EPA fines as well as criminal charges.

Analyzer measurements qualitative and quantitative.

MECHANICAL INTEGRITY OF FIXED EQUIPMENT Permanent damage could occur to process equipment when the process is cloudy, contaminated, too acidic, or too basic. Fixed equipment such as exchangers, pumps, valves, vessels, pipes, and so on can become corroded, eroded, or plugged if the process goes unchecked. Analyzing instruments can check for these problems or the potential for these problems to occur, thus preventing equipment failure.

ECONOMICS Online product analysis gives technicians a real time view of the product. This allows adjustments while the product is being made and helps to eliminate off specification product. Online analysis is more expensive to install, but it is less expensive over the long haul because of reduction in off specification product.

PRODUCT QUALITY ASSURANCE Online analyzers are the first step in ensuring product quality, but analyzers in control laboratories are also used to run finished product samples to ensure all specifications are met before a product is sold to a customer. These analyzers are located in controlled environments and are calibrated under strict conditions. The measurements they make should be highly reliable, accurate, and repeatable.

Common Types of Analyzers

Online refers to a process measurement that is updated continuously or any measurement that does not involve a person physically going out to get a sample or taking a measurement–like a grab sample for laboratory analysis.

Example:

A gas chromatograph or a stack monitor that has a sample brought to the analyzer on a continuous basis is considered to be an online analyzer.

Inline usually refers to a measurement taken directly in a process pipe. Inline measurements might include a retractable pH probe, a retractable conductivity sensor, or a temperature element where the probe is located in the process stream.

Laboratory analysis involves a person physically going out to get a sample for industrial processing and quality control for lab analysis in a dedicated facility staffed by trained chemists, material scientists, and technicians with years of industry knowledge and expertise.

The more common types of analyzers used in the process industry include the following:

- pH and ORP meters
- Conductivity meter
- Optical analyzer (color, turbidity, and/or opacity)
- Gas chromatograph (GC) and GC mass spectrometer
- Spectrometer (UV/VIS and/or infrared light)
- Total carbon analyzer.

pH AND ORP METERS pH and ORP meters are among a group of instruments known as electrochemical analyzers. The pH analyzer, for example, is a specific ion analyzer that is capable only of measuring hydrogen ion concentration in a solution. A pH meter does this by measuring a small electromotive force (EMF) across a hydrogen-ion-sensitive glass bulb. The ORP meter also measures a small voltage across the electrode within the meter, but it registers all oxidizing or reducing ions in the solution. Both of these analyzers play an important role in controlling the chemistry of industrial processes.

pH Meter. The **pH meter** measures the acidity and alkalinity of a solution by measuring the hydrogen ion concentration. pH is defined mathematically as the negative logarithm of the hydrogen ion concentration: $pH = -\log_{10}[H^+]$. When an acid or base is mixed into water, the acid/base disassociates either partially or fully into positive and negative ions. The pH analyzer then measures the concentration of hydrogen ions (H^+) in solution by comparing the hydrogen ions to a buffered solution (exactly 7 pH) located inside the measuring

pH meter an electrochemical instrument that measures the acidity and alkalinity of a solution; an analyzer that measures hydrogen concentration in a solution.

electrode. The millivolt potential developed across the glass membrane is proportional to the pH of the solution. pH meters (Figure 6.2) are calibrated with standard pH solutions at several pH levels (typically 4.0, 7.0, and 10.0).

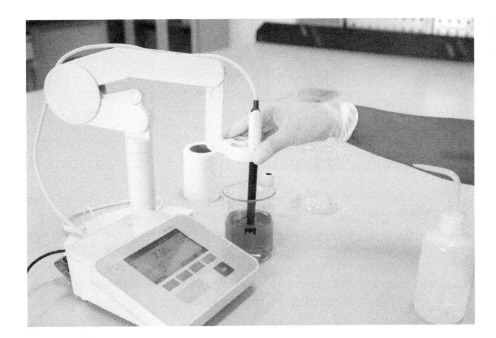

Figure 6.2 pH meter.
CREDIT: Choksawatdikorn/Shutterstock.

The pH scale ranges from 0 to 14, with 7 considered to be neutral. A pH of less than seven is considered to be acidic and a pH greater than seven is basic (caustic). Every technician needs to understand that each pH number constitutes a tenfold change in concentration. For example, a pH of 4 is 10 times more concentrated than a pH of 5; a pH of 4 is 1,000 times more concentrated with hydrogen ions than a pH of 7. See Table 6.1.

Table 6.1 pH Relationship to H Ion Concentration

	pH Range	H Ion Concentration
acid	pH 0–6	10^{-0}–10^{-6}
neutral	pH 7	10^{-7}
base	pH 8–14	10^{-8}–10^{-14}

Oxidation Reduction Potential (ORP) Meters. Another type of electrochemical analyzer is the **ORP analyzer**, which stands for **oxidation reduction potential (ORP)** or *redox* (short for reduction oxidation). An ORP measurement indicates how much *free electron* potential exists in the sample. In the chemical world, oxidation means the loss of electrons, whereas reduction means the gain of electrons. This potential is similar to pH in that they both are measuring ion potential based on concentration. ORP analyzers are used to monitor the electrochemical properties of a reaction as it is taking place. ORP analyzers let a process technician see into the chemical reaction of a process in a way unlike any other analytical method. ORP analyzers are used in control loops in the same manner as pH meters. ORP meters (Figure 6.3) are calibrated in millivolts. For the ORP scale, positive millivolts indicate oxidation and negative millivolts indicate reduction.

ORP analyzer an electrochemical analyzer that measures a small electromotive force (EMF) across a hydrogen-ion-sensitive glass bulb; an ORP (oxidation reduction potential) meter also measures a small voltage (EMF) across its electrode and has the capability of detecting all oxidizing and reducing ions in the solution.

Oxidation reduction potential (ORP) a measure of redox potential created by the ratio of reducing agents to oxidizing agents present in the sample.

CONDUCTIVITY METERS Conductivity is the reciprocal of resistance. Conductivity is an electrochemical property measured in micromhos and in the metric unit siemens per square centimeter (siemens/cm^2).

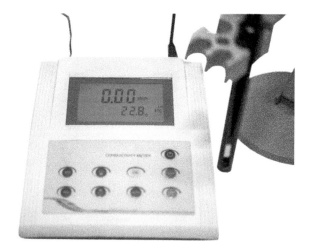

Figure 6.3 ORP meter.
CREDIT: Rabbitmindphoto/Shutterstock.

Electrolyte a nonmetallic substance that is an ionic conductor; the greater a substance conducts an electrical charge, the more ionic the substance becomes.

Conductivity analyzer a device that measures all ions in an aqueous solution.

When a salt such as table salt (sodium chloride [NaCl]), or another strong electrolyte is dissolved in water (Figure 6.4), the salt disassociates into cations carrying a positive (+) charge and anions carrying a negative (−) charge. Once this happens, the solution acts as an **electrolyte** (a nonmetallic substance that becomes an ionic conductor). Table salt is just one example of an electrolyte that acts in this manner. Other electrolytes can be either metallic or nonmetallic and act in a similar manner. The more ionic the solution becomes, the greater its ability to conduct an electrical charge. A **conductivity analyzer** measures all ions in an aqueous solution.

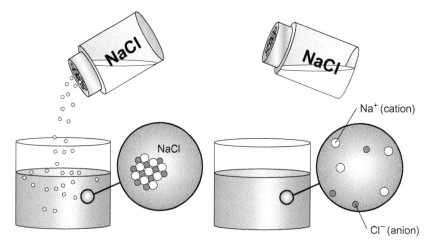

Figure 6.4 Salt (NaCl) introduced into water.

Conductivity meter a device that measures the conductivity of a process sample by comparing it to a known value or standard cell.

Optical meters any or all of the following: turbidity meters, opacity meters, and color meters.

Color/optical analyzers photometers that operate in the visible light spectrum (400–800 mμ); two types include either the visual color analyzer and the photometer, or spectral color analyzer.

Photometers color analyzers that operate in the visible light spectrum (400–800 μm).

One important industrial conductivity application is in measuring for dissolved solids in water. Pure water is nonconductive. There are various levels of conductivity in boiler water, condensate, and cooling tower water.

A **conductivity meter** (see Figure 6.3) uses a standard cell (1 cm^2 in size) to quantify the ability of a process material to conduct an electrical current. Numerous styles or types of conductivity probes are used to insert in the process. Conductivity meters use standards to check calibration at regular intervals, and a reference (or control) solution is used for that purpose.

OPTICAL ANALYZERS (COLOR, TURBIDITY, AND OPACITY) Optical measurements are those measurements that determine how much reflection, refraction, or absorption of visible light takes place through a fixed distance in a sample. Generally, **optical meters** include color meters, turbidity meters, and opacity meters.

Color. Color measurement implies the measurement of visible light. **Color/optical analyzers** are **photometers** that operate in the visible light spectrum (400–800 μm) (Figure 6.5). A color analyzer is used to determine the color of a process sample. Some off color products are impossible to sell. Take for example, bottled water. Most consumers would not buy or drink water if it had a tinge of yellow coloring. The same is true for other mass-produced products.

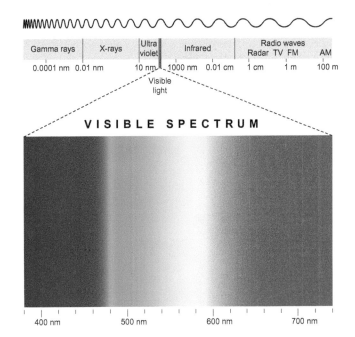

Figure 6.5 Visible light spectrum.

CREDIT: Peter Hermes Furian/Shutterstock.

Color can also indicate process problems. If a chemical reaction goes too far or is heated beyond a certain point, a color change may result. Some products that are off specification with color can still be sold if they can be reworked. Color analyzers (Figure 6.6) are found in the process areas of the plant as well as in the lab.

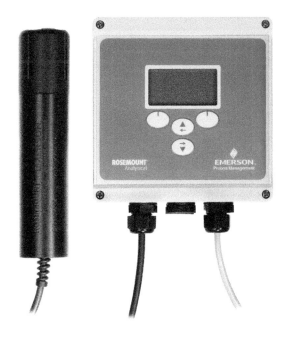

Figure 6.6 Optical analyzer—color.

CREDIT: © Emerson 2019.

There are two types of color analyzers: the **visual color analyzers** and photometers, or spectral color analyzers. In the visual type, a process technician or lab analyst visually compares the sample to a standard. A visible light is shone through the sample, is detected by the human eye, and is analyzed by the human brain. This method is obviously the older of the two and is highly subjective, since it depends on an individual's color perception and discretion. The other method is photometric, using a **photometry system** to analyze the color of the sample. Although the photometry system is automated, it may use a standard sample for comparison.

Turbidity. **Turbidity and opacity meters** essentially test the same thing; they measure the transmittance or absorption of light passing through a sample. The **turbidity analyzer** (Figure 6.7)

Visual color analyzers compare a physical sample with a standard by shining visible light through a sample; the color comparison is made by the human eye/brain.

Photometry system an automated system used to determine the color of a sample.

Turbidity and opacity meters optical meters; measure the transmittance or absorption of light passing through a sample.

Turbidity analyzer an optical analyzer used to determine the cloudiness of a liquid.

is used to determine the cloudiness of a *liquid*. Cloudiness is due to suspended particles. To detect the particles, the photometer is located across from a light source so that it detects the result of the scattering or absorption of the light beam by the particulate matter in the sample.

Figure 6.7 Optical analyzer—turbidity.
CREDIT: © Emerson 2019.

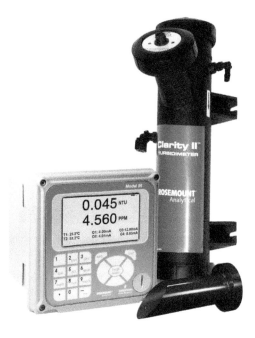

Types of electromagnetic waves used for the light source within these meters include the following:

- Gamma rays
- X-rays
- Ultraviolet waves
- Visible
- Near infrared (NIR)
- Thermal infrared
- Microwave
- TV/radio

Opacity analyzers optical analyzers used to determine how much particulate matter is in a gas sample.

Opacity. **Opacity analyzers** (Figure 6.8) are used to determine how much particulate matter is in a *gas* sample. Opacity analyzers work essentially the same as turbidity analyzers. Again, a beam of light projecting from a source located on one side of a furnace or boiler

Figure 6.8 Optical analyzer—opacity.
CREDIT: © Emerson 2019.

stack penetrates the gases rising through the stack. A photodetector across from the source provides an optical measurement indicative of the absorption and scattering of the light beam as it strikes particulate matter in the stack.

OXYGEN ANALYZER Oxygen (O_2) level measurements must be accurate and are essential in preventing injury or death in situations where safe levels are required. The most common use of oxygen gas detection is in confined spaces or closed areas not designed to be permanently occupied. Oxygen level detection instruments trigger an alarm when the oxygen level drops below 19.5 percent volume, which is the OSHA mandated level. It is critical that O_2 gas samples be taken prior to entering closed spaces and that oxygen levels continue to be monitored even after entry.

Lower Explosive Limit (LEL)–Upper Explosive Limit (UEL)

Lower explosive limit (LEL) refers to the lowest concentration of gases, fumes, and vapors required to produce an explosion in the presence of an ignition source such as flame or heat. When the vapors of a flammable or combustible liquid are mixed with air in the right proportion, they can lead to explosion or combustion if some form of ignition is present. Generally, there is a minimum concentration of gas below which a flammable or combustible gas will not ignite. There is also an upper concentration above which the air will not ignite; it is called the upper explosive limit (UEL). These are usually measured as percentages. The Occupational Safety and Health Administration (OSHA) has guideline limits for the presence of explosive concentrations, as well as for the safe storage of these to prevent fire and explosion.

Chromatography

Gas chromatography is an analytical method that can provide a molecular separation of one or more individual components in a sample. This method identifies the chemical composition of the molecules within the separated sample. A **gas chromatograph (GC) analyzer** provides the necessary means to accomplish the chromatographic separation and analysis. A GC takes a mixture and separates the mixture into its individual components with each being identified and quantified.

In continuous processes, an online GC can actually provide feedback (closed-loop) to an instrument control loop or just give information that can be used in an informative manner (open-loop). A GC can monitor as few as one component in the sample and produce a standard instrument signal that can be used in the same way as any other transmitter in a control loop. All chromatographs operate in the same manner. The GC (Figure 6.9A) consists of an injection port that vaporizes the sample, a column (a hollow tube filled

Gas chromatography an analytical method that can provide a molecular separation of one or more individual components in a sample.

Gas chromatograph (GC) analyzer an analyzer that provides the necessary means to accomplish the chromatographic separation and analysis.

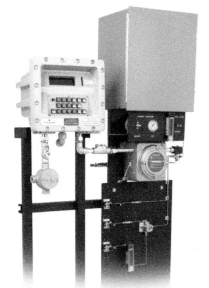
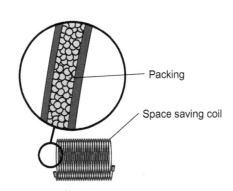

A. B.

Figure 6.9 A. Gas chromatograph. **B.** Gas chromatograph coiled column configuration.

CREDIT: A. © Emerson 2019.

with a special material called packing), a detector that senses the separated gases, and an output device that generates a chromatogram. The GC may also include a microprocessor that takes the signal generated by the detector and converts it into a chart and/or reports the output as a weight percent, as parts per million (ppm) volume, and so on. The column may be configured in coils (see Figure 6.9B) to conserve space within the GC. See the chromatogram depicting the separated components as peaks on a graph (Figure 6.10).

Figure 6.10 Gas chromatogram.

CREDIT: Chromatos/Shutterstock.

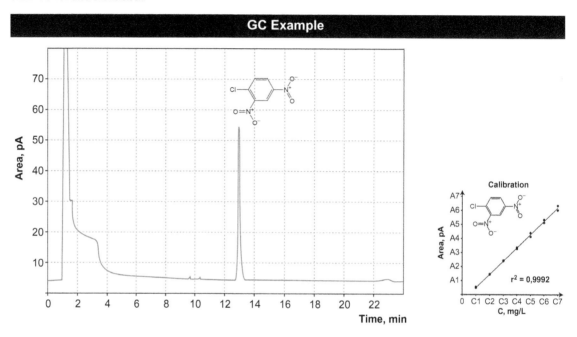

Mass Spectrometers

Mass spectrometer (MS) a device capable of separating a gaseous stream into a spectrum according to mass and charge.

A **mass spectrometer (MS)** (Figure 6.11) is a device capable of separating a gaseous stream into a spectrum according to mass and charge. There are chromatograph applications that measure the composition of liquids without converting to a vapor state. These may be referred to as liquid chromatographs.

Figure 6.11 Mass spectrometer.

CREDIT: Cergios/Shutterstock.

In one type of mass spectrometer analyzer, a sample of process gas is converted into an accelerated stream of highly charged ions. The sample is then sent through a magnetic filter (field) and then into a detector. The more massive particles tend to go in a straight line through the magnetic prism; the less massive particles tend to move away from the center toward the negatively or positively charged rods in the prism. The detector located at the exit of the magnetic prism senses both the mass and charge of each particle at their position of impact on the detector. Different molecules tend to have unique identifying patterns represented by the impact positions, which are registered on the detector.

Mass spectrometers may be used in many of the same applications as gas chromatographs. However, GC output is on a volume percent data, while mass spectrometers provide weight percent data. This can be useful in identification of components. Mass spectrometers are faster than GCs, but they are also a lot more expensive. The most common place to find a mass spectrometer analyzer is in a quality control lab.

A mass spectrometer analyzer is sometimes used in combination with a gas chromatograph. Effectively, the mass spectrometer analyzer becomes the detector for the chromatograph. There are several reasons why these two analyzers are used in combination. One is that the chromatograph separates the sample gases into partitioned components (or groups of components) so that they can have a more definitive analysis when they enter the MS. Another is that the gas chromatograph can provide a separation, then backflush the unwanted components in the sample, leaving only the components of interest to be analyzed. Of course, when the GC and MS are used in parallel with each other, the analysis time is longer. These units can be a combination gas chromatograph mass spectrometer (GC-MS) unit.

SPECTROMETERS A **spectrometer analyzer** (Figure 6.12) is used to detect and/or quantify chemical components in a process sample by measuring variations in transmittance (or absorption) of a spectrum of light passed through the sample. A spectrometer may use visible light (VIS), ultraviolet light (UV), or infrared light (IR and NIR) as a source and detection measurement.

Spectrometer analyzer an analyzer used to detect chemical components in a process sample by measuring variations in transmittance (or absorption) of a spectrum of light passed through the sample; may use visible light (VIS), ultraviolet light (UV), or infrared light (IR and NIR) as a source and detection measurement.

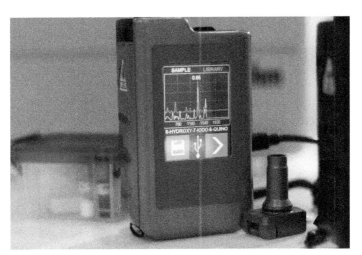

Figure 6.12 Spectrometer.
CREDIT: Warut Sintapanon/Alamy Stock Photo.

The word spectrum implies a continuum of wavelength bands of light. When a beam of white light is passed through a prism, a spectrum of colors can be seen as a result. A spectrophotometer is capable of detecting a spectrum of wavelengths rather than a single frequency band. Scientists have determined that unique and reliable fingerprints for specific chemical components in a mixture of process fluids can be generated when exposed to a spectrum of light analysis. For example, CO_2 is best analyzed with the near infrared (NIR) spectrum.

Total Carbon Analyzers

Total carbon analyzers (Figure 6.13) are used to determine how much carbon is in a sample. They are used to detect carbon based contaminates in steam condensate and wastewater. Carbon analyzers can be used to determine how much organic as well as inorganic carbon is

Total carbon analyzers analyzers used to determine how much carbon is in a sample; used to detect carbon based contaminants in steam condensate and wastewater.

Figure 6.13 Total organic carbon analyzer.

CREDIT: Engineer story/Shutterstock.

present in the sample. Carbon analyzers can test for total carbon (TC), total inorganic carbon (TIC), or total organic carbon (TOC).

The most common type of total carbon analyzers ultimately converts all carbon based compounds into carbon dioxide (CO_2) and then determines the concentration of total carbon based on a CO_2 analysis with an NIR analyzer.

Miscellaneous Monitors

Besides the many obvious types of process analyzers that process technicians encounter, there are still many other types worth mentioning that play a very important role in day-to-day operations. These monitors are sensitive instruments that are usually only maintained by instrument personnel. However, process technicians should observe these monitors on their daily rounds in their work areas to ensure appropriate housekeeping is performed so that airflow around the monitors is maintained. Process technicians may come in contact with these monitors in certain work sites or on special occasions.

There are other monitors that analyze pipe welds, pipe corrosion, scrubber and column tray placement, and more. These monitors may use radiation or ultrasonic types of detection methods. There are also portable types of nuclear analyzers that give proper chemical composition of pipes or valve parts that are capable of giving a printout listing the percent composition by common name.

PERSONNEL MONITORS Plant workers sometimes wear a monitor to detect potential exposure to harmful substances. A ring badge or a pocket dosimeter may be used that is capable of indicating exposure to nuclear radiation, and both badge and pocket types respond to x-rays or gamma rays. Chemical badges react by changing colors to represent exposure levels in ppm, while electronic pocket devices detect exposure and respond by various means. Process technicians must ensure that these devices are kept in good operating condition and must discard exposed badges appropriately.

AREA MONITORS Specific types of monitors (e.g., Cl_2, CO, H_2S) are placed around loading and unloading stations, compressor stations, specific vessels, or piping manifolds where dangerous chemicals could potentially leak. These monitors work in conjunction with wind direction and speed or weather stations.

UNIT OR PLANT BOUNDARY MONITORS Unit or plant boundary monitors are similar to area monitors, but they address a broader scale. They may be installed at or near unit boundaries or on property fence lines. They may employ laser beams or passive IR coupled with computers to analyze for several types of gases. Laser beams may be shot down property lines, using mirrors at various points to return the final beam to a computer to be analyzed for escaped gases.

6.3 Miscellaneous Devices

Some devices do not fall into the main categories or types of instrumentation used in process facilities but that are, in fact, commonly used. Two types that require discussion here are devices that measure vibration and those that measure speed. Several applications and variations of these miscellaneous measuring devices are also discussed.

Vibration

Vibration (Figure 6.14) in an object, device, or system is the random or periodic change in velocity, acceleration, or displacement from a predetermined point. A **vibration meter** (Figure 6.15) is a device used to measure velocity, acceleration, or displacement due to vibration. Vibration meters consist of a pickup device, an electronic amplification circuit, and an output meter.

Vibration the periodic motion of an object.

Vibration meter a device used to measure displacement, velocity, or acceleration due to vibration; consists of a pickup device, an electronic amplification circuit, and an output meter.

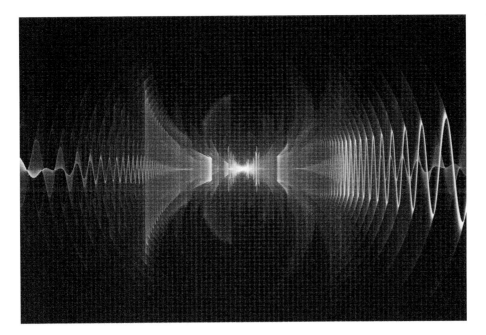

Figure 6.14 Vibration.
CREDIT: Pixelparticle/Shutterstock.

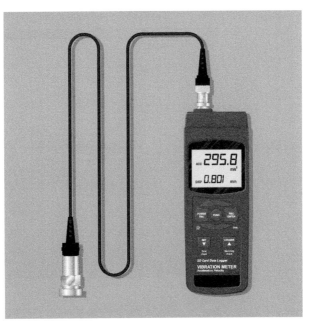

Figure 6.15 Vibration meter.
CREDIT: Rumruay/Shutterstock.

Acceptable limits the operating range within which a piece of rotating equipment can operate without causing excessive wear to the bearings, or other types of catastrophic failure.

Vibration sensors or monitors a device used to sense the effects of vibration and send a signal to a meter or monitor, or to shut down a device if operating limits are exceeded.

Accelerometer a vibration measuring device (e.g., piezoelectric type).

There are **acceptable limits** within which rotating equipment can operate without causing excessive wear to the bearings or otherwise causing the equipment to be subjected to high stress forces. If operating limits are exceeded, severe and potentially catastrophic damage can occur very quickly. To alleviate this problem, the vibration in large and/or high speed rotating equipment must be monitored and controlled.

For example, most compressors have high vibration alarms as well as shutdowns associated with them. The two types of vibration encountered are either axial or radial thrust on the shaft. If a vibration monitor works properly, the compressor shuts down before major damage is caused, thereby avoiding a safety problem or excessive downtime in the process. Having **vibration sensors or monitors** on high energy rotating equipment makes good sense from a safety standpoint as well as from an economic standpoint.

Vibration can be measured with an **accelerometer** (Figure 6.16). One type of accelerometer sensor is the piezoelectric type. The piezoelectric sensor is self generating. It generates a small voltage when strained or pushed on by the operating equipment. This characteristic makes this sensor a good choice for measuring vibration. The piezoelectric crystal (a type of

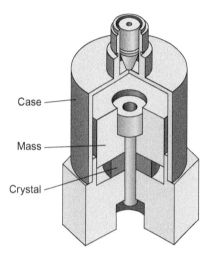

Figure 6.16 Accelerometer sensor (cutaway).

transducer) is attached to a sensing mass (Figure 6.17). The sensing mass is a small but relatively heavy piece of metal that provides enough mass so that when the rotating equipment vibrates, the transducer can be strained and subsequently creates an output signal. As the sensing mass changes directions, vibrating back and forth with the rotating equipment, the

Figure 6.17 Accelerometer sensor (function).

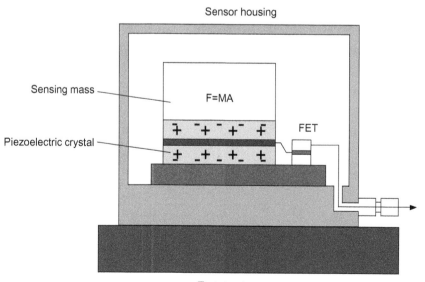

mass presses or pulls on the crystal that in turn responds with a voltage output. This voltage can be measured and transduced into an instrument signal. The signal drives an electronic circuit, providing a readout, high vibration alarm, and/or shutdown point.

Speed

Speed is the distance traveled per unit of time irrespective of direction. **Velocity** is speed with direction (Figure 6.18). A **speed monitor** (see Figure 6.15) is a device that measures speed. As with other monitors, a speed monitor consists of a speed sensor and a readout/receiving device.

Speed the distance traveled per unit of time irrespective of direction (e.g., feet per second).

Velocity the distance traveled over time or change in position over time.

Speed monitor a device that measures speed; comprises a speed sensor and a readout/receiving device.

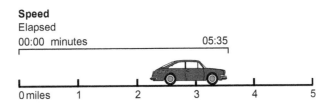
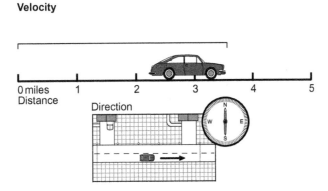

Figure 6.18 Speed and velocity.

The speed sensor or pickup device can be one of many different designs, such as a tachometer generator or an induction or magnetic proximity sensor. Another type of sensor is an optical device that can count transparent slits or reflective patches in a rotating wheel. All these sensors detect rotation by counting (one or more) markers located on the rotating component. For example, the proximity sensors count the teeth on a rotating wheel while the optical sensor counts transparent slits in a rotating wheel.

Speed can be monitored as **rectilinear speed** (as in linear feet/second) and **rotational speed** (as in revolutions/minute or RPMS). Generally, the direction of motion of a monitored process can be identified, making the determination a velocity measurement instead. If direction of motion is not important, then speed is a sufficient descriptor. In any case, monitoring rate of motion can be extremely important.

Overspeed, for example, is one of the most dangerous conditions occurring in a turbine or other type of rotating equipment. If a turbine is left to spin out of control, one of two things will happen. First, the bearings will eventually seize, causing a lot of damage and bringing the turbine to a stop. Second, the turbine will accelerate to the point where the centrifugal forces cause the machine to explode, resulting in injury to personnel and/or equipment. In either case, speed monitoring and control is vitally important with rotating process equipment. Most high speed rotating equipment has overspeed protection installed in the form of mechanical and/or electronic trip devices.

Another application for speed control is with rectilinear speed (or just linear speed). One way to measure the flow of solid materials is to observe the material on a moving belt. If the conveyor belt moves over a weigh scale, then a **mass flow rate** can be calculated as well as a total flow. To do this accurately, the instantaneous weight measurement must be accurate and the rate of motion must be known. The rectilinear speed of a conveyor belt, or

Rectilinear speed linear speed expressed in distance per unit of time (e.g., feet per second).

Rotational speed number of revolutions per unit of time (e.g., revolutions per minute, or RPMS).

Overspeed a dangerous condition that can occur in a turbine or other type of equipment that moves too fast.

Mass flow rate amount of mass passing through a plane per unit of time.

the speedometer in a car, can be measured by converting the linear (straight line) motion into rotational motion and then measuring it accordingly.

6.4 Analytical Instruments and the Role of the Process Technician

Analytical instrumentation provides process technicians with a unique means of *looking* into the process. Analysis gives them an insight into the chemical composition and/or physical attributes of the process fluid missed by the more common measurements of temperature, pressure, level, and flow.

The role of the process technician has changed significantly from expectations in the past. Today's role includes operator assisted maintenance, in which process technicians initiate the first steps in troubleshooting for problems in the unit. Operations personnel are also expected to provide assistance by doing routine tasks and checks that relate to instrumentation and analytical equipment. For example, they are expected to monitor carrier gas bottles and/or to check for analytical sample flows, temperatures, and pressures. Process technicians must have a basic understanding of the theories and operations of instruments, analytical devices, and their sample systems.

Summary

Process facilities have many types of analytical instruments that are designed to monitor various chemical and physical properties of process streams. Analyzers, sometimes called meters in the field, are specially designed instruments that are used in more harsh environments than their counterparts in controlled atmospheres, such as in laboratories. These analytical instruments are also designed to be left unattended and continuously monitor process streams.

Sample systems are responsible for obtaining and transporting samples before they go into some analyzers. After the sample is analyzed, the sample system either returns the sample to the process or suitably discards the sample. Many analyzers are installed in line (inline) or have a probe that eliminates the need for a sample system. These innovations help to reduce problems and expenses associated with sampling systems.

Analyzers measure quantitative (proportion) and qualitative (composition) properties of process streams. With hundreds of types of properties that can be analyzed, there are a few that are most common within the industry.

The more common types of analyzers include the following:

- *pH or ORP meter:*
  Used in the control of the chemistry of industrial processes; electrochemical; pH meters only measure hydrogen ion concentrations in solutions; ORP meters measure a small voltage across an electrode and also see all the oxidizing or reducing ions in a solution

- *Conductivity meter:*
  Measures all ions in an aqueous solution with the more ions in the solution giving the solution a greater ability to conduct an electrical charge; commonly used to detect contaminants in condensate or spent steam

- *Optical analyzer (color, turbidity, and/or opacity):*
  Optics such as reflection, refraction, or absorption of visible light observed through a fixed distance in a sample; color meters (visual and photometer or spectral color) operate in the visible light spectrum with results in a simple on spec or off spec (off color) result; turbidity analyzers determine the cloudiness of a liquid using various forms of light with a photometer for detection of either the scattering or absorption of the beam; opacity analyzers are used to determine how much particulate matter is in a gas by using a beam of light to penetrate rising gases and detect by a photo detector like the turbidity meter

- *Gas chromatograph (GC) and a combination gas chromatograph mass spectrometer (GC-MS) unit:*
  GCs provide a molecular separation of one or more individual components in a sample to identify the chemical composition (proportionally); GCs can be stand-alone

units or included into a loop by transmitting the measured data to control an operation; GC-MSs separate gas streams into spectrums according to mass and charge; GC-MSs are more expensive and mostly found only in laboratory environments

- *Spectrophotometer (UV/VIS and/or infrared):*
Used to detect chemical components in a process sample by measuring variations in transmittance (or absorption) of a spectrum of light passed through a sample; each specific chemical component has its own fingerprint when exposed to a spectrum of light analysis with the transmittance plotted (spectrogram) with respect to wavelength

- *Total carbon analyzer:*
Used to determine how much carbon (organic and inorganic) is in a sample in steam condensate and wastewater

Through the use of analytical instrumentation, process technicians have a way to look into various processes that may be missed by the more common measurements of pressure, temperature, level, and flow. Process technicians are expected to understand the theories and operations of these instruments so they can troubleshoot process problems.

Vibration is the amount of change in velocity, acceleration, or displacement from a predetermined point as applied to an object, a device, or a system. To monitor this change, a meter may be installed. The meter is composed of a pickup device sensor, an electronic amplification circuit, and an output meter that may or may not have a transducer to communicate a signal of vibration indicators that are out of an operating limit.

Monitoring vibration in rotating equipment is vitally important since vibration amounts exceeding limits could cause catastrophic damage to equipment and personnel.

Another type of vibration measuring device is an accelerometer (piezoelectric type) that has a sensor that self-generates an electrical voltage when the sensor is either strained or pushed on by the operating equipment. The resulting voltage can be measured and transduced into an instrument signal that drives an electronic circuit providing a readout, high vibration alarm, and/or a shutdown point.

Speed is the distance traveled per unit of time irrespective of direction while velocity is speed with direction. Speed monitors measure speed and comprise a speed sensor and a readout/receiving device. Speed sensors (pickup devices) can be made in several different designs such as a tachometer generator, induction or magnetic proximity sensor, or an optical type device that counts transparent slits or reflective patches in a rotating wheel.

Speed can be measured rectilinearly (feet per second) or in rotational speed (revolutions per minute). The direction of motion of the velocity may or may not be important, but monitoring the rate-of-motion can be extremely important.

Machines have an overspeed trip to prevent the centrifugal forces from tearing the rotating assembly apart. Excessive overspeed for a prolonged period of time could also cause the bearings to seize or malfunction. This could cause major damage to the machine or even an explosion if the machine contained flammable material and this was allowed to escape through damaged seals.

The speed of solids across a conveyor path can be measured rectilinearly (linear speed versus rotating speed). A mass flow rate is established through an instantaneous weighing scale measurement and the rate of motion of the conveyor belt itself. Rotational motion can be converted to linear motion and vice versa depending on what type of speed needs to be measured. A car odometer is an example of converting the linear motion of driving a certain number of miles and measuring them by how many times they rotate fully (number of rotations times the circumference of the tire). The speedometer, by comparison, takes the amount of time (hour) and the amount of miles covered (as in the odometer) and determines the speed (miles per hour) of the car.

Checking Your Knowledge

1. Match the terms below with their correct definitions.

Term	Definition
I. pH	a. Measurement of the consumption of fuel
II. ORP	b. Measurement of the ratio of reducing agents to oxidizing agents
III. Conductivity	c. Measurement of hydrogen ion concentration
IV. Chromatography	d. Measurement of the molecular components of a liquid or gas
V. Combustion	e. Measurement of the ability of a solution to conduct electricity

2. Which of the following instrument capabilities would you use to measure pH?
   a. Measuring reflection, refraction, or absorption of light
   b. Measuring the level of ionic concentration
   c. Measuring variations in the transmittance of light through a sample
   d. Measuring the molecular composition
   e. Measuring free electron potential

3. In the chemical world, oxidation means the loss of _____.
   a. pH
   b. hydrogen
   c. electrons
   d. protons

4. Which type of analyzer would be used to find out what components and their respective percentage are in a sample stream?
   a. pH meter
   b. Gas chromatograph
   c. ORP meter
   d. Optical analyzer

5. Which type of analyzer is best suited to measure cloudiness in process water?
   a. pH meter
   b. ORP meter
   c. Turbidity meter
   d. Opacity analyzer

6. At what percentage point does an oxygen analyzer trigger an alarm in confined spaces?
   a. 19.5%
   b. 20%
   c. 35%
   d. 85%

7. _____ in an object, a device, or a system is the random or periodic change in velocity, acceleration, or displacement from a predetermined point.
   a. Voltage
   b. Variance
   c. Vibration
   d. None of the above

8. _____ is speed with direction.
   a. Variance
   b. Vibration
   c. Velocity
   d. Viscosity

9. A piezoelectric sensor is used in what type of vibration device?
   a. Magnetic proximity
   b. Optical
   c. Tachometer generator
   d. Accelerometer

NOTE: Answers to Checking Your Knowledge questions are in the Appendix.

Student Activities

1. Draw a horizontal linear scale from 0 to 14 using each increasing number (2, 3, 4, etc.) as a point on the scale. Label the diagram for pH. Perform research (library, internet, etc.) on common household substances to determine their pH levels. List the verified items (at least 20) by their scale number.
   NOTE: Many commercial substances (e.g., shampoo) list pH levels on the label.

2. Construct a simple conductivity tester as per the description that follows. Determine the conductivity of various metal and nonmetal objects by performing the following steps:
   a. Put a flashlight bulb in a socket and mount them onto a piece of wood.
   b. Connect one wire from the socket to one terminal of the battery.
   c. Connect another wire to the other side of the socket (NOT to the battery).
   d. Connect a third wire to the other battery terminal (NOT to the socket).
   e. Wrap (at least six times) each of the loose wire ends to clean head thumbtacks.
   f. Place the tacks (after wire wrapping) firmly into the erasers of two pencils that are used as the conductivity probes.
   g. Now that you have constructed a conductivity tester (a simple conductivity analyzer), test a variety of objects and record the results in chart form.

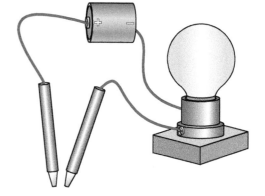

3. Identify analytical devices in the lab.

4. Use a pH meter to measure pH. Make a 5 percent, 10 percent, and 15 percent solution using Peter's fertilizer and distilled water. Use the solutions to define clarity and concentration. Determine the concentration mathematically, then use a meter to determine the pH of each solution.

5. Use conductivity meter to show the concentration of dissolved solids in cooling towers and boilers.

6. Use oxygen analyzers to check oxygen concentration in air and other gas streams.

7. Use gas test meters to show flammability of gasoline and diesel.

8. **ANALYTICAL (pH) MEASUREMENT: Laboratory Procedure**

Background Information

pH is the hydrogen ion concentration of an aqueous (water based) solution. Pure water ionizes to a very small extent to produce a few hydrogen and hydroxyl ions in equilibrium with the water molecules. Pure water always contains equal amounts of each of these ions and is considered neutral. An acid is a substance that yields hydrogen ions when dissolved in water. Bases are substances that yield hydroxyl ions when dissolved in water. The relative strength of an acid or base is found by comparing the concentration of hydrogen ions in solution with that of pure water. Any solution that contains equal concentrations of hydrogen and hydroxyl ions is neutral. Any solution that contains an excess of hydrogen ions is acidic, and any solution that contains an excess of hydroxyl ions is basic.

The pH probe is designed to selectively measure a hydrogen ion concentration difference between the inside of the pH sensitive glass probe and the solution it is immersed in. The small voltage (millivolt) produced by this probe is usually indicated on a meter marked 0 to14 with 7 at midscale. Seven is considered to be neutral having equal amounts of hydrogen ions and hydroxyl ions.

Finally, each whole number on the pH scale represents a tenfold change in concentration. That means that a pH of 5 is ten times more acidic than a pH of 6.

Materials Needed

- a pH meter and pH probe(s)

    NOTE: The pH probe(s) may be a combination probe (singular) or it may be two separate probes. In this lab, assume that the probe is a singular combination probe.

- five 250-mL beakers
- buffer solutions: 4 pH and 10 pH (Alternatively, rather than a solution of one material in different concentrations, use different liquids. Try fruit juices, milk, Coca-Cola®, and liquid detergent.)
- squirt bottle filled with distilled water for rinsing the probe
- distilled water or tap water to make up fertilizer solutions
- fertilizers such as Peters® or Miracle-Gro®

Safety Requirements

Safety glasses are required in the lab.

Precautions

Glass electrodes are very fragile. Electrodes should be left in liquid.

Procedure

a. Check the calibration of the pH meter by immersing the probe in the two buffer solutions (first the 4 pH buffer solution and then the 10 pH buffer solution). Rinse the probe with distilled water between separate immersions. Always rinse the probe before placing it in another beaker.

b. If the calibration is NOT correct, follow the instructions provided by the instructor or read and perform the calibration procedure found in the instruction manual. Once the meter is correct, go on to the next step.

c. Add distilled water or tap water to a separate beaker. Place the probe into the water and record the reading.

    pH = _____

d. Make up a 5 percent solution of fertilizer in the beaker. Record the new pH.

    pH = _____

    NOTE: To make a 5 percent solution, add 5 grams of fertilizer to the beaker and enough distilled water to bring the solution level to the 100-mL mark. Use this procedure to make up the other solutions. Another option is to add the additional 5-gram increments to the previous solution.

e. Make up a 10 percent solution of common fertilizer in the same or another beaker. Record the new pH.

    pH = _____

f. Finally, make up a 20 percent solution of common fertilizer in the same or another beaker. Record the new pH.

    pH = _____

g. Correlate the percent solution concentration to the pH recorded.

	Percent Solution	pH
1	5%	
2	10%	
3	20%	

h. Draw a graph representing the percent versus the pH.

Additional Information

The actual solution concentration (chemical to water weight concentration) will probably not have a linear relationship with the resulting pH. Specific chemicals will have different effects on the water and its resulting pH. If you were careful while conducting the lab procedure, your data should support a reasonable conclusion.

Findings

Compile your observations and readings into a report that includes your recorded data, an explanation of the results of the experiment, and a conclusion. Include a paragraph explaining correlation of the observed pH for each concentration level.

9. Draw a scale and label it from 0 to 100 with 10 (0, 10, 20, 30, etc.) evenly spaced and marked increments along the line. Look for things that produce vibrations in your home, at work, at school, or any other place you happen to be going. Make sure to have a full range of vibration intensities. On the scale you have drawn, try to visually indicate how each vibration that you feel corresponds to the other vibrations. The less strong ones should be closer to the smaller numbers and the very strong ones should be charted closer to the higher numbers. Write a 300+ word description of your observations in relation to vibration.

    NOTE: Electromagnetic vibrations go from zero (0) to $+\infty$.

10. Research vibration via the internet of other library sources to find normal vibration measurements for various things (e.g., guitar strings, fan motors) and plot them on a similar scale as drawn in the activity above. Increase the scale increments to match the lowest and highest vibration measurements. Be sure to ONLY use the same measurement scales on one graph—if using more than one type of measurement scale, be sure to plot the *like* measurements on different graphs.

11. Identify rotational vibration analytical sensing and measurement devices in the lab.

12. Use bench units to show flow, level, pressure, and temperature loops.

13. Purge a tank with nitrogen and measure the reduction in oxygen content.

14. Use cutaways of common primary sensors placed in their respective positions in the control loop. Use a show and tell type of demonstration.

PART 2 Symbology, Hardware, and Instrumentation Communication

Chapter 7
Process Diagrams and Instrumentation Symbology

Objectives

After completing this chapter, you will be able to:

7.1 Describe the types of drawings that contain instrumentation an operator might use. (NAPTA Symbology: Process Diagrams Part I 1; Process Diagrams Part II 1; Instrument Sketching 1, 2, 3*) p. 130

7.2 Explain the lettering and numbering standards based on ISA instrumentation symbols. (NAPTA Symbology: Process Diagrams Part I 2) p. 133

7.3 Demonstrate how to determine the instrument type or signal line type from the symbol information. (NAPTA Symbology: Process Diagrams Part I 3, 4) p. 136

7.4 Using a legend, locate and identify instrumentation on a PFD or P&ID drawing. (NAPTA Symbology: Process Diagrams Part I 5; Process Diagrams Part II 2, 3) p. 137

Key Terms

Balloon—a basic instrumentation symbol, to represent one of many functions **p. 137.**
Basic equipment symbol—common equipment such as pumps, towers, furnaces are basic pieces of equipment for most processing facilities and have commonly recognizable, or basic, equipment symbols, **p. 136.**
Block flow diagram (BFD)—a simple illustration that shows a general overview of a process, indicating its parts and their interrelationships, **p. 130.**

*North American Process Technology Alliance (NAPTA) developed curriculum to ensure that Process Technology courses will produce knowledgeable graduates to become entry-level employees in process technology. Objectives from that curriculum are named here in abbreviated form. For example, "(NAPTA Symbology: Process Diagrams Part I, 1; Process Diagrams Part II, 1; Instrument Sketching 1, 2, 3)" means that this chapter's objective addresses objectives under NAPTA's course content for symbology and diagrams and also for instrument sketching.

130 Chapter 7

ISA—the International Society of Automation (originally known as the Instrument Society of America); a global, nonprofit technical society that develops standards for automation, instrumentation, control, and measurement, **p. 133.**

Legend—a section of a drawing that explains or defines the information or symbols contained within the drawing, **p. 131.**

Line symbols—connectors between the basic pieces of equipment without which process streams could not be moved, **p. 136.**

Piping and instrumentation diagram (P&ID)—detailed illustration that graphically represents the relationship of equipment, piping, instrumentation, and flows contained within a process in the facility, **p. 131.**

Process flow diagram (PFD)—basic illustration that uses symbols and direction arrows to show the primary flow path of material through a process, **p. 131.**

Symbology—various graphical representations used to identify equipment, lines, instrumentation, or process configurations, **p. 133.**

7.1 Introduction to Process Diagrams

The three drawings that a process technician is most likely to use are the following:

- Block flow diagram (BFD)
- Process flow diagram (PFD)
- Piping and instrumentation diagram (P&ID)

Block flow diagram (BFD)
a simple illustration that shows a general overview of a process, indicating its parts and their interrelationships.

A **block flow diagram (BFD)** (Figure 7.1) shows the flow scheme in a simple sequential block form. Not all, but most, block flow diagrams show flow from left to right and tend not to cross over lines.

Figure 7.1 Block flow diagram.

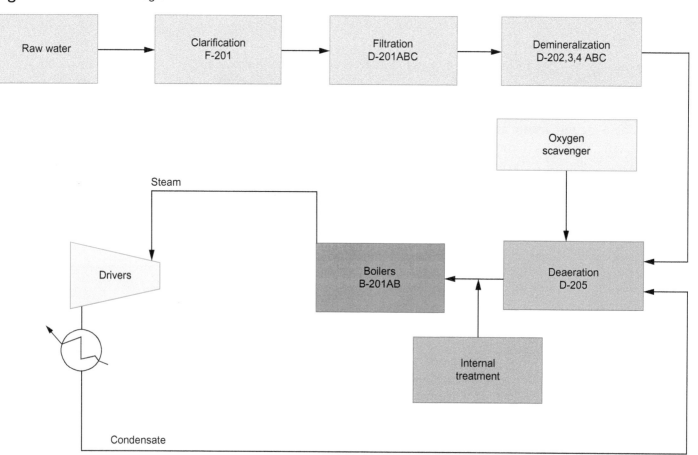

A **process flow diagram (PFD)** (Figure 7.2) describes the actual process pictorially. It includes the major process equipment and may provide process variables as well as heat and material balance information. This is one of the first documents developed when initiating the design of a new plant. The material balance is used in all further flow calculations including main process pumps and compressors, vessel sizing, and so on. The PFD contains major controls but only those block valves necessary for understanding flow control.

Process flow diagram (PFD) basic illustration that uses symbols and direction arrows to show the primary flow path of material through a process.

Figure 7.2 Process flow diagram.

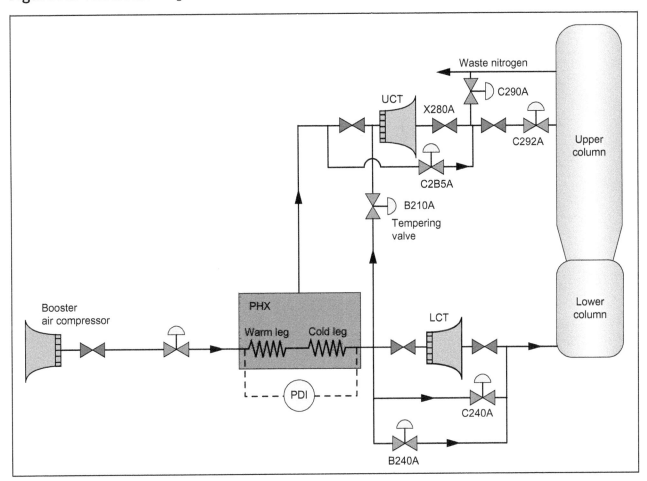

A **piping and instrumentation diagram (P&ID)** (Figure 7.3) is similar to a PFD. However, it contains no process information but much more detail, including piping, instrumentation, and the entire control system. These drawings provide the basic mechanical design details and operating philosophy for the plant.

P&IDs are usually filed within a drawing package when a plant or new process is designed. Other drawings that come in the complete drawing package that may be useful to the process technician are the site plan and utility drawings, as well as the more complete instrument drawings called the loop diagrams.

All PFDs and P&IDs should have an associated legend. A **legend** is an explanation of what the symbols and codes represent. The legend may be located in a small box (Figure 7.4), in the margin of the drawing, or, if it is very large, on an entire separate page.

Different drawing types exist to provide relevant information to the user. That is why both PFDs and P&IDs are useful. The PFD primarily illustrates the flow of materials through the process. To do this, a PFD must include process equipment and piping symbols. PFDs may also include process flow notations and even some instrumentation. Generally, any piece of equipment that moves fluids or comes in direct contact with the flowing process is on the PFD. The intent of the PFD is to help the technician understand and troubleshoot process flow problems.

Piping and instrumentation diagram (P&ID) detailed illustration that graphically represents the relationship of equipment, piping, instrumentation, and flows contained within a process in the facility.

Legend a section of a drawing that explains or defines the information or symbols contained within the drawing.

Figure 7.3 Piping and instrumentation diagram (P&ID).

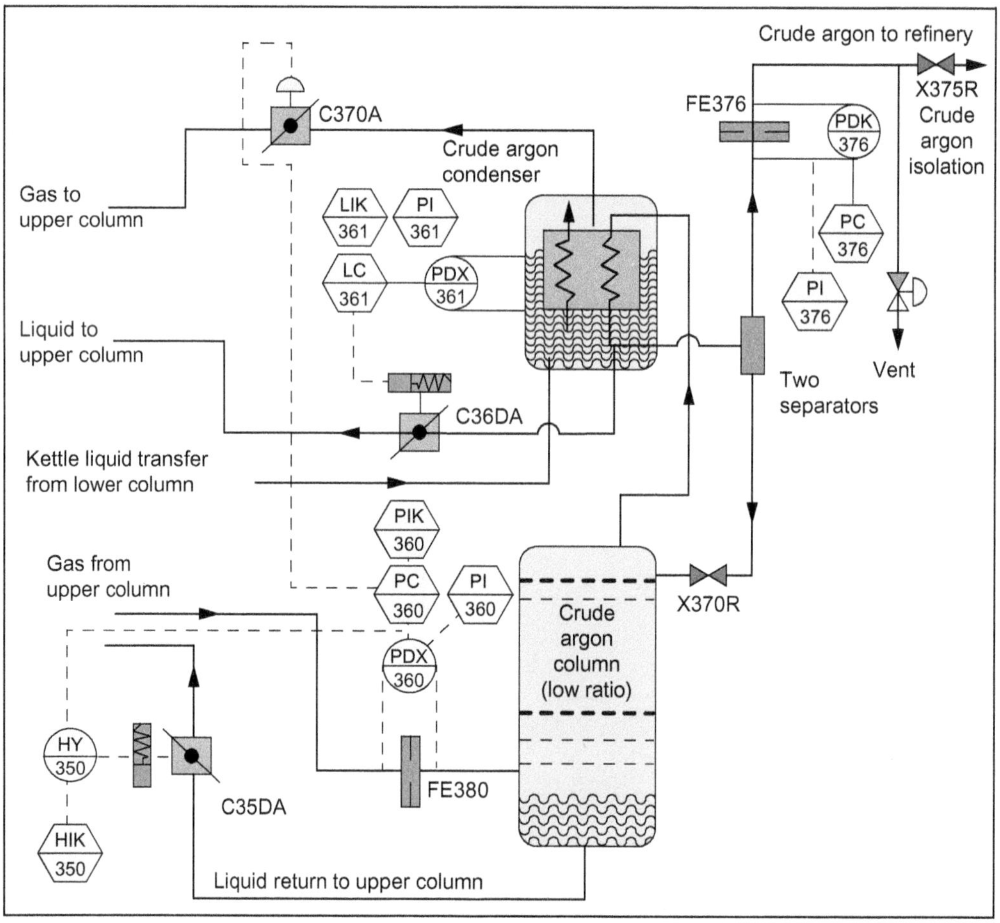

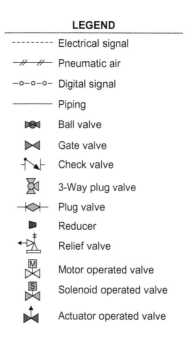

Figure 7.4 Legend or symbols chart.

LEGEND

- ---------- Electrical signal
- —//—//— Pneumatic air
- —o—o—o— Digital signal
- ————— Piping
- Ball valve
- Gate valve
- Check valve
- 3-Way plug valve
- Plug valve
- Reducer
- Relief valve
- Motor operated valve
- Solenoid operated valve
- Actuator operated valve

The P&ID, by comparison, has most of the same items as a PFD with the addition of the control instrumentation and considerable mechanical details. The instrumentation is drawn in a schematic form to illustrate functionality. A P&ID shows the entire control loop in proximity to the field instrumentation. Again, this is a schematic representation of the loop, not a drawing. A P&ID does not represent the actual physical placement of the components as they are situated within a plant or unit. P&IDs help technicians understand how the control instruments are interconnected and function together with respect to the process. By understanding how the control loop(s) function(s), a technician becomes a safer, more capable operator and troubleshooter.

7.2 Introduction to Symbology

Throughout the years of modern processing design, a system of symbols has been utilized to streamline the various drawings that are used to describe how both equipment and its associated instrumentation are interconnected. While various engineering and/or chemical companies have striven to make these symbols standardized within their own facilities or companies, there have been others that have formed organizations or societies to address issues across various process industries. A few of these organizations are utilized in the process industry to define symbology. The International Organization for Standardization (ISO) is an international standard setting body made up of representatives from various national standards organizations. ISO creates standards that provide requirements, specifications, guidelines, or characteristics that can be used consistently to ensure that materials, products, processes, and services are fit for their purpose. The Process Industry Practices (PIP) is a member consortium of process industry owners and engineering construction contractors. Members collaborate to harmonize internal company standards and "best practices" around design, procurement, construction, and maintenance into industry wide PIP standards. For instrumentation, **ISA** (International Society of Automation) is the dominant source for industry symbology, specifically its **symbology** Standard 5.1. The ISA symbol is shown in Figure 7.5.

ISA the International Society of Automation (originally known as the Instrument Society of America); a global, nonprofit technical society that develops standards for automation, instrumentation, control, and measurement.

Symbology various graphical representations used to identify equipment, lines, instrumentation, or process configurations.

Figure 7.5 ISA—International Society of Automation.

CREDIT: Courtesy of ISA.

ISA
67 Alexander Drive
Research Triangle Park, NC 27709
www.isa.org

ISA S5.1

This standard consists of both specific symbols denoting functionality and a coded system built on the letters of the alphabet that depicts functionality of the instrument. Although many, if not most, large companies have moved toward adopting the ISA S5.1 standard in its entirety, vestiges of proprietary or locally preferred symbols may be kept in their inventory. Therefore, anyone who uses a drawing should not assume that the ISA standard is the only standard being employed. Regardless of which standard is used, all standard and nonstandard symbols should be identified in the legend of each drawing.

ISA Instrument Tag Number

An instrument tag number (Figure 7.6) should identify the measured variable, the function of the specific instrument, and the loop number. Accordingly, the ISA instrument tag number is described with both letters and numbers and should be unique since most plants now use a global database to identify devices.

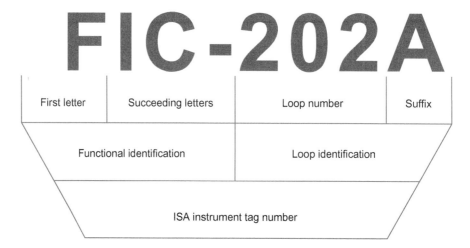

Figure 7.6 Instrument tag number.

The first letter identifies the measured or initiating variable and the following or succeeding letters describe the function of the instrument. For example, in Figure 7.6 above, "F" stands for flow, "I" for indicate, and "C" for control. That makes this instrument a *flow indicating controller*, or a controller that is controlling flow and has an indicator on its faceplate.

Loop numbers are unique numbers assigned locally by the plant engineering group or by the engineering firm that has been hired to produce the drawings. In either case, every instrument in the loop exclusively shares the loop number. If a loop has more than one instrument with the same functional identification, then a suffix is added to the end of the loop number. For example, a split range control loop has two control valves. One should be designated with the suffix "A" (TV-202A) and the other with the suffix "B" (TV-202B). A multipoint temperature recorder (a temperature recording device that records numerous temperature trends using a printed number and/or color) would instead have numerical suffix identifications, such as TE-43-1, TE-43-2, and so on. Suffix identifiers are the exception rather than the rule. When you encounter them, you need to know what they are expressing in terms of the physical loop instrumentation.

The ISA Functional Identification Table (Table 1 in the ISA S5.1 standard) lists the first letter with its possible modifiers and the succeeding letters including the passive or readout function column, the output function column, and possible modifiers associated with the succeeding letters. The following information in Table 7.1 is taken from that table and should

Table 7.1 ISA Table 1—Identification Letters

	First Letter		Succeeding Letters		
	Measured or Initiating Variable	Modifier	Readout or Passive Function	Output Function	Modifier
A	Analysis		Alarm		
B	Burner, combustion		User's choice	User's choice	User's choice
C	User's choice			Control	
D	User's choice	Differential			
E	Voltage		Sensor (primary element)		
F	Flow rate	Ratio (fraction)			
G	User's choice		Glass, viewing device		

	First Letter		Succeeding Letters		
	Measured or Initiating Variable	Modifier	Readout or Passive Function	Output Function	Modifier
H	Hand				High
I	Current (electric)		Indicate		
J	Power	Scan			
K	Time, time schedule	Time rate of change		Control station	
L	Level		Light		Low
M	User's choice	Momentary			Middle, intermediate
N	User's choice		User's choice	User's choice	User's choice
O	User's choice		Orifice, restriction		
P	Pressure, vacuum		Point (test) connection		
Q	Quantity	Integrate, totalize			
R	Radiation		Record		
S	Speed, frequency	Safety		Switch	
T	Temperature			Transmit	
U	Multivariable		Multifunction	Multifunction	Multifunction
V	Vibration, mechanical analysis			Valve, damper, louver	
W	Weight, force		Well		
X	Unclassified	X axis	Unclassified	Unclassified	Unclassified
Y	Event, state, or presence	Y axis		Relay, compute, convert	
Z	Position, dimension	Z axis		Driver, actuator, unclassified final control element	

always be referenced back to the original ISA table for verification. The original table in the ISA publication also lists reference numbers where additional explanatory information may be found within Section 5.1.

Instrument tag examples can be interpreted by using the functional identification table shown in Table 7.2.

Table 7.2 Instrument Tag Examples

Letters	Functional Interpretation
P	Pressure
T	Temperature
F	Flow
L	Level
E	Element
I	Indicator
C	Controller
CV	Control Valve
Y	Transmitter/transducer
R	Recorder
PT	Pressure Transmitter
TT	Temperature Transmitter
FRC	Flow Recording Controller
PIC	Pressure Indicating Controller
LV	Level Valve (preferred way of identifying a control valve in a loop; may also be expressed as PV, FV, TV)

Letters	Functional Interpretation
PY	Pressure relay or compute (convert) (e.g., could be an I/P transducer in a pressure loop)
TE	Temperature Element (e.g., could be a thermocouple, RTD, or filled thermal system)
LI	Level Indicator
PC	Pressure Controller (since this controller does not have an indicator or recorder function, it would probably be behind the panel out of the sight of the operator)
FFIC	Flow Fraction (ratio) Indicating Controller

7.3 Instrumentation Symbols

Technicians should be familiar with each **basic equipment symbols** commonly used on P&IDs (Table 7.3). Instruments shown on P&IDs are associated with various pieces of equipment and measure process variables such as pressure, temperature, level, flow, speed, and position. Process analyzers associated with equipment are also shown on the P&ID.

Connecting each of the basic pieces of processing equipment are the various piping arrangements and/or signal paths that communicate between the instruments controlling the processes. **Line symbols** are more important than most technicians realize. Not only do they show how loop instrumentation is related and connected, but together they also identify the type of energy the instrument uses.

Basic equipment symbol common equipment such as pumps, towers, furnaces are basic pieces of equipment for most processing facilities and have commonly recognizable, or basic, equipment symbols.

Line symbols connectors between the basic pieces of equipment without which process streams could not be moved.

Table 7.3 Basic Equipment Symbols

Equipment Symbol		Equipment Name
PFD	P&ID	
(tank image)	(tank image)	Tank
(heat exchanger image)	(heat exchanger image)	Heat exchanger
(motor image)	Motor	Motor
(pump and motor image)	(pump and motor image)	Pump and motor
(tower image)	(tower image)	Tower or column
(compressor image)	(compressor image)	Compressor

An analog pneumatic transmitter produces a pneumatic signal, and an analog electronic transmitter produces an electrical signal. Most analog control systems are fairly consistent with their signal forms. Pneumatic loops tend to be pneumatic from one end to the other. In contrast, analog electronic usually switches to pneumatic at the control valve. If a digital control system is used, then another line symbol, the software link, may be introduced into an electronic analog loop. One possible scenario is this: An electronic signal can be accepted into a distributed control system (DCS) by first converting the signal into a binary number. This is done with an analog-to-digital (A/D) converter. Once the signal value has been converted into a binary number, it moves through the computer programming as a software link. The output of the computer leaves through a digital-to-analog (D/A) converter and reenters the loop as an analog signal before manipulating the process. Without a system of symbols depicting these instrument signal lines, it would be very difficult to understand how a controlled variable signal moves through the computer system.

Other line symbols include instrument connections to the process. The electromagnetic or sonic symbol includes radar, radiation, and even a video camera focused on a flare. The capillary tubing usually indicates a liquid filled system, and the hydraulic line symbol may be any line that carries a fluid signal, such as a piston actuator.

Standards for line symbols, as indicated on most symbol charts, include those shown in Table 7.4.

Table 7.4 Standard Line Symbols

Line Symbol	Line Type
—//—//—	Pneumatic signal
— — — — —	Electrical signal
—∽—∽—	Electromagnetic or sonic signal
—✕—✕—	Capillary tubing
—L—L—	Hydraulic signal
—o—o—	Digital (software) link
—————	Connection to process; secondary line; utility line
━━━━━	Process line
—●—●—	Mechanical link

7.4 Instrumentation Symbol Interpretation

Unlike process equipment symbols, line symbols and instrumentation symbols may or may not look like the physical device they represent. For example, a 7/16 inch diameter circle, called a **balloon**, is commonly used to represent any number of functionally different instruments. The only distinguishing difference from one balloon to another is its unique alphanumeric tag number. Considering the complexity of many control systems, this schematic approach works very well. The tag number, covered earlier in this chapter, is the primary key to defining the functionality of the instrument, whereas slight modifications of the balloon depict where the instrument is physically located.

Balloon a basic instrumentation symbol to represent one of many functions.

In Figure 7.7, a typical legend may show how the various instrument balloons are represented for a particular drawing. One balloon (Figure 7.8) is further explained as to what each of the letters or numbers that may be found on the tag may represent. Also indicated in Figure 7.8 are two variations of the balloon.

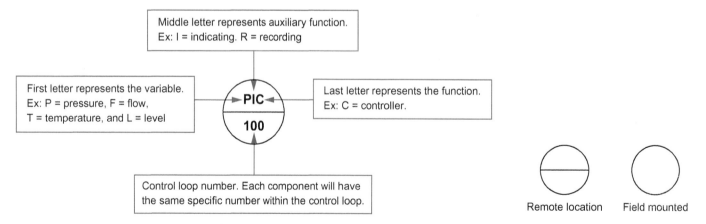

Figure 7.7 Legend example for instrument balloon interpretation.

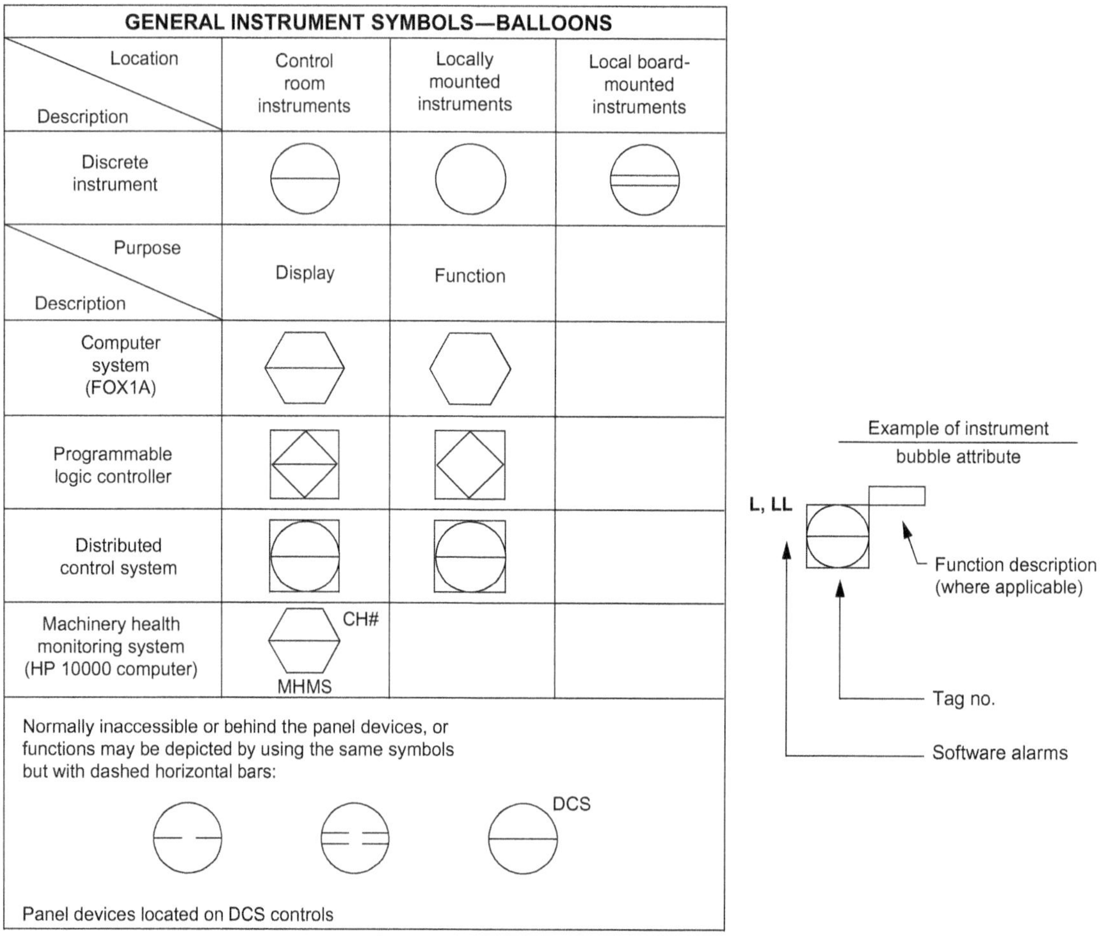

Figure 7.8 Instrument symbol interpretation key.

After putting the two items (basic balloon representation and symbol interpretation) together, use the following tables to learn the many variations of instrumentation combinations that may be found on the drawings you may encounter.

Use the Instrument Symbol Interpretation Key as you study the balloon symbols in Table 7.5 to ensure understanding of how the symbol was developed for each of the main types of process variables where instrumentation is used in Table 7.6.

Table 7.5 Instrument Balloon Symbols

Balloon Variation	Interpretation
No lines in balloon (FT 1280) (LT 1024)	A field-mounted or local instrument
A solid line through it (FRC 288) (PT 768) (PIC 121)	A board-mounted or remote instrument
Two parallel lines through it (TT 72)	Located in an auxiliary location, usually on a control panel in the processing area
A broken line through it (FY 360)	Located behind the panel board, or at least is not readily accessible
A box around it [TC 144]	Digital control systems (DCS) or computer interface

Table 7.6 Instrument Balloon Symbol Examples

Balloon Variation	Interpretation
FLOW	
(FC)	Flow Controller
(FE)	Flow Element
(FI)	Flow Indicator
(FR)	Flow Recorder
(FT)	Flow Transmitter
LEVEL	
(LAH 15)	Level Alarm High (LAH) ("H" for High)
(LIC 30)	Level Indicator Controller
(LG)	Level Gauge, field mounted
(LI)	Level Indicator, field mounted
(LR 20)	Level Recorder, board mounted
(LT 25)	Level Transmitter

Balloon Variation	Interpretation
PRESSURE	
PAH	Pressure Alarm High (PAH or PHA), panel mounted
PIC 60	Pressure Indicating Controller, board mounted
PI	Pressure Indicator, field mounted
PIC 100	Pressure Indicating Controller, board mounted
PR 50	Pressure Recorder, board mounted
PCR 90	Pressure Recording Controller, board mounted
PT 45	Pressure Transmitter, board mounted
TEMPERATURE	
TC	Temperature Controller
TE	Temperature Element
TI	Temperature Indicator
TR	Temperature Recorder
TT	Temperature Transmitter
MISCELLANEOUS	
AT pH	pH Analyzer Transmitter
I/P	I/P and P/I converters change a current signal to a pneumatic signal so a distributed control system (DCS), programmable logic control (PLC), or personal computer (PC) can control a valve or actuator; it may also convert a pneumatic signal to current so remote pneumatic devices can interface with electronic instruments and computer based monitoring systems
E/P	E/P transducers convert electrical signals to equivalent pneumatic signals; they are commonly used in the field to supply instrument air to field control elements
TY E/P	Temperature transducer—transducer electronic to pneumatic symbol used to activate a pneumatic positioner on a valve for a temperature loop
LY I/P	Level transducer—transducer electronic to pneumatic symbol used to activate a pneumatic positioner on a valve for a pressure loop
PY I/P	Pressure transducer—transducer to pneumatic symbol used to activate a pneumatic positioner on a valve for a pressure loop
FY I/E	Flow transducer—transducer, isolator and converter; changes current to voltage

Using information from the previous chapters on pressure, temperature, level, and flow, the following tables (Tables 7.7 through 7.12) are provided to show how the instrumentation balloon symbols, line types, and equipment symbols are used together to represent process components. In addition, a table of final control elements is shown with corresponding instrument symbol configurations.

Table 7.7 Pressure

Symbol Variation	Interpretation
	A Pressure Indicator directly connected to a tank/vessel
	A Pressure Indicator connected to process piping
	A Pressure Transmitter connected via piping to a low pressure lead coming off a flow transmitter
	A Pressure Indicator connected to a special type of chemical seal to protect the instrument from the process fluid (capillary)
	A Pressure Indicator connected to process piping where a siphon is installed
	Pressure element, strain gauge type, connected to pressure indicating transmitter (TAG strain gauge PE-33)
	Back-pressure regulator, self-contained with handwheel adjustable set point
	Pressure reducing regulator, self-contained
	Back-pressure regulator with external pressure tap

Table 7.8 Temperature

Symbol Variation	Interpretation
	A local Temperature Recorder, filled thermal system, and thermowell connected to process piping
	A local Temperature Recorder, thermocouple, or RTD and thermowell connected to process piping
	A Temperature Indicating Transmitter and filled thermal system connected to a tank or vessel and thermowell

Symbol Variation	Interpretation
	A Temperature Transmitter of the thermal radiation type using an optical pyrometer connected to a furnace fire box
	A bimetallic thermometer (Temperature Indicator) inserted into a thermowell in process piping
	A Temperature Indicating Controller
	Temperature Differential Indicator
	Temperature regulator capillary, filled-system type

Table 7.9 Level

Symbol Variation	Interpretation
	Level Gauge (gauge glass) connected to a tank/vessel and read visually
	Level Indicator connected to a tank/vessel and read locally
	Level Transmitter connected to a tank/vessel and read remotely
	Level Transmitter (low side vented) connected to a tank/vessel and read remotely
	Level Indicator (gauge board—float actuated) connected to a tank/vessel and read locally
	Level Recorder/Level Electronic (bubble tube direct connect to final device) connected to a tank/vessel and read remotely
	Local Controller (piped direct) connected to a tank/vessel and read remotely
	Level regulator with mechanical linkage

Table 7.10 Flow

Symbol Variation	Interpretation
FE 10	Flow Element (orifice plate with flange/corner taps) installed in piping
FI 11	Flow Indicator (orifice plate with flow indicator) installed in piping
FE 12	Flow Element (orifice plate with vena contracta radius or pipe taps) installed in piping
FE 13	Flow Element (orifice plate in quick change fitting) installed in piping
FE 14	Flow Element (pitot tube) installed in piping
FE 15	Flow Element (venturi or flow nozzle) installed in piping
FT 22	Flow Transmitter installed in piping
FQI 23	Flow Quantity Indicator
FE 17	Flow Element (weir) installed in piping
FE 18	Flow Element (flume) installed in piping
FE 24	Flow Element (turbine or propeller type primary element) installed in piping
FE / FT 25	Flow Target (meter) installed in piping
FI 26	Rotameter (variable area flow indicator) installed in piping
FR A / FFC 544 / FR B / FT A / FT B	Flow-ratio controller with two pens to record flow
FC 427 SP / LC 428 / FT	Cascade control

Table 7.11 Final Control Elements (Valve Bodies)

Symbol Variation	Interpretation
	General symbol for valve (ON/OFF only)
	General symbol for angle valve NOTE: ISA uses connected triangles.
	General symbol for butterfly valve
	General symbol for globe valve
	Ball (Rotary) Valve
	General symbol for three-way valve NOTE: ISA uses connected triangles.
	3-Way Valve (Fails to Bottom)
	3-Way Valve (Fails Straight)
	General symbol for four-way valve NOTE: ISA uses connected triangles.
	General symbol for diaphragm valve
	General symbol for a motor-operated valve

Table 7.12 Actuator Symbols

Symbol Variation	Interpretation
	Hand actuator or handwheel
	Diaphragm, spring-opposed, or unspecified actuator
	Control Valve (Fail Open) (straight through, diaphragm vs. spring actuator)
	Control Valve (Alternate) (Fails Open) (push down to open)
	Control Valve (Fail Closed) (straight through, diaphragm vs spring actuator)
	Control Valve (Alternate) (Fails Closed) (push down to open)

Symbol Variation	Interpretation
FO	Butterfly Control Valve (Fails Open)
FC	Butterfly Control Valve (Fails Closed)
	Butterfly Control Valve (Alternate) (Fails Closed) (push down to open)
	Butterfly Control Valve (Alternate) (Fails Open) (push down to close)
	Control Valve Actuator with Positioner
	Actuator (Diaphragm versus diaphragm)
	Actuator (Piston) NOTE: ISA draws the piston from side to side on the box.
	Actuator (Double-acting Piston) NOTE: ISA draws the piston from side to side on the box.
	Three-way Solenoid Valve

There are many other types of symbols used on drawings frequently or infrequently. For example, between updates, hand-drawn clouds (Figure 7.9) may be used to signify a revision of some type, such as an addition.

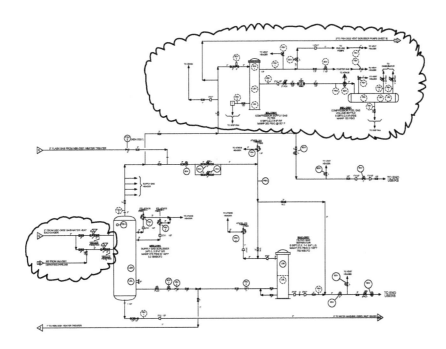

Figure 7.9 Revision or addition—clouds.

Equipment that has been abandoned or dismantled is bordered and filled with crosshatched lines (Figure 7.10) indicating abandonment in place or dismantlement.

Figure 7.10 Crosshatching.

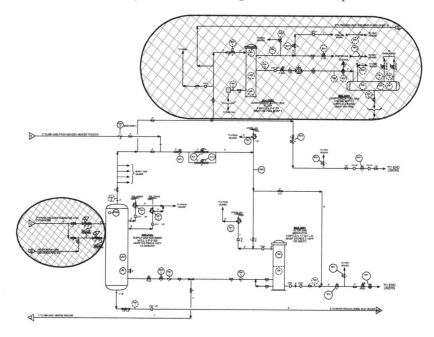

Still other miscellaneous symbols are common to drawings that have continuations (e.g., connector lines) similar to key maps used for driving across large distances where you have to connect one map drawing to the next.

Summary

There are several types of drawings that process technicians come in contact with often. Of those, the three most common are the block flow diagram (BFD), the process flow diagram (PFD), and the piping and instrumentation diagram (P&ID). Block flow diagrams show simple sequential flow from left to right while process flow diagrams include major process equipment and provide heat and material balances of the chemical processes represented within the drawing. Piping and instrumentation diagrams are similar to process flow diagrams but contain much greater amounts of detail about mechanical design, piping, instrumentation, and primary control systems. More detailed drawings such as PFDs and P&IDs usually have legends that explain how various symbols and codes are used. When process technicians understand the interrelationships of the various pieces of equipment, piping, and instruments that monitor and maintain the processes, they can relate to their job responsibilities such as troubleshooting and quality control with safer, more qualified judgment capability.

People have always used symbols to communicate; however, standardization is the key to understanding between facilities and location-to-location. The International Organization for Standardization (ISO), Process Industry Practices (PIP), and the International Society of Automation (ISA) have attempted to bring together across these boundaries a set of drawing symbols to standardize instrumentation and other basic symbols so that engineering companies and facility design and implementation personnel are able to understand and replicate drawing parameters. Even so, there are still variations being used. To alleviate some of this problem, legends are used on drawings for explanations of symbols used. These legends should be carefully studied when observing and/or studying PFDs and P&IDs.

Instrument tag numbers are similarly standardized by the ISA and contain an alphanumeric code to include the instrument functionality and a specific numeric identifier specific to a particular drawing or control loop. The ISA has a Functional Identification Table to help identify what the specific letters indicate and what their modifiers may mean.

Symbols can be quickly divided into several types to include the following:

- Basic equipment
- Lines
- Instrument balloons
- Combinations by process variables for pressure, temperature, flow, level, and so on.

Checking Your Knowledge

1. Which of the following types of diagrams provides heat and material balances of the chemical process?
   a. BFD
   b. PFD
   c. PRD
   d. P&ID

2. Which of the following types of drawings contains the most detail to include instrumentation and the entire primary control system?
   a. BFD
   b. PFD
   c. PRD
   d. P&ID

3. What is the part of a drawing called that contains a compilation of the symbols used as well as any codes or other notes?
   a. Margin
   b. Legend
   c. Perimeter
   d. Cloud

4. What does ISA stand for?
   a. Instrumentation Society of America
   b. International Society of Automation
   c. International Standards and Automation Society
   d. Instrumentation Standards Association

5. The ISA standard for specific drawing symbols denoting functionality and a coded system built on the letters of the alphabet are in which of the following standards?
   a. ISA-S5.1
   b. ISA-S5.2
   c. ISA-S5.3
   d. ISA-S5.4

6. What does the first letter of an ISA instrument tag number identify?
   a. Function of the instrument
   b. Name of the instrument
   c. The measured or initiating variable
   d. The type of instrument

7. Why are line symbols so important?
   a. They identify the type of energy an instrument uses.
   b. They identify safety concerns when using the instrument.
   c. They identify the level of training required for using the instrument.
   d. All of the above

8. What is the interpretation of a symbol with no lines in the balloon?
   a. A board-mounted or remote instrument
   b. Digital control systems
   c. A field mounted or local instrument
   d. Equipment located in an auxiliary location

9. What is the interpretation of a symbol with a box around it?
   a. A board-mounted or remote instrument
   b. Digital control systems
   c. A field mounted or local instrument
   d. Equipment located in an auxiliary location

NOTE: Answers to Checking Your Knowledge questions are in the Appendix.

Student Activities

1. Photocopy symbol pages or draw symbols on repositionable sticky paper found at most craft supply stores. Cut out symbols and temporarily mount them on plastic such as transparency film or plastic protective page covers. Print Drawing 1 in this section, scaling to 11 × 17, if possible. Place symbol stickers in appropriate places on the drawing where instrumentation would be found. When complete, bring to class and present your work to your instructor and classmates.

2. Repeat the above procedure for Drawing 2.

3. Given a simplified flow diagram of a section of your pilot plant or table top model, use the symbols legend to insert instrumentation on the diagram.

4. Create an original PFD of the pilot plant section or tabletop model located at your facility. Add control loops to the diagram and write a short paragraph for each control loop added describing why the loop was added and how the process would be affected by having the control system in place.

5. Using the symbols in this chapter, create a simple diagram with process equipment and its associated instrumentation. If trainer resources are available, then these drawings may be created there. Drawings and printouts should be passed around to other groups for discussion and improvement.

6. Complete the following:
   - Given a P&ID, explain the relationship of one piece of instrumentation to another.
   - Given a drawing that has major system equipment, add control loops:

     flow

     level

     temperature

     pressure
   - Mark-up instrumentation control loops on available trainer resources.

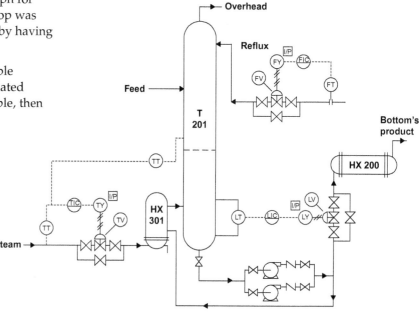

7. Diagrams, Symbols, and Mark-ups

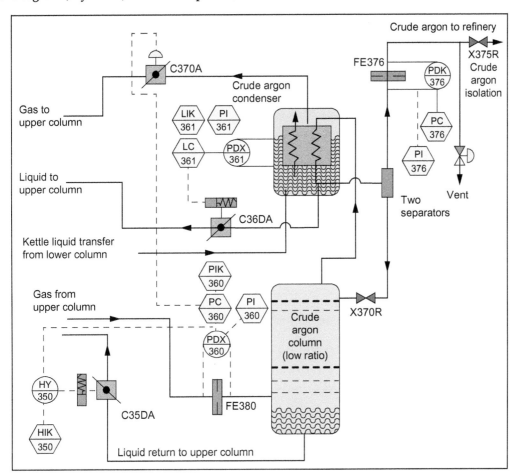

Instructions

a. Review the P&ID.

b. Visually trace the control loop from the steam input to the heat exchanger on the left side of the P&ID.

c. Write a description of all the control elements for the loop and the purpose of the loop.

d. Follow your specific instructor guidelines as to submission of your work.

8. USING LEGENDS

Instructions

a. Review the P&ID and symbols legend.

b. Identify all the listed control elements.

c. Write a description for each control element.

d. Follow your specific instructor guidelines as to submission of your work.

Symbols Legend

	LEGEND
--------	Electrical signal
—//—//—	Instrument air
–o–o–o–	Digital signal
————	Piping
▶◀	Ball valve
▷◁	Gate valve
↳◁	Check valve
⧗	3-Way plug valve
—◇—	Plug valve
▬	Reducer
⚐	Relief valve
M	Motor operated valve
S	Solenoid operated valve
↑	Actuator operated valve

Chapter 8
Switches, Relays, and Alarms

 Objectives

After completing this chapter, you will be able to:

8.1 Define the purpose and function of a switch:

energizes alarms, interlocks, safety systems, equipment, or other devices when a process condition meets a preset value.

may be operated by hand, actuated by a mechanical signal, or actuated by a process or electrical signal.

alarms

equipment operation (sequential)

shutdown

autostart. (NAPTA Switches, Relays, Alarms 1*) p. 151

8.2 Define the following terms associated with switches used in process control:

autostart

bypass

on/off/auto (HOA—hand [on]/off/auto)

limit

proximity

vibration

process variable switch. (NAPTA Switches, Relays, Alarms 3) p. 152

---

*North American Process Technology Alliance (NAPTA) developed curriculum to ensure that Process Technology courses will produce knowledgeable graduates to become entry-level employees in process technology. Objectives from that curriculum are named here in abbreviated form. For example, "(NAPTA Switches, Relays, Alarms 1)" means that this chapter's objective relates to objective 1 of NAPTA's course content on switches, relays, and alarms.

8.3 Given a drawing, picture, or actual device, identify and describe basic switch devices used in process control. (NAPTA Switches, Relays, Alarms 3, 5, 6, 7) p. 158

8.4 Explain how relays are used in the process industry. (NAPTA Switches, Relays, Alarms 2, 4, 5) p. 158

Key Terms

Alarm switch—a switch used to notify an operator when a process variable enters an abnormal range (e.g., high or low) by triggering an alarm (in the form of a light, a horn, or both), **p. 153**.

Alarm system—a series of alarms associated with one process unit or equipment system, **p. 160**.

Bypass switch—a switch used to override the normal operation of a system or device for equipment startup operations or for routine check of alarms, **p. 154**.

Limit switch—an electromechanical device used to limit travel of a component beyond a predetermined point, **p. 155**.

Process variable switch—a type of switch that actuates when a predetermined value of a process variable (e.g., pressure, temperature, level, flow, or analytics) is present, **p. 156**.

Proximity switch—a type of switch operation requiring the presence, or closeness, of an object or device to facilitate its operation, **p. 155**.

Relay—an electromechanical device that boosts, maintains, or controls the electrical flow of a signal through a separate circuit, **p. 158**.

Shutdown switch—a switch used to actuate a circuit that shuts down a process system or equipment unit, **p. 153**.

Switch—an electrical device used to start, stop, or otherwise reconfigure the flow of electricity in a circuit, **p. 151**.

Vibration sensor—a device used to determine the velocity, acceleration, displacement, or any combination of these characteristics for the purpose of predicting wear or impending failure. High vibration would activate a switch for an alarm or equipment shutdown, **p. 156**.

8.1 Introduction

Switches, relays, and alarms are important components in process operation units. These components are interdependent in their relationship to each other.

Switches allow electrical circuits to operate. For example, switches can be used to initiate a startup, a shutdown, or some other change of components or equipment and can even trigger alarms. The passing of information to a switch is often facilitated by a relay. One function of relays is to help pass or "boost" signals so they are received properly. An audible or visual alarm is received when a signal is outside of normal range. These alarms reside on panelboards in the distributed control system (DCS) to let the process technician know there is a problem.

Switch Categories

A **switch** is a binary mechanical or electrical device that is used to operate, energize, or de-energize mechanical or electrical circuits for alarm, shutdown, or control purposes using a predetermined operating point, or set point.

Switch types may be categorized as hand operated (commonly called a hand switch) or pushbutton (actuated by a mechanical signal such as a limit or proximity switch, by a process

Switch an electrical device used to start, stop, or otherwise reconfigure the flow of electricity in a circuit.

signal, or by an electrical signal). Switches may be used to monitor the process variables of pressure, temperature, level, flow, or analytical values. Switches may be used in units that follow an automated sequence of steps, such as air dryers, water polishing units, or dual filter applications.

Process field switches can operate with normally open (N.O.) contacts, normally closed (N.C.) contacts, or both. The N.O. switch closes its contacts when powered, while an N.C. switch opens its contacts when powered. When a specific process condition is met, the contacts open and create an open circuit. Switches are commonly used in safety systems to energize alarms or to shut down the process when electrical power is lost or when the circuit is cut. Examples include:

- Hand switch (also called a toggle switch)—hand, off, or autostart (HOA)
- Pushbutton switch—start/stop equipment, open/close valves
- Selector switch—moves between circuits and can be used to bypass alarms temporarily when checking shutdowns or to satisfy a shutdown condition temporarily during equipment startup (e.g., a pump placing liquid into a tank where the shutdown switch is located on the pump motor)
- Limit switch—initiated by physical contact of the switch to the process component
- Proximity switch—initiated by nearness of the switch contacts to the process component

8.2 Switch Nomenclature and Identification

Switches' nomenclature often reflects their use within the process unit or piece of equipment. The following are some of the more common uses in process areas:

- Alarm—process high or low condition
- Shutdown—process high high or low low condition
- Autostart
- Bypass
- Vibration

These types of switches fall into two broad categories. One category describes how the switch is physically constructed and what it senses; the other describes how the switch is integrated into the system (what the system response to that switch will be). If these switches were to be divided logically, they would be categorized as shown in Table 8.1.

Table 8.1 Switch Categories

Physical Hand Switch Type	Switch System Response
Hand switch	Start, stop, auto (HOA)
Pushbutton switch	Start/stop, open/close
Push switch	Hand switch closes circuit only when pushed or activated by hand Circuit opens when released
Selector switch	Bypass, select one of two or more positions
Limit switch	Response initiated by physical contact of the switch to the process component
Proximity switch	Response initiated by nearness of the switch contacts to the process component

The switch types listed here are not mutually exclusive. Any given physical switch type may be used to initiate a given system response. For example, a limit switch sensing mechanical motion may be used to sound an alarm, initiate a shutdown, autostart an equipment component, and/or bypass a process condition. As another example, a shutdown switch may have any number of different physical forms: shutdown on high vibration or shutdown on extreme variations of a process variable (e.g., level, temperature, pressure, flow).

There are many different applications of switches used in process control, including autostart, bypass, limit, proximity, vibration, and process variable condition; three major applications are alarm, shutdown, and equipment autostart.

Toggle or Hand-Off-Auto (HOA) Switches

On/off/auto switches may also be referred to as a toggle, rotary, or HOA switches (Figure 8.1). These types of switches may be set to one of three positions:

- **Hand (manual)**—The switch places the process equipment in the on position unless interrupted by safety interlocks.
- **Off**—The switch prevents the process equipment from running.
- **Auto**—The switch allows automatic start of the process equipment at a preset condition.

Figure 8.1 A. Process technician at switch location. **B.** Toggle and HOA switch examples.
CREDIT: **A**. Oil and Gas Photographer/Shutterstock. **B**. Eakasarn/Shutterstock.

A.

B.

Switches for Alarm Application

An **alarm switch**, like the one shown in Figure 8.2, is used to notify when a process variable enters an abnormal range (high or low). The switch triggers an alarm (lights and/or a buzzer) that informs the process technician of the condition. For example, a low-pressure switch may be calibrated to alarm when pressure decreases to 140 PSIG and clear at 145 PSIG. The 5-PSIG difference is called the *deadband* of the pressure switch.

Alarm switch a switch used to notify an operator when a process variable enters an abnormal range (e.g., high or low) by triggering an alarm (in the form of a light, a horn, or both).

Switches for Shutdown Application

Shutdown switches can be used to actuate a circuit that shuts down a process. In fail-safe situations they can be used to deactivate a circuit or detect an open circuit to shut down part or all of the process unit or equipment. Figure 8.3 shows an example of a shutdown switch during a tank level low low process condition.

Shutdown switch a switch used to actuate a circuit that shuts down a process system or an equipment unit.

Figure 8.2 Alarm switch diagram.

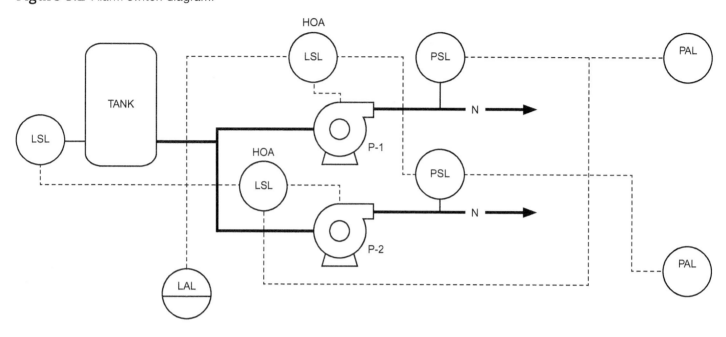

Figure 8.3 Example of a shutdown switch.

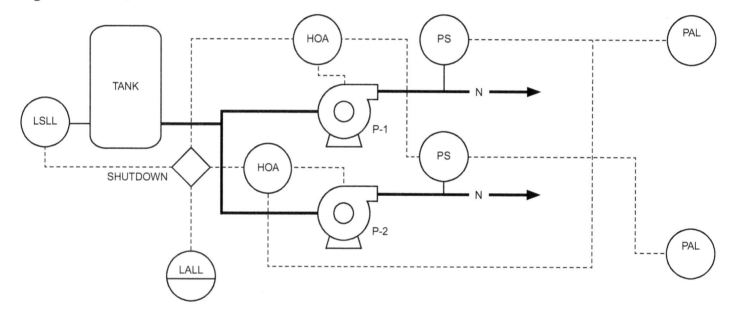

Switches for Autostart Application

Autostart switches are used to initiate an action based on its response to a predetermined process condition. These switches are generally located at the process equipment in the field. An example would be a series of two pumps, pump A and its spare, pump B (offline). Pump A is running; its switch is in the hand position; pump B is set to auto with its switch set in the auto position. The predetermined condition could be low pump discharge pressure or low process flow. Figure 8.4 shows an example of an autostart switch.

Switches for Bypass Application

Bypass switches are used to override (or bypass) different parameters in the operation of a system or device. Figure 8.5 shows an example of a bypass switch.

Bypass switch a switch used to override the normal operation of a system or device for equipment startup operations or for routine check of alarms.

Switches, Relays, and Alarms **155**

Figure 8.4 Example of an autostart switch.

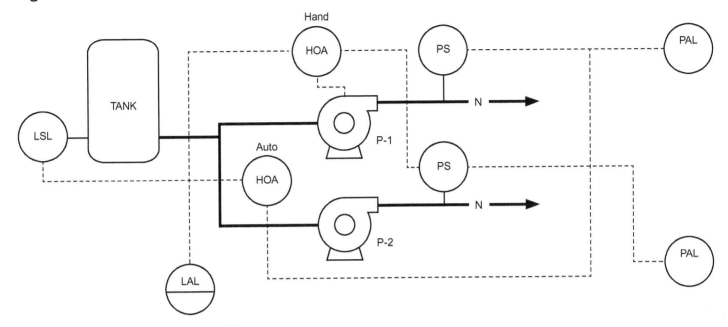

Figure 8.5 Example of a bypass switch.

CREDIT: Engineer story/Shutterstock.

Limit Switches

A **limit switch** is an electromagnetic device that is normally used to verify the state or presence of a condition that exists in the process (e.g., the position of a valve to be fully open or fully closed). Figure 8.6 shows an example of a limit switch.

Limit switch an electromechanical device used to limit travel of a component beyond a predetermined point.

Proximity Switches

Proximity switches sense the closeness of an object without needing to make physical contact as a limit switch does. Some proximity switches use magnetic fields to sense the position of an object, while others use beams of light. These switches are typically sealed against environmental and process conditions (liquid, dust, gases), as are their connecting cables. Figures 8.7 shows an example of a proximity switch.

Proximity switch a type of switch operation requiring the presence, or closeness, of an object or device to facilitate its operation.

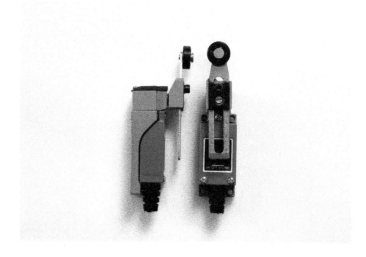

Figure 8.6 Mechanical limit switch.

CREDIT: Nattakit.K/Shutterstock.

Figure 8.7 Proximity switch.

CREDIT: Molpix/Shutterstock.

Vibration Sensors

Vibration sensor a device used to determine the velocity, acceleration, displacement, or any combination of these characteristics for the purpose of predicting wear or impending failure. High vibration would activate a switch for an alarm or equipment shutdown.

Vibration sensors may be used to determine vibration for the purpose of predicting wear or impending failure of the equipment in which the process is taking place. At a predetermined signal from the sensor, a switch may activate to send an alarm or shutdown signal, depending on the severity of the condition. Vibration sensors are commonly used on rotating equipment such as compressors, pumps, centrifuges, steam turbines, and blowers. Figure 8.8 A, B, and C show examples of vibration sensors.

In some machines, such as turbines, centrifugal pumps, and compressors, the rotor is contained within a heavy casing by rigid bearings. Owing to the weight and rigidity of the casing, vibrations in the moving rotor may not be felt on the frame of the machine, so vibration detectors attached to the outside of the machine may not properly detect the condition. In such cases, it is better to get a direct measurement of shaft vibration relative to the casing to indicate when seal and bearing clearances are in danger. This may be accomplished using special proximity detectors, or pickups, that indicate motion (displacement) between the shaft and the machine casing. Figure 8.9 shows the basic operation of a typical noncontact pickup.

Process Variable Switches

Process variable switch a type of switch that actuates when a predetermined value of a process variable (e.g., pressure, temperature, level, flow, or analytics) is present.

Process variable switches actuate when a predetermined value of a process variable is exceeded. A process variable is any physical or chemical property of a process that can be measured and is used for information or control purposes. Variables include: pressure, temperature, level, flow, and analytical variables.

Figure 8.8 **A.** Vibration sensor. **B.** Side view of vibration sensor. **C.** Front view of vibration sensor.

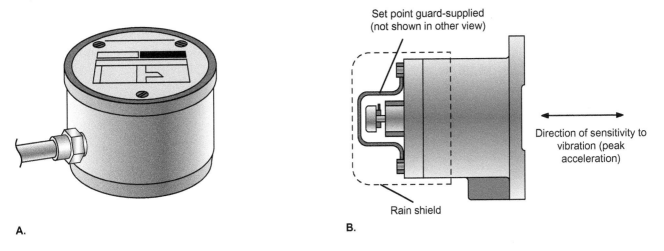

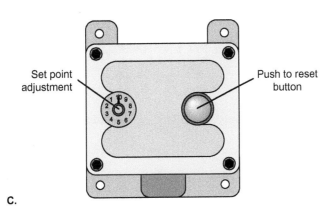

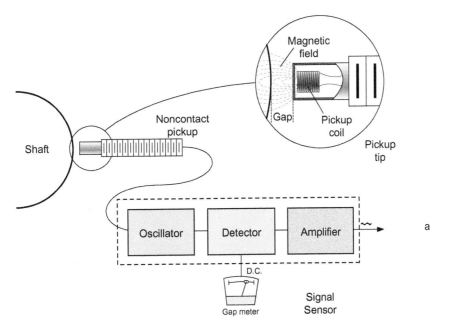

Figure 8.9 Example of noncontact pickup.

8.3 Switch Symbols

The piping and instrumentation diagram (P&ID) of a flow control loop shown in Figure 8.10 contains several examples of switches and other piping and instrument details (for more information on symbology, review symbols for switches presented in Chapter 7: *Process Diagrams and Instrument Symbology*).

Figure 8.10 Piping and instrumentation diagram (P&ID) example.

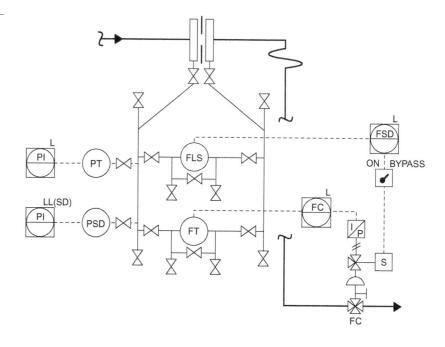

Switches are represented on process flow diagrams (PFDs) and P&IDs using standardized symbols and should be noted in the legend of the drawing, if used. The images shown in Figure 8.11 A, B, and C are representative of commonly used switches discussed in this chapter.

Figure 8.11 A. Toggle, or HOA switch. **B.** Bypass switch. **C.** Start or stop switch.

CREDIT: **A.** Bankrx/Shutterstock. **B.** Surawit Klanliang/Shutterstock. **C.** Ampol sonthong/Shutterstock.

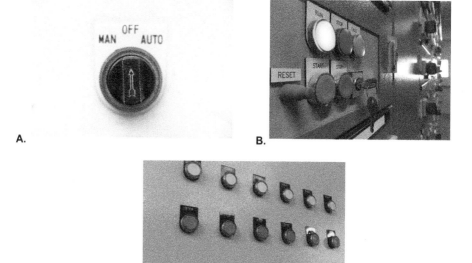

8.4 Relays

Relay an electromechanical device that boosts, maintains, or controls the electrical flow of a signal through a separate circuit.

Relays, when the term is applied to pneumatic (computational) devices, are analog instruments that handle continuously variable signals. When applied to electrical devices, relays are discrete, on/off devices.

A relay is an electrical, electronic, pneumatic, or hydraulic device whose primary function is to pass information, unchanged or in some modified form, to an external circuit (e.g., a low voltage or low current control circuit can be used to control the starting and stopping of high powered motors). In addition, a relay may be used as a computing device (e.g., a square root extractor, or a high or low selecting relay that chooses between two incoming signals based on their relative values).

The term relay is more specifically applied in this chapter to an electric, pneumatic, or hydraulic switch that is actuated by a signal from a control circuit. Generally, however, a relay describes an electromechanical device that is operated by a relatively low power signal and is used to control a higher power source through one or more sets of electrical contacts. This allows a control system to turn on high power devices remotely.

Relay Functionality

There are several types of relays. These include pneumatic, hydraulic, electrical, electronic (on/off control), timing, pneumatic booster, and selection. Table 8.2 lists these relay types and a brief description of their function.

Table 8.2 Relay Types and Functions

Relay Type	Function
Electromechanical (pneumatic and hydraulic)	Controlled by a coil and mechanized parts, application determines its size. Used to perform mathematical operations, signal conditioning, timing, signal boosting, and selection or modification operations
Solid state/semiconductor [electronic (on/off control)]	Electrically operated, used in electronic controls and circuits. Used to control one or more circuits that are not normally part of the control circuit
Thermally operated	Used in temperature applications
Timing	Electrical, mechanical, or electromechanical; used to determine the elapsed time between the start and stop of related operations or events (e.g., a timing relay may be used to start an agitator 15 seconds after the pump starts)
Pneumatic booster	Containing a high pressure or volume source that increases the signal value or volume
Selection	Can take two or more inputs and choose a predetermined value (e.g., high, medium, or low)

Relay Applications

Relays can be used for many applications. Some of the most common are power automation systems, control and drive systems, and protection systems. These applications would include areas in maintenance and operations for the purpose of modification, selection, and computation.

The advent of programmable logic controllers (PLCs) has greatly broadened the use of relays in regard to process plant equipment operation. These discrete controllers have the option of multiple inputs and outputs, making them much more viable for use in quality and production control. PLCs are discussed in depth in Chapter 18, *Programmable Logic Controls*. Electromechanical relays are used to control the flow of an electrical signal and are used to carry a process through a logical sequence to ensure safety and/or product quality.

Although a current to pneumatic (I/P) transducer is called a converter (4–20 mA to 3–15 PSI), it is also considered to be a relay (ISA identifies it as a relay or compute function). A modifying relay can easily accomplish this task.

Alarm Systems

Alarms are used in process units on process instrumentation parameters and on equipment components as a means of alerting personnel to situations that are outside the realm of normal operation. Alarms are signaled by activating a light and a horn. Types include deviation warning (H/L), warning of extreme deviation conditions (HH/LL), and warning of the shutdown of a process unit or piece of equipment. Alarms can be found locally in the

Alarm system a series of alarms associated with one process unit or equipment system.

process area, on a panelboard, or in the DCS. These alarm points can be singular or be a part of a system. Switches activate alarms at a predetermined point in accordance with plant process operation and procedures. Alarms must be acknowledged and reset by the process technician, which is a part of the process technician's job responsibility. The horn should cease and not be audible; the light may flash or go to a steady state in color. Different process units can have distinct alarm sounds to distinguish them from each other. They also may change sound depending on the severity of the situation.

An **alarm system** is a series of alarms on a panelboard relating to a unit or operating equipment component. It provides audible and visual reference to some measured variable. These alarms are associated with the shutdown condition of a unit or piece of operating equipment. The system can be configured as the alarm package of a DCS such as a computer. These systems are often programmable, so they can reside on the Ethernet or process variable input signal and use LCD displays. Figure 8.12 shows an example of an alarm system. Table 8.3 describes the alarms and how they are intended to operate.

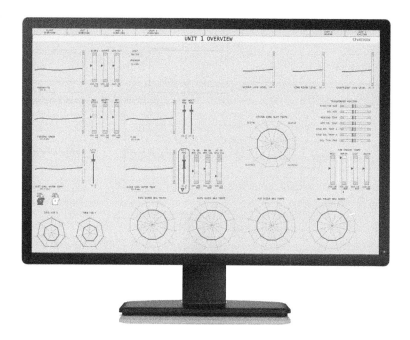

Figure 8.12 Computer interface type of alarm.
CREDIT: © Emerson 2019.

Table 8.3 Condition and Description of Alarms in an Alarm System

Operating Condition	Description
Normal operating condition	Alarm circuits are not activated (no lights), and the horn circuit is not activated (no audible signal).
Alarm condition	Alarm circuit is de-energized; flashing lights and horn sounds.
First out (first shutdown)	A first-out alarm is the one denoting which process condition caused the shutdown. This is very useful for troubleshooting. *Note*: A number of alarm points have a first-out provision to indicate which of several conditions occurred first to cause a machine to trip off. When a first alarm occurs, these points are audible and they display an intermittent fast flash. Subsequent first-out alarm points are audible and display a steady fast flash.
Acknowledge	Audible is silenced. The first-out alarm light continues to flash slowly and all subsequent alarms remain steady on. *Note*: Redundant alarms can have two windows. The signal comes from a local panel. Both lights come on and the audible sounds in the alarm condition. "Acknowledge" silences audible in the control room, but both lights continue to show until local acknowledgment is made. Once acknowledgment is made in the control room, the first window in the control room will go off, while the second window remains steady on.
Reset	Used to put the panel indicator back in a ready position after a problem has been corrected.
Test	On panel board systems this functional test, which is conducted per operating procedure, triggers an audible signal and intermittent fast flash. *Note*: On a DCS or PC system the *test* functionality will differ.

When working with alarm panels like the one in Figure 8.13, it is important to remember that the backlit panels on the display will light up, flash, and sound an audible alarm when a condition reaches alarm status. Pushing the Acknowledge (ACK) button shuts off the audible alarm and triggers the flashing lights to remain lit in a solid display.

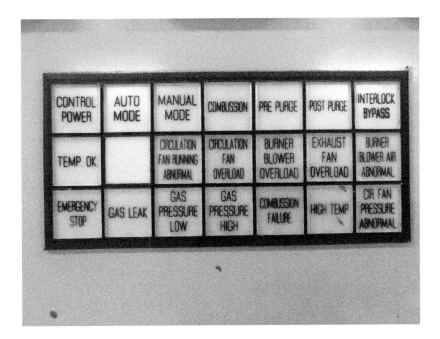

Figure 8.13 Example of alarm panel.

CREDIT: NongphomanMTN/Shutterstock.

Summary

Switches, relays, and alarms are critical components in the operation of process plant units and equipment.

Switches, which may be operated by hand, actuated by a mechanical signal, or actuated by a process or electrical signal, may be used to energize many things, including alarms, interlocks, safety systems, equipment, or other devices when a process condition meets a preset value.

There are many different types of switches in process control (e.g., autostart, bypass, HOA, limit, proximity, vibration, and process variable). The most common in regard to interfacing with the process technician are alarm, shutdown, and HOA.

In process plants, there are many different kinds of relays, including pneumatic, hydraulic, electrical, electronic, timing, pneumatic booster, and selection. Relays assist in signal transfer by increasing (boosting), passing, or maintaining a signal.

If a signal is sent that a process condition or equipment status is out of an acceptable range or that meets preset variable limits, an alarm may be triggered. Alarms are instruments that provide audible and/or visual warnings in reference to some measured variable for a process unit or equipment component. Alarms can be standalone panelboard devices or configured as part of a DCS.

A typical alarm system is designed to notify the process technician that a potential shutdown condition is present. If an alarm condition is present, the *first-out* feature lets process technicians know which alarm came on first and caused the shutdown condition. From there they can acknowledge the alarm and/or reset the system if the problem has been corrected. If process technicians wish to verify the system is functioning properly during nonalarm conditions, they may conduct a system test.

Checking Your Knowledge

1. Which switch closes its contacts when powered?
   a. Normally open
   b. Normally closed
   c. Both normally open and normally closed
   d. Neither normally open nor normally closed

2. What type of switch is initiated by physical contact of the switch to the process component?
   a. Hand switch
   b. Selector switch
   c. Limit switch
   d. Proximity switch

3. Which of the following is a position that a toggle switch can be placed in?
   a. Hand
   b. Off
   c. Auto
   d. All of the above

4. What type of switch is used to override different parameters in the operation of a system or device?
   a. Bypass switch
   b. Proximity switch
   c. Limit switch
   d. Vibration sensor

5. Which of the following defines relays when applied to pneumatic devices?
   a. Discrete devices that handle continuous signals
   b. On/off devices that handle continuous signals
   c. Analog instruments that handle continuously variable signals
   d. Digital instruments that handle sporadic signals

6. What type of relay contains a high pressure or volume source that increase the signal value or volume?
   a. Timing
   b. Selection
   c. Thermally operated
   d. Pneumatic booster

NOTE: Answers to Checking Your Knowledge questions are in the Appendix.

Student Activities

1. Explore your home and educational institution for applications of switches. List the applications and explain what types of switches are used in each.

2. Explore your home and educational institution for applications of relays. List the applications and explain what types of relays are used in each.

3. Explore your home and educational institution for applications of alarms. List the applications and explain what types of alarms are used in each.

4. Given a set of actual switches, identify each type and explain how it is used.

5. Given a set of actual relays, identify each type and explain how it is used.

6. Given a control panel with a simulated alarm situation, do the following:
   a. Identify the alarm condition.
   b. Tell which system item was the first out.
   c. Acknowledge the alarm.
   d. Reset the system.
   e. Conduct a test of the system.

7. Given a PFD or P&ID, locate all the switches and identify each type of switch. Describe the role each switch plays in controlling the processes.

Chapter 9
Signal Transmission and Conversion

Objectives

After completing this chapter, you will be able to:

9.1 Explain the purpose and operation of transmitters. (NAPTA Control Loops: Primary Sensors, Transmitters, and Transducers 2*) p. 164

9.2 Describe methods for protecting integrity and reliability of signal transmission:

shielding

insulation

materials of construction

other safety concerns. (NAPTA Control Loops: Primary Sensors, Transmitters, and Transducers 3, 4*) p. 166

9.3 Recall types of common transmissions:

mechanical

pneumatic

electrical

digital. (NAPTA Introduction to Control Loops, Simple Loop Theory 3) p. 168

9.4 Discuss the purpose and operation of signal converter equipment. (NAPTA Introduction to Control Loops, Simple Loop Theory 4; Control Loops: Primary Sensors, Transmitters, and Transducers 5) p. 169

9.5 Describe wet leg and dry leg instrument lines. (NAPTA Control Loops: Primary Sensors, Transmitters, and Transducers 4) p. 170

---

*North American Process Technology Alliance (NAPTA) developed curriculum to ensure that Process Technology courses will produce knowledgeable graduates to become entry-level employees in process technology. Objectives from that curriculum are named here. For example, "(NAPTA Control Loops: Primary Sensors, Transmitters, and Transducers 2)" means that this chapter's objective relates to objective 2 of the NAPTA curriculum about control loops, the section on primary sensors, transmitters, and transducers.

9.6 Perform scaling calculations:

linear signal

I/P conversion

E/P conversion

I/E conversion

nonlinear signal

square root to linear signal (NAPTA Introduction to Control Loops, Simple Loop Theory 1, 3, 4) p. 171

9.7 Explore wireless technologies—Industry 4.0—and open internet protocol transmission/addressing. p. 173

Key Terms

Analog transmission—a signal transmission in which process variable measurements change in a continuous manner mirroring the actual process variable, **p. 169.**

Digital transmission—a signal transmission method that uses digital devices to create a binary signal, **p. 167.**

E/I conversion—the conversion of a signal from voltage to current, **p. 170.**

E/P conversion—the conversion of a signal from voltage to analog pneumatic, **p. 170.**

Electrical transmission—a signal transmission method that uses electrical devices to create a signal, **p. 165.**

I/E conversion—the conversion of a signal from current to voltage, **p. 170.**

I/P conversion—the conversion of an analog signal from current to pneumatic, **p. 170.**

Insulation—any substance that prevents the passage of heat, light, electricity, or sound from one medium to another, **p. 165.**

Mechanical transmission—a signal transmission method that uses mechanical methods to create a signal, **p. 168.**

Noise—any unwanted or excessive sound, **p. 166.**

Optical transmission—a signal transmission method that uses light frequencies to create a signal, **p. 168.**

P/I conversion—the conversion of an analog signal from pneumatic to current, **p. 170.**

Pneumatic transmission—an analog signal transmission method that uses compressed air/gas to create a signal, **p. 168.**

Shielding—a technique used to control external electromagnetic interference (EMI) by preventing transmission of noise or static signals from the source to the receiver; consists of foil, mesh, or woven wire, **p. 165.**

Zeroing out—adjusting a measuring instrument to the proper output value for a zero measurement signal, **p. 171.**

9.1 Introduction

Transmitters and conversion equipment are integral parts of signal transmission. They send process information or data in different signal formats to displays, controllers, or other transmitters. The sent information is commonly used to control a process or part of a process. Sometimes it just conveys information for human decisions or verifications, such as on/off or safety data.

There are many different types of transmitters. Most are categorized by the type of information they convey. When transmitting information, it is important to protect the integrity and reliability of signal transmission since the accuracy of this signal and its transmission determine how well the final control elements perform their intended function.

Signal integrity can be improved or protected using several methods, including **shielding**, **insulation**, error correction, signal conditioning, and proper construction material selection.

Transmitted information can be in a mechanical, an analog pneumatic, an analog or a digital electrical, or an optical format. When sending information between two or more devices, it is sometimes necessary to convert the signal(s) from one format to another. This type of conversion is performed through a piece of equipment called a signal converter.

Transducers are devices that convert energy from one form to another—for example, sound to electrical. Whether the variable is measured and converted or just simply measured, the information or data will need to be transmitted in order to be useful for process control.

Shielding a technique used to control external electromagnetic interference (EMI) by preventing transmission of noise or static signals from the source to the receiver; consists of foil, mesh, or woven wire.

Insulation any substance that prevents the passage of heat, light, electricity, or sound from one medium to another.

Signal Transmission

Signal transmission is an important part of data transfer. In order to transmit signals properly, it is important to ensure the integrity and reliability of the signal through such methods as grounding and shielding, insulation, error correction, signal conditioning, and construction material selection.

Transmitter Purpose and Operation

Transmitters are devices that encode information (data) in a format that is suitable for a transmission medium. In process control applications, this information corresponds to process variable data (e.g., flow, pressure, and level). The type of transmitter and transmission medium used are determined by the kind of process variable and its sensor type, physical conditions or environment, and the degree of precision required.

The most common of all transmission modes is the electrical signal. In **electrical transmission**, information or data is sampled or collected by a sensor or transducer, then mixed with or encoded onto an electrical carrier signal. (NOTE: Sometimes the carrier signal may simply be a power supply.)

Electrical transmission a signal transmission method that uses electrical devices to create a signal.

In Figure 9.1, the microphone is the sensor-transducer. Sound waves are sampled or collected by the microphone diaphragm, which presses on small crystals, causing a voltage change in proportion to the magnitude and frequency of the sound waves hitting the microphone diaphragm. This small voltage signal is the information data or process variable. This signal is applied to a voltage source (the battery) and together they act as a transmitter to convey the data to the receiver or final control element (the speaker).

Figure 9.1 Electrical signal transmission. **A.** Illustration of a microphone and speaker system. **B.** Microphone.
CREDIT: B. Mipan/Shutterstock.

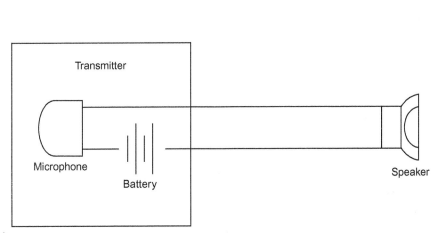

A.
B.

9.2 Protecting Integrity and Reliability of Signal Transmission

During a process, it is important to maintain the integrity and reliability of signal transmission. To accomplish this, several methods may be employed. These include shielding, insulation, error correction, signal conditioning, and proper construction materials.

Grounding and Shielding

Two of the problems associated with signal transmission are radio frequency interference (RFI) and electromagnetic interference (EMI). RFI and EMI are induced, radiated, or conducted electrical disturbances. They can interfere with other signals by raising or lowering the signals themselves. This can cause undesirable responses or malfunctions in electrical or electronic equipment. Failure to reduce or eliminate these RFI and EMI interferences can lead to erroneous signal transmission or signal **noise**. One way to prevent signal noise is through proper grounding and shielding.

Noise any unwanted or excessive sound.

Grounding is the establishment of a conductive connection, so harmful current (from an electrical circuit or equipment) is diverted to the earth or some other conducting body that serves in place of the earth. Proper grounding redirects rogue signals away from the intended transmission, thereby reducing interference. Grounding is typically accomplished through a single grounding wire connected directly to the housing of the unit being grounded. The grounded unit commonly has a grounding wire or a plant grounding grid to which it is attached. Figure 9.2 shows an example of grounding.

Figure 9.2 Grounding.
CREDIT: Somkhuanfoto/Shutterstock.

Shielding is another technique for controlling external RFI and/or EMI in signal transmission. In shielding, a wire mesh or foil is used to encase the transmission wire. This mesh or foil is then connected to a grounding wire which diverts erroneous signals and reduces interference. Figures 9.3 and 9.4 show examples of shielding, and how it is incorporated into a transmission wire.

Although we incorporate grounding techniques in shielding, the actual grounding of equipment is primarily for safety purposes for both people and equipment. It allows for conductivity to earth, thereby protecting equipment personnel from high voltage shorts. Grounding equipment may also help shield it from signal noise.

Figure 9.3 Insulation and shielding. **A.** Illustration of insulation and shielding layers. **B.** Cutaway showing insulation and shielding.
CREDIT: B. Nordroden/Shutterstock.

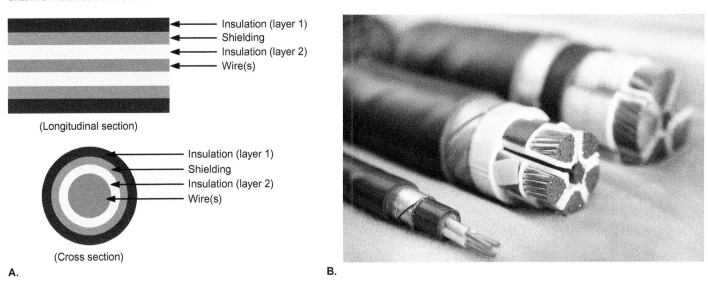

Figure 9.4 Shielding.

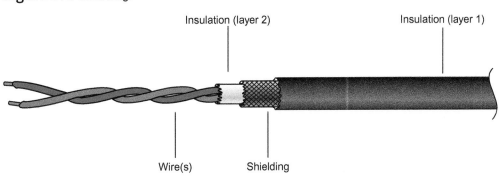

Ground loops occur when two points of a circuit, both intended to be at ground reference potential, have a potential between them. This can be caused, for example, in a signal circuit referenced to ground, if enough current is flowing in the ground to cause two points to be at different potentials. To prevent ground loops, instrument DC shielded cable should have the shield or ground wire grounded in the control room or the junction box, not in the field. Shielding ensures that the control signal is transmitted without interference (e.g., static).

Insulation

There are varieties of wires and cables as well as wire and cable insulations. Insulated wire or cable consists of nonconductive material or some other kind of material that is resistant to an electric current. It surrounds and protects the wire and cable inside.

Dielectric is a term that is often used in relation to radio frequency cables. It describes cable and wire insulation that prevents the insulated wire's current from coming into contact with other conductors. It preserves the wire material against environmental threats and resists electrical leakage. There are three major categories of wire insulation: plastic, fluoropolymers, and rubber, each with a variety of styles.

Error Correction

In **digital transmission**, information is sampled and converted to binary data. That is, the signal exists in two states, either "On" (a one) or "Off" (a zero). This on-off signal is usually very reliable and does not require any correction to the data. If, however, there are large

Digital transmission a signal transmission method that uses digital devices to create a binary signal.

noise sources, error correction can be employed using two methods: (1) binary signal filtering transmission, or (2) the addition of error parity bits (check bits) to the transmission itself. (*Parity bits* are the simplest form of error detecting code.) Parity checking at the receiver can detect the presence of an error if the parity of the receiver signal is different from the expected parity.

Signal Conditioning

In gathering data with various sensors and transducers, the converted energy or signal may be suitable for the receiver (the controller or final control element). However, it may need amplification (making it larger), filtering (separating something out of the signal), or signal conversion (change from one kind of data or signal to another). (Signal conversion will be discussed later in this chapter.)

Amplification involves increasing the magnitude of a signal voltage or current level increase to a level that allows the transmission to arrive unchanged from its original transmission point to its destination. Signals that have considerable amounts of noise may require filtering to eliminate the noise and enhance data reliability. And finally, to ensure reliable data transmission, the signal may need to be converted to another type of signal that is more suitable to its environment.

Construction Material Selection

One of the greatest enemies of signal transmission is friction or heat. Friction or heat may be due to the resistive conductive properties of poorly formed attachment of wire joints, splices, or connections. It may also be due to the two physical surfaces rubbing or sliding against each other as in liquid or gas flow. In any case, material selection is crucial to maintaining proper signal conductivity or propagation by reducing heat loss or friction. The quality of physical joints, splices, and connections is also critical for ensuring proper conductivity.

Other Safety Concerns

It is very important that proper safety precautions be implemented and equipment be kept in good working order (e.g., insulation must not be torn or removed, and proper construction materials must have been selected). Failure to use or maintain proper safety precautions could result in human injury and/or equipment loss or damage.

The following are just a few safety issues that could arise as a result of improper safety precautions:

- Equipment fires (e.g., grounding problems)
- Electrical thermal burns (personal)
- Sensor damage (loss of measurement and control signals causing process upsets and/or safety issues)
- Process piping damage (erosion and corrosion)

Mechanical transmission a signal transmission method that uses mechanical methods to create a signal.

Pneumatic transmission an analog signal transmission method that uses compressed air/gas to create a signal.

Optical transmission a signal transmission method that uses light frequencies to create a signal.

9.3 Common Types of Transmissions

Transmissions can be in a variety of forms. The most common, however, are **mechanical transmissions**, **pneumatic transmissions**, and electrical transmissions (including **optical transmissions**). Table 9.1 lists these transmission types and provides a brief description of each.

Analog Transmission

Analog signals have an infinite continuous number of values and can vary both in magnitude and frequency, depending on the mode. Amplitude-varying signals (the usual process control type) contain the information in the magnitude of the signal.

Table 9.1 Transmission Types and Descriptions

Transmission Type	Description
Mechanical transmissions	Including levers, arms, torque tubes, cables, pulleys, and more
Pneumatic transmissions	Analog—involving use of compressed gas (air or some other inert gas) to send signals; normally 3 to 15 PSIG
Electrical transmissions	May be any of several types: • Analog—Varying signals from VLF (very low, normally 4 to 20 mA) to gamma ray frequency • Digital—Discrete or binary signals (having only two states: 1 for open and 0 for closed, for example) • Optical—Light frequency signals of either analog or digital modes

Simple **analog transmission** in which the voltage or current amount (magnitude) is transmitted or conducted to a controller or receiver is the usual method for process control signals. The method is reliable and inexpensive and satisfactory for low noise, nonhostile environments. It uses simple electrical circuits. The main disadvantages are the length of transmission (typically 2,500 feet [762 meters]) and its susceptibility to noisy environments and power loss (therefore signal weakening). Standard 4–20 mA, 0–10 V, and 0–50 mV loops utilize this method of transmission but usually also use some kind of noise filtering, signal amplification, or buffering.

Analog transmission a signal transmission in which process variable measurements change in a continuous manner mirroring the actual process variable.

Digital Transmission

Digital transmission consists of either (1) retransmitting digital pulses that are already produced by the field sensor (rotary encoders, etc.) or (2) converting the analog signal containing the process control variable data or information into a digital representation that is transmitted. The latter is normally achieved by an analog-to-digital or A/D converter and is most commonly used in the process industries.

The advantages of converting signals to binary data for transmission are that the signal has very good noise immunity and that simple buffering filters noise. The disadvantage is that it is more expensive and generates its own noise, which is harmful to other signals. Essentially, analog, current to pneumatic (I/P), or pneumatic to current (P/I) communication is used most at the controller or sensor levels in a process. The digital signal is used most from the process area or equipment back to the control room or station.

There are standards (IEEE-488, RS-232, Ethernet, fieldbus, etc.) for the various digital transmission methods. However, compatibility between the transmitting field device and the controller, final control element, or receiver must be considered.

Optical Transmission

Fiberoptic transmission is becoming more and more common in process control because of cost reductions, noise immunity, and intrinsically safe operating characteristics. Light transmitters (LEDs or laser diodes) and receivers (photo diodes, photo transistors, or photoresistive cells) used with fiberoptic cabling make up a complete circuit. The transmitter is modulated with the process variable information that is conveyed over the optical cable to the receiver, which senses or detects the modulated signal contained in the light signal. The process variable data may either be analog or digital but is typically digital.

A big advantage of the digital signals in fiberoptic lines is that it is almost immune to outside interferences. They also do not have the degradation of the signal over distance that metal wires do.

9.4 Signal Conversion

In order for a signal to be transmitted and interpreted properly, it is sometimes necessary to convert the signal from its original form to something the receiving device can understand. Consider a typical cell phone call. The caller's voice is converted into an analog electrical signal by a microphone, and the analog signal is converted to a digital stream using an analog to digital converter (ADC). Then a digital to analog converter (DAC) converts this back into an analog electrical signal, which drives an audio amplifier, which in turn drives a speaker, which finally produces sound.

Transducers are the devices used to perform this type of conversion. There are generally two forms of transducers: active and passive. An active transducer has the ability to convert one form of energy into another form without using an external source of power. A passive transducer converts one form of energy into another but can only do so by making use of an external source of power.

Purpose and Operation of Signal Converter Equipment

The purpose of signal conversion is to change the form of one signal value into another signal value to ensure device or signal compatibility. Transducers perform this type of conversion.

The following are common signal conversions:

- **A/D conversion** = analog to digital
- **D/A conversion** = digital to analog
- **I/P conversion** = current to pneumatic
- **P/I conversion** = pneumatic to current
- **I/E conversion** = current to voltage
- **E/I conversion** = voltage to current
- **E/P conversion** = voltage to pneumatic
- **I/F conversion** = current to frequency
- **F/I conversion** = frequency to current
- **E/F conversion** = voltage to frequency
- **F/E conversion** = frequency to voltage

> **I/P conversion** the conversion of an analog signal from current to pneumatic.
>
> **P/I conversion** the conversion of an analog signal from pneumatic to current.
>
> **I/E conversion** the conversion of a signal from current to voltage.
>
> **E/I conversion** the conversion of a signal from voltage to current.
>
> **E/P conversion** the conversion of a signal from voltage to analog pneumatic.

9.5 "Zeroing Out" Wet Leg Instruments

A differential pressure level transmitter is located below the tank with impulse tubing connecting the high side (tank hydrostatic pressure) and the low side (ullage space in the tank, or atmospheric pressure). These impulse tubes can be in a dry leg or a wet leg condition. Dry leg is a low side transmitter piping which will remain empty if gas above the liquid does not condense. This is a dry leg condition. Range determination calculations for dry leg conditions are the same as those described for bottom-mounted differential pressure transmitters in an open vessel (Figure 9.5).

If the low side leg might get condensate in the impulse tubing, the technician will create a wet leg as a reference leg containing a liquid. Slow condensation of the gas above the liquid could cause the low side of the transmitter piping to fill with liquid, resulting in errors in the

Figure 9.5 Dry leg, wet leg, and open tank.

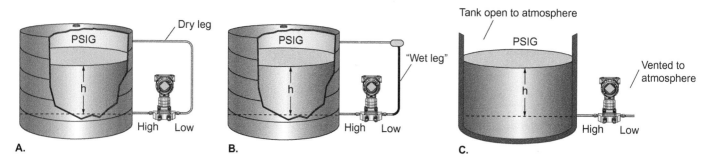

A.   B.   C.

differential pressure readings. The impulse tubing is intentionally filled with a suitable reference fluid to eliminate this potential error. This is a wet leg condition. The wet leg condition is a reverse transmission instrument, in that the low side will be higher than the high side.

Because of this reverse action, an instrumentation technician needs to "zero out" the wet leg. **Zeroing out** means adjusting a measuring instrument to the proper output value for a zero measurement signal.

Zeroing out adjusting a measuring instrument to the proper output value for a zero measurement signal.

9.6 Scaling Calculations

Scaling calculations convert process signals so that they are compatible with instrument and control systems. Through scaling, it is possible to equate the numerical value of one scale to its mathematically proportional value on another scale. For example, measurements applied to a standard analog electronic temperature transmitter are represented on an appropriate temperature scale, while the output signal is represented on a milliampere scale.

Figure 9.6 shows an example of various scale types and how they relate to one another.

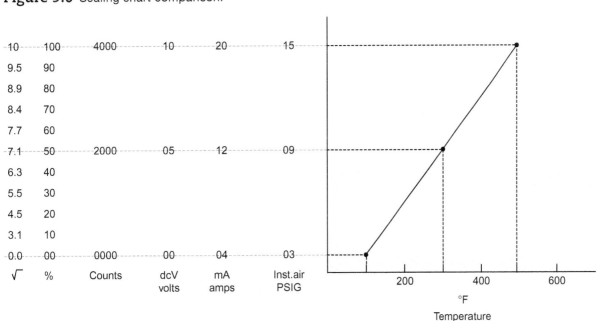

Figure 9.6 Scaling chart comparison.

Instrument Scale Definitions

In order to interpret and discuss various scales properly, it is important to know a few terms—specifically, upper range value (URV), lower range value (LRV), range, and span (Figure 9.7).

- *Upper range value* (URV) is the number at the top of the scale, expressed as one number.
- *Lower range value* (LRV) is the number at the bottom of the scale, expressed as one number.
- *Total range* is the set of values that exist between the URV and LRV of a scale. A *range* (as in the correct range for running a piece of equipment) is expressed as two numbers (e.g., a calibration range of 50–150 PSIG or see the range from 40 to 70 in Figure 9.7).
- *Span* is the algebraic difference between the upper range value of a scale minus the lower range value of a scale expressed as one number (e.g., in Figure 9.7, span = 100 PSIG).

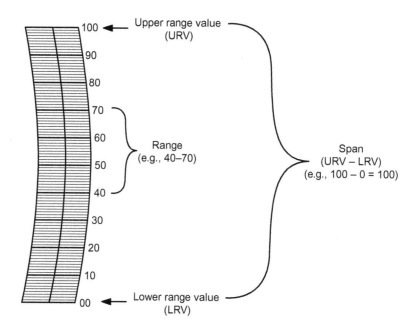

Figure 9.7 Example of URV, LRV, range, and span.

Common Linear Signal Conversions

Linear scaling, or linear signal conversion, is used when there is a linear relationship between two scales (e.g., transmitter input versus transmitter output). Certain types of conversions are the most common:

- I/P conversion
- E/P conversion
- I/E conversion

The following is a formula for conducting a linear conversion:

$$\text{VALUE}_B = \frac{\text{VALUE}_A - \text{LRV}_A}{\text{SPAN}_A} \times \text{SPAN}_B + \text{LRV}_B$$

Nonlinear Signal Conversion

The flow rate through an orifice meter piping run is proportional to the square root of the differential pressure created by the orifice. Consequently, the need to calculate flow rate as a result of a square root number is required.

Percent D/P is a measurement of the actual differential pressure drop compared to the D/P at 100 percent flow. If a D/P transmitter is calibrated to measure 0 to 100 inches water column (in. w.c.), then a measurement of 25 in. w.c. would also be 25 percent D/P. In order to convert the differential pressure into percent flow rate, use the following table (Table 9.2).

$$\text{Percent flow rate} = \sqrt{\text{D/P\%}}$$

Table 9.2 Square Root to Linear Signal Conversion

Step	Equation
Insert the % D/P into the equation.	Percent flow rate = $\sqrt{25\%}$
Change percent into a decimal form.	Percent flow rate = $\sqrt{0.25}$
Take the square root.	Percent flow rate = 0.50
Make the decimal a percent again.	Percent flow rate = 0.50 × 100
Record your answer.	Percent flow rate = 50%

9.7 Wireless Technologies—Industry 4.0

The Industrial Internet of Things (IIoT) is migrating into industrial Ethernet transmissions. Industry 4.0 refers to the transition of the operational technology (OT) into the world of open systems architecture and advance transmission systems based on the IIoT. These technologies provide higher speeds, increased performance, and interoperability among components throughout the process.

One of the Industry 4.0 technologies is wireless transmission and communication. Many concerns have been raised around the security of wireless communications, but with the integration of information technology (IT) and OT, secure networks are achievable. We will cover security in depth in later chapters.

Development of wireless devices (Figure 9.8), gateways, access points, and software provides potentially seamless integration across all devices. Control-ready wireless networks result in lower installation costs and greater reliability than traditional systems.

Figure 9.8 Wireless transmitter.
CREDIT: © Emerson 2019.

When deployed in remote areas or harsh environments, wireless technologies can prevent accidents and increase productivity. The operations workforce will become more mobile, working with handheld interfaces. Wireless connectivity will be critical for control, data analytics, and communications.

Staying current with these and future technologies will require every operations technician to study the systems of Industry 4.0 and learn new approaches as industry continues to change.

Open Internet Protocol Transmission/Addressing

Scalable, secure technologies and services in the IIoT are advancing into process industry. Just as the transition was made from pneumatic control systems to distributed control systems (DCSs), so now process automation is moving from proprietary systems to open internet protocol transmission/addressing and open source software. Interoperability of software, controllers, instrumentation, and control elements will change the landscape of process operations.

These systems will include distributed control nodes (DCNs). DCNs have an input/output (I/O) interface, are addressable, and are connected to switches and gateways within the network. A good comparison is the cell phone, which contains a unique physical address and can communicate to many other devices within the network, using an array of software applications. Interconnected devices allow us to adjust our home thermostat with our smartphone from anywhere. Similarly, with DCNs, we can control field devices and get feedback from those devices from almost anywhere.

Of course, a major concern with such open transmissions and control systems is security. We will cover this in depth in later chapters, but security needs to be addressed at every level of control. Communications and applications will be based on open protocol and transmission, but the network itself will need to be completely secured.

Interoperability of open system architecture will foster innovation, reduce life cycle costs, lower barriers for competition, and allow sharing of both services and data. Interoperability relies on unique address identification of individual parts. The IPv4 address numbering system, which allowed 4.3 billion unique addresses with the 32-bit address length, has already begun to run out of addresses. The IPv6 address numbering system increases the address length to 128 bits. This results in 340 undecillion unique addresses (i.e., 340 with 36 zeros). This capacity will allow every piece of instrumentation and equipment in the plant to carry a unique identification.

As these Industry 4.0 technologies work their way into the operations of the process industry, operators will need to continue to educate themselves and be prepared to embrace these changes.

Summary

Transmitters and conversion equipment are integral parts of signal transmission. Transmission is the conveying of information from one place (the sample point) to another (the endpoint) via some kind of transmission medium (transmitter).

Information from the real world (physical data) is imperative in process control. Transmitters send process information or data in different signal formats to displays, controllers, or other transmitters to utilize this information to control a process or part of a process, or sometimes just to convey information for human decisions or verifications.

There are many different types of transmitters, and most are categorized by the type of information they convey (e.g., temperature, pressure, level, flow, and computer, machine, or controller status). The operation of these various process control transmitters is relatively simple in that they all convey the sampled data, either in the same format it is gathered or possibly the format signal is converted to a more appropriate one because of the environment the transmission is occurring in.

When transmitting information, it is important to protect the integrity and reliability of signal transmission. The accuracy of this signal and its transmission determine how well the final control elements perform their intended function. The integrity of these signals or data must be maintained by transmitters and converters, or additional error compensation is required. This may be accomplished using several methods including shielding, insulation, error correction, signal conditioning, and proper construction material selection. Transmitted information can be in either a mechanical, a pneumatic, an electrical (analog or digital), or an optical format. When sending information between two or more devices, it is sometimes necessary to convert the signal(s) from one format to another (e.g., from pneumatic to electrical). This type of conversion is performed through a piece of equipment called a signal converter. Some of the most common conversions are P/I, I/P, and I/E (or E/I).

The actual sampling or measurement of a process variable is performed by a sensor. This sensor is specifically designed for the type of variable to be measured and has operating characteristics unique to that design and is usually of two categories, contacting or noncontacting. These variables include temperature, color, light, density, chemical properties, force, weight, motion, position, flow, level, pressure, sound, particle radiation, electromagnetic radiation, and electrical.

Transducers are devices that convert energy from one form to another. Whether the variable is measured and converted or just simply measured, the information or data will need to be transmitted in order to be useful for process control.

Process technicians must be on a lifelong learning curve in order to stay current with technological developments and changes. Wireless technologies are becoming part of process industries. Open internet protocol transmission/addressing is just one example of information the new process technician will encounter when entering the workplace.

Checking Your Knowledge

1. Which of the following types of devices is used to convert control signal from one form to another?
   a. Transducer
   b. Signal converter
   c. Heat tracing
   d. Sensor shielder

2. Which of the following can cause interference with a control signal?
   a. Overhead airplanes
   b. Electrical interference from a thunderstorm
   c. Power outages within a plant
   d. Loud noises within the plant

3. Which of the following can be used to maintain the integrity and reliability of a signal transmission? (Select all that apply.)
   a. Grounding
   b. Error correction
   c. Transducers
   d. Signal conditioning

4. You are working in a process facility that is a nonhostile environment with a low level of noise. You need to identify an electrical transmission type that is inexpensive and reliable. Which transmission type should you use?
   a. Analog
   b. Digital
   c. Optical
   d. Mechanical

5. You are working in a process facility that needs to identify a transmission method that is immune to outside interferences. Your facility also needs to use a signal that can travel over long distances without degradation. Which transmission type should you use?
   a. Analog
   b. Digital
   c. Optical
   d. Mechanical

6. Match the abbreviations below to the type of signal conversion they represent.

Abbreviation	Conversion Type
I. P/I	a. Current to pneumatic
II. I/P	b. Voltage to current
III. E/I	c. Pneumatic to current
IV. I/E	d. Current to voltage

7. With a wet leg condition as in a level transmitter or differential pressure instrument, the low side will be _____ than the high side.
   a. lower
   b. higher

8. Calculate the percent flow rate for 14% D/P.

9. Which of the following are advantages of open systems architecture in the process industries?
   a. Reduced life-cycle costs
   b. Lower barriers for competition
   c. Sharing of both services and data
   d. All of the above

10. For the image below, record the numerical values for each:
    a. Span _____
    b. Lower range value _____
    c. Range _____
    d. Upper range value _____

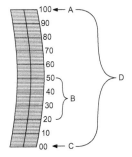

NOTE: Answers to Checking Your Knowledge questions are in the Appendix.

Student Activities

1. Perform a zero-based calibration of a differential pressure, gauge pressure, or absolute pressure transmitter. It is recommended that a calibration graph, calibration report, and a lab report be included with this exercise. If a lab exercise has not been developed, use the calibration and documentation procedures in the manufacturer's instruction manual.

2. Perform a zero suppression calibration of a differential pressure, gauge pressure, or absolute pressure transmitter. It is recommended that a calibration graph, calibration report, and a lab report be included with this exercise. If a lab exercise has not been developed, use the calibration and documentation procedures in the manufacturer's instruction manual.

3. Perform a zero elevation calibration of a differential pressure, gauge pressure, or absolute pressure transmitter. It is recommended that a calibration graph, calibration report, and a lab report be included with this exercise. If a lab exercise has not been developed, use the calibration and documentation procedures in the manufacturer's instruction manual.

4. Using a table with an mA input column and a PSIG-output column (corresponding to an I/P transducer). Determine the input values based on 10 percent, 30 percent, 50 percent, 70 percent, and 90 percent of the analog input signal (4–20 mA). Then calculate the output values based on a 3–15 PSIG signal. Finally, set up an I/P converter (transducer) in the lab and verify the answers by applying the mA input values to the instrument and reading its output.

5. Set up an I/P converter (transducer) in the lab and check its calibration. To do this, you will have to understand its I/O relationship. Then calculate several additional output values based on random input values verifying each result with the I/P transducer.

6. Using examples of signal transmission lines, locate and identify the various transmission signal types. It is recommended that this activity be done independently.

PART 3 Control Loops, Controllers, Control Schemes, and Programmable Logic

Chapter 10
Introduction to Control Loops: Simple Loop Theory

 Objectives

After completing this chapter, you will be able to:

10.1 Explain the function and elements of process control loops. (NAPTA Introduction to Control Loops, Simple Loop Theory 1, 2*) p. 179

10.2 Explain the differences between open and closed control loops. (NAPTA Introduction to Control Loops, Simple Loop Theory 5) p. 180

10.3 Classify the functions of a control scheme:
sensing
measuring
comparing
controlling
transducer (converter) (NAPTA Introduction to Control Loops, Simple Loop Theory 4) p. 182

10.4 Explain signal transmission:
pneumatic (analog)
electronic
digital (discrete)
mechanical (NAPTA Introduction to Control Loops, Simple Loop Theory 3) p. 186

*North American Process Technology Alliance (NAPTA) developed curriculum to ensure that Process Technology courses will produce knowledgeable graduates to become entry-level employees in process technology. Objectives from that curriculum are named here in abbreviated form. For example, "(NAPTA Introduction to Control Loops, Simple Loop Theory 1, 2)" means that this chapter's objective 1 relates to objectives 1 and 2 of NAPTA's curriculum on simple loop theory.

Key Terms

Accuracy—how close a measurement corresponds to its true value, **p. 184.**

Algorithm—preset mathematical function calculated in a controller that can be mechanical, analog, or digital. The three most common output functions deal with proportional (P), integral (I), and derivative (D) tuning, **p. 185.**

Basic control functions—sensing, measuring, comparing, calculating, correcting, and manipulating, **p. 180.**

Closed control loop—when a control loop has feedback (e.g., controller in automatic mode), **p. 180.**

Comparator—a component of a controller that compares the measurement to a predetermined set point, **p. 185.**

Comparing, calculating, and correcting element—the control loop component that receives the appropriate signal from the transmitter and compares the signal to a desired value (set point); if there is a difference, then the output of the comparison causes a calculation to be performed to cause a corrective response by the controller output signal to the final control element, **p. 184.**

Controlled variable—a process variable that is sensed to initiate the control signal, **p. 180.**

Controller—an instrument that receives a signal from the transmitter, compares it to a set point, and produces an output to a final control element, **p. 180.**

Converting and transmitting element—the control loop component that converts the sensed process variable and transmits the measured signal, **p. 183.**

Converting device—a device that receives information in one form of an instrument signal and changes it into another form of an instrument signal, **p. 185.**

Digital signal—data that is represented as coded information in the form of binary numbers; used to transmit data to and from field transmitters on a twisted pair of wires; may also be between computers and computer components, **p. 188.**

Electronic signal—analog or digital signal; current or voltage signal, **p. 187.**

Feedback loop—the most common type of control loop where the change caused by the output of the controller is fed back to the process, providing a self-regulating action, **p. 181.**

Final control element—the last active device in an instrument control loop; it directly controls the manipulated variable; usually a control valve, a louver, or an electric motor, **p. 180.**

Live zero—a standard bias added to the instrument signal (e.g., pneumatic 3 to 15 PSIG or electronic 4 to 20 mA); instead of reading zero, 0 percent of scale would have a reading of 3 PSIG (pneumatic) or 4 mA (electronic), **p. 189.**

Loop error—the accumulated error of each device in the loop; calculated as the square root of the sum of the squares of individual device accuracy, **p. 189.**

Manipulating element—the final control element (e.g., control valve) manipulated by the corrective response of the controller output to maintain the process variable at the correct set point value, **p. 186.**

Measured variable—a process variable that is measured, **p. 187.**

Mechanical link—a way of mechanically transmitting the motion of a primary sensor to a controlling mechanism; conveys linear or rotary motion by using a pivoting crank, **p. 188.**

Open control loop—when a control loop does NOT have feedback (e.g., controller is in manual mode), **p. 180.**

Parallel data communication—one wire per bit or 64+ wires for a 64-bit binary word; used primarily in short distances (a few feet), **p. 188.**

Pneumatic signal—an instrument communication with a range of 3–15 PSIG; must have an instrument air supply; has a lag time associated with the signal; relatively short transmission distances, **p. 187.**

Precision—how close repeated measurements are versus the action; reproducibility; the closeness of repeated measurements of the same quantity, **p. 184.**

Process control—the act of regulating one or more process variables so that a product of a desired quality can be produced, **p. 179.**

Process error—the difference between set point and process variable (SP − PV). **p. 182.**

Sensing—the act of detecting, **p. 183.**

Sensing element—the control loop component that detects, or senses, the process variable, **p. 183.**

Serial data communication—two data wires; the most common means of communication used between plant equipment, **p. 188.**

Set point—the desired RPM indication where a controller is set for optimum operation, **p. 180.**

Standard instrument signal—any kind of detectable quantity used to communicate information in a standard format for control, **p. 183.**

Transducer—a device that receives information in one form of instrument signal and changes it into another form of instrument signal, **p. 180.**

Transmitter—the control loop component that receives a signal from the sensor, measures that signal, and sends it forward to the controller or indicator, **p. 180.**

10.1 Introduction

The information found in this chapter introduces the concept of **process control** that occurs when various instruments are combined to function together in a control loop. These instruments control the process by measuring, controlling, and manipulating it by communicating with each other. Loops may be either open or closed, depending on the type of feedback communicated within the particular loop. The various components found in a control loop may sense, measure, compare, control, or convert signals between the various components. Signal transmission between the instruments may be pneumatic, electronic, digital, or mechanical. Each of these characteristics of a control loop is discussed below.

Process control the act of regulating one or more process variables so that a product of a desired quality can be produced.

Process Control and Control Loop Elements

A process variable (PV) is usually a quantity that is measured or controlled, such as pressure, temperature, level, flow, or pH. Process control (Figure 10.1) is the act of regulating one or more process variables so that a product of a desired quality can be produced. To control any process, a variable must first be sensed and measured. Then the PV must be compared

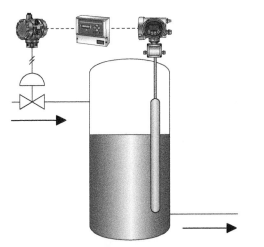

Figure 10.1 Process control.

Set point the desired RPM indication where a controller is set for optimum operation.

Transmitter the control loop component that receives a signal from the sensor, measures that signal, and sends it forward to the controller or indicator.

Controller an instrument that receives a signal from the transmitter, compares it to a set point, and produces an output to a final control element.

Transducer a device that receives information in one form of instrument signal and changes it into another form of instrument signal.

Final control element the last active device in an instrument control loop; it directly controls the manipulated variable; usually a control valve, a louver, or an electric motor.

Basic control functions sensing, measuring, comparing, calculating, correcting, and manipulating.

Controlled variable a process variable that is sensed to initiate the control signal.

to a desired value, or **set point** (SP—the target value); next, a calculation (some algorithm) is performed, and then a counter response (corrective output) must be produced that enables the measurement to move toward set point.

The most common instruments (Figure 10.2) that are connected together to form a loop are the sensor, the **transmitter**, the **controller**, the **transducer**, and the **final control element**. All the **basic control functions** (e.g., sensing, transmitting, measuring, converting, etc.) are identified and associated with one of these instruments in a control loop. The arrows indicate the direction of information flow.

Figure 10.2 Most common control elements.

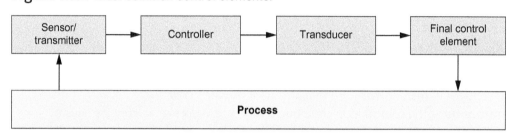

The PV itself is any measured property of a process other than an instrument signal. In a control loop, these measured properties are referred to as:

- **Controlled variables**: Process variables are sensed to initiate the feedback signal.
- Transduced variables: A quantity such as temperature, pressure, air humidity, sound pressure, or light is transformed into normalized signals (e.g., 4 to 20 mA).
- Manipulated variable: The quantity or condition is adjusted as a function of the actuating error signal.

In many instances, the measured and controlled variables are the same process variable. In some situations, all of these variables can be the same PV. For example, in a simple flow control loop, the flow is measured (PV), controlled, and manipulated.

10.2 Closed and Open Control Loops

A simple control loop is a group of instruments working together to control a single process variable such as pressure, temperature, level, flow, or some analytical or physical property. A typical instrument loop consists of the individual instruments necessary to control a process variable, along with any additional instruments used to either convert signals or mathematically compute process variable measurements so that a signal can be manipulated.

There are different types of control loops in a processing plant. All process technicians should understand the fundamental concept of open loop control versus closed loop control. A **closed control loop** is a combination of control units in which a process variable is measured and compared to set point. If there is a difference, a corrective signal is sent to the final control element to bring the process variable back to the desired value (SP). An **open control loop** is a system of control that provides information necessary for control without comparing a measured value of a process variable to the desired value of a process variable. The difference between the two types of control loops is based upon feedback. A control loop is closed if it has feedback (Figure 10.3) and open if it does NOT have feedback. Most control loops in process industries are closed loop.

Closed control loop when a control loop has feedback (e.g., controller in automatic mode).

Open control loop when a control loop does NOT have feedback (e.g., controller is in manual mode).

Figure 10.3 Function block diagram of a feedback control loop (closed).

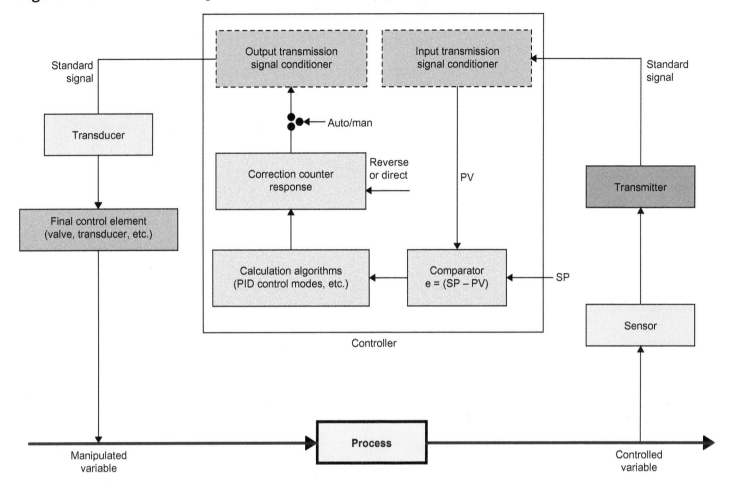

Closed Loop Control (Feedback)

The most common control loop is the closed, or **feedback loop** (Figure 10.4). A feedback system measures a value and reacts to changes in that value. As an example, a thermostat measures the ambient temperature in a home, and if the temperature falls below its minimum setting, the thermostat activates the furnace to heat the home back to the appropriate temperature.

Feedback loop the most common type of control loop where the change caused by the output of the controller is fed back to the process, providing a self-regulating action.

Figure 10.4 Closed control loop (feedback).

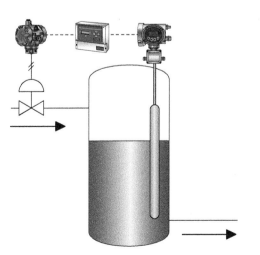

Process error the difference between set point and process variable (SP − PV).

The thermostat measures the temperature, but it also feeds that value back into its control scheme to maintain the desired temperature. As the error diminishes, so does the controller output. A small amount of **process error**, the difference between set point and process variable (SP − PV), is always present in a feedback control loop since an error must exist before the error can be eliminated.

An air conditioner thermostat in a house acts as an ON/OFF controller. The air conditioner cools the air, and the temperature measurement (PV) is the feedback signal compared to the set point. The air conditioner does not come on until the temperature has risen to a point above the set point. In a simple controller, this corresponds to the automatic mode. Feedback control loops always work to eliminate error, not prevent it.

Another example of closed loop control is controlling the speed of a vehicle with cruise control. A set point is entered and the cruise control mechanism automatically adjusts the fuel to the engine in order to adjust to the desired speed. In a simple controller, this corresponds to the automatic (auto) mode.

Open Loop Control

As stated earlier, open loop control (Figure 10.5) is defined as a signal path without feedback. Open loop control exists when a vehicle operator turns off the cruise control and physically manipulates the accelerator while monitoring the speedometer to achieve the desired speed. In a simple controller, this would correspond to the *manual* mode.

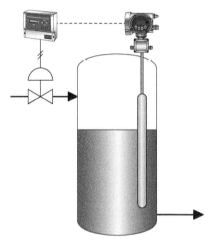

Figure 10.5 Open loop control.

10.3 Control Loop Functions and Signals

This section discusses the main components of a control loop in more detail, and how they transmit signals between themselves.

All closed control loops must contain the following basic control components:

- Sensor
- Transmitter
- Controller
- Transducer
- Final control element

The components perform numerous functions, as shown in Table 10.1.

Table 10.1 Control Loop Components

Component	Function
Sensor	Sensing
Transmitter	Converting
	Transmitting
Controller	Comparing (set point)
	Calculating (algorithms)
	Correcting (counter response)
Transducer	Converting
Final control element	Manipulating

Sensor: Sensing Element

Sensing is the act of detecting, and a sensor, or **sensing element** (Figure 10.6), is the first device in a control loop. Obviously, the sensing element must be capable, in one way or another, of detecting the process variable. For example, a metal diaphragm can sense pressure when it moves as force is applied. A resistance temperature device (RTD) can sense temperature through a change in electrical resistance. Sensing the process variable is possibly the most important action taken by a control system. If the sensor is incapable of responding accurately, precisely (repeatable), and quickly enough to the process variable, then controlling the process variable may be difficult or even impossible.

Sensing the act of detecting.

Sensing element the control loop component that detects, or senses, the process variable.

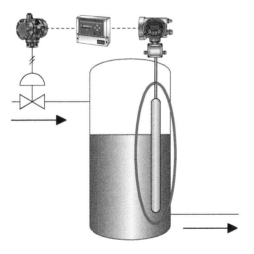

Figure 10.6 Level sensing element (displacer type).

Transmitter: Converting and Transmitting Element

Once the process variable has been sensed, it must be converted to a measurement. The measurement operation in an instrument is generally an assessment of the response of the sensor to the process variable rather than a direct measurement of the process variable itself. The device that transmits the measured signal is a transmitter (Figure 10.7). For example, a metal diaphragm responds to a change in position. The measurement transducer then responds electrically to the motion by producing an output. (Normally, such quantities as temperature, pressure, air humidity, sound pressure, or light are transformed into normalized signals [i.e., 4 to 20 mA].) Effectively, the process variable is transduced (changed) into a **standard instrument signal** such as 3 to 15 PSI, 4 to 20 mA, or digital.

Prior to the use of transmitters, controllers were directly connected to the process, which meant that they had to be locally mounted in the field close to the process. The introduction

Converting and transmitting element the control loop component that converts the sensed process variable and transmits the measured signal.

Standard instrument signal any kind of detectable quantity used to communicate information in a standard format for control.

Figure 10.7 Level transmitter.

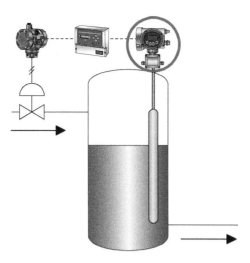

Accuracy how close a measurement corresponds to its true value.

Precision how close repeated measurements are versus the action; reproducibility; the closeness of repeated measurements of the same quantity.

Comparing, calculating, and correcting element the control loop component that receives the appropriate signal from the transmitter and compares the signal to a desired value (set point); if there is a difference, then the output of the comparison causes a calculation to be performed to cause a corrective response by the controller output signal to the final control element.

Figure 10.8 Controlling element (controller).

of the transmitter into industry ushered in the small (6" × 6" [15 cm × 15 cm]) and eventually the more modern (3" × 6" [7.5 cm × 15 cm]) board-mounted controllers that allowed for the centralization of controllers (remote) into the more modern control rooms of today. The transmitter produces a standard instrument signal that can carry the measurement information to a remote location, such as a control room. This was a very significant improvement in field instrumentation and a true mile marker on the evolutionary path to modern automatic process control, which today includes computers.

A measuring transducer must respond in an accurate and precise manner, just as the sensing element must. **Accuracy** is determined by how close the measurement corresponds to its true value (the process variable). **Precision** describes how repeatable the measurement is each time the same value is applied. Reproducing the same measurement results each time a specific value is applied to the sensor is as important, if not more important, than the actual accuracy of the measurement itself. Precision is the agreement between the numerical values of two or more measurements made in the same way and expressed in terms of deviation.

For example, a person shooting at the bullseye on a target may hit the target every time. Precision is how close each of the holes is to each other, or the tightness of the pattern. Accuracy would depend on how close (bias error) the holes are to the centermost point on the target. The possibility exists for there to be great precision and very poor accuracy.

Controller: Comparing, Calculating, and Correcting Element

Once the PV has been measured and the appropriate signal has been transmitted to the controller (Figure 10.8), the PV must then be assessed. The **comparing, calculating, and correcting element** receives the appropriate signal from the transmitter and compares the signal to a desired value

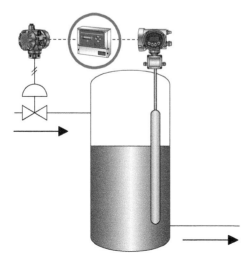

(set point). If there is a difference, then the output of the comparison causes a calculation to be performed to cause a corrective response by the controller output signal to the final control element. The **comparator** part of the controller determines the difference between the SP and the incoming PV signal. If the measurement and SP are the same, the output from the comparator is zero and the controller output stays the same. If there is a difference between the SP and the PV, then the output of the comparator causes a change to the calculating function of the controller.

The next function is the calculating portion of the controller performed by some type of preset **algorithm**, which can be mechanical, analog, digital, or mathematical. The calculation element must respond to the magnitude and direction of the error (signal difference) received from the comparator. This means that as the difference generated by the comparator increases, the response of the calculation element increases as well. Also, the direction that the corrective element drives the final control element is directly related to whether the measurement is above or below the set point. In either case, an appropriate response is produced to reduce or eliminate the error; this is the corrective output.

The controller then acts upon this deviation by calculating a corrective output signal. These calculations are performed using one or more algorithms. The three main controller algorithms are usually referred to as proportional (P), integral (I), and derivative (D) modes. Other common accepted names for these functions are gain (G), reset (R), and rate (Rt). Digital controllers may be programmed with any combination of mathematical functions to produce a desired output. A controller usually contains the comparing, calculating, and corrective output elements. The controller action can be chosen direct (+) or reverse (−) acting to counteract the process direction.

Comparator a component of a controller that compares the measurement to a predetermined set point.

Algorithm preset mathematical function calculated in a controller that can be mechanical, analog, or digital. The three most common output functions deal with proportional (P), integral (I), and derivative (D) tuning.

Transducer: Converting Element

While not always present in a control loop, **converting devices** (transducers) (Figure 10.9) may be used to receive information in one form of an instrument signal and change it into another. These devices may also be called converters or relays. Converting instruments are placed between the main control loop elements to convert one element signal type to another type of signal that must be read or understood by the next element in the loop.

Converting device a device that receives information in one form of an instrument signal and changes it into another form of an instrument signal.

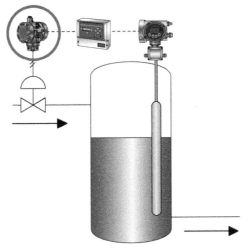

Figure 10.9 Converting element I/P transducer.

The most prominent converting device is the current-to-pneumatic converter (I/P). An I/P converter is used in most electronic control loops because the final control element is usually a pneumatically actuated control valve. Many final control elements have several components: an I/P transducer, positioner, actuator, and valve integrally mounted as a single unit. Examples of signal converters include:

- Current-to-pneumatic transducer (I/P)
- Pneumatic-to-voltage converter (P/E)
- Voltage-to-pneumatic converter (E/P)
- Analog-to-digital converters (A/D)
- Digital-to-analog converters (D/A)

Computing relays are those devices that are capable of computing values based on instrument signals. A square root extractor, for example, is a computing device that takes the linear output signal from a D/P transmitter that measures the pressure drop across an orifice plate and changes it to a square root output that is proportional to the flow rate. There are numerous types of *computing relays* and mathematical functions used in applications within control loops. They include the following:

- Addition and subtraction
- Multiplication and division
- Inversion
- Scaling and proportioning
- High and low select
- Mathematical function (e.g., $\sqrt{}$)

Final Control Element: Manipulating Element

Sometimes the manipulated process stream (**manipulating element**) (Figure 10.10) is the same as the controlled process stream. For example, in a flow control loop, the manipulated variable is also the controlled variable. In this situation, the control valve is placed into the same process stream as the flow transmitter.

Manipulating element the final control element (e.g., control valve) manipulated by the corrective response of the controller output to maintain the process variable at the correct set point value

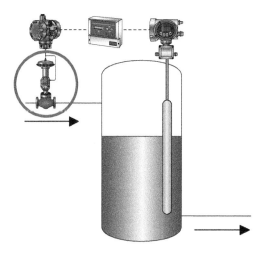

Figure 10.10 Manipulating element (final control element).

For other situations, one PV may have a controlling effect on another PV, so the manipulating element is not the same as the controlled process stream. For example, the flow of steam into the tube side of a heat exchanger is used to raise (control) the temperature of the product flowing through the shell side of the heat exchanger—two totally separate streams—one containing steam and the other containing the process fluid. In this case, the product temperature is the important process variable, and this temperature is controlled by a steam valve opening and closing in the steam.

10.4 Signal Transmissions, Loop Error, Live Zero

All instrument signals can be thought of as a language that instruments use to communicate information between one another. There are four basic types of signal transmission used for instrumentation communication within processing facilities:

- Pneumatic (analog)
- Electronic (analog)
- Digital (discrete)
- Mechanical

Pneumatic (Analog)

In the United States, the standard **pneumatic signal** has a range of 3–15 PSIG. Instrument signals always represent 0–100 percent of the information moving between loop devices. For example, a 3–15 PSIG signal leaving a pneumatic transmitter represents 0–100 percent of the **measured variable**; whereas the 3–15 PSIG signal leaving the controller represents 0–100 percent of the response of a controller to error (the output).

Any instrument capable of producing a pneumatic signal (Figure 10.11) must have an instrument air supply connected. The instrument air is usually about 20 PSIG for standard 3–15 PSIG signal devices. The instrument air of a pneumatic instrument can be equated to a power supply for a piece of electronic equipment. After leaving an instrument, the signal travels through stainless steel, copper, or plastic tubing. The tubing connects the individual instruments together in a loop.

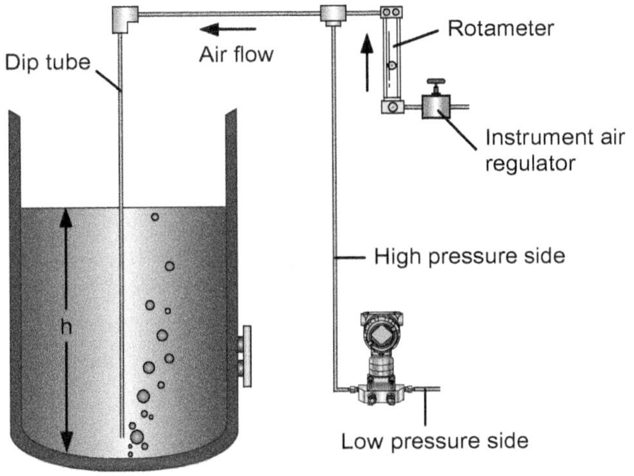

Pneumatic signal an instrument communication with a range of 3–15 PSIG; must have an instrument air supply; has a lag time associated with the signal; relatively short transmission distances.

Measured variable a process variable that is measured.

Figure 10.11 Pneumatic signal.

Compared with **electronic signals** (discussed later in this chapter), pneumatic signals are slower. Imagine the changing pressure propagating down a one-quarter to one-half inch (0.64 to 1.27 cm) tube that may be several hundred feet long. Pneumatic signal lines are kept to relatively short distances, usually no more than a couple of hundred feet (the longest are up to 500–600 feet [152.4 to 183 meters]) because of this potential transmission lag problem. Even though a pneumatic signal travels at about the speed of sound, the resistance to flow in the tubing and the size of the signal producing element may cause the signal to take several seconds to reach its full pressure, even in a short run of tubing. For long distances, and for large volume runs, a device called a booster relay may be added.

Electronic signal analog or digital signal; current or voltage signal.

Electronic (Analog)

There are two types of electronic signals (Figure 10.12) used in industry: analog electronic and digital electronic. An analog electronic signal is a continuously variable representation of a process variable. Analog electronic signals usually have a range of 4 to 20 mA. Earlier analog signals had a range of 10 to 50 mA and 1 to 5 VDC.

Current signals are preferred over voltage signals because of their higher immunity to electrical and electromagnetic interference. Voltage signals are decreased by the resistance in the wires and also lose voltage to every device in the loop. Thus, the voltage signal sent is not the same as the one that is received. A current loop is much better because it is similar to water flowing in a pipe (water in = water out; current sent = current received).

Most field transmitters have two wires connecting them to the controller. Power is supplied to the transmitter through the same two wires that carry the signal. In a sense, the *current-generating circuit* of the transmitter acts more like a *current-limiting circuit* in that it controls the current flow through the two wires to a range of between 4 mA and 20 mA. The actual output circuit is usually designed with operational amplifiers and other discrete components.

Figure 10.12 Electronic signal.

Digital signal data that is represented as coded information in the form of binary numbers; used to transmit data to and from field transmitters on a twisted pair of wires; may also be between computers and computer components.

Parallel data communication one wire per bit or 64+ wires for a 64-bit binary word; used primarily in short distances (a few feet).

Serial data communication two data wires; the most common means of communication used between plant equipment.

Electronic (Digital)

A **digital signal** (Figure 10.13) is characterized by data that is represented as coded information in the form of binary numbers. Digital signals are used to transmit data to and from field transmitters on a twisted pair of wires, between computers and computer components on a data highway, and in a new generation of instruments systems called *fieldbuses*. Digital information is transmitted as either parallel or serial data. **Parallel data communication** requires one wire per bit, or 64+ wires for a 64-bit binary word, whereas **serial data communication** requires only two wires (data). Parallel communication becomes impractical to use beyond a few feet. That is why serial data communication (serial bus) is the most common means of transmission used between plant equipment.

Figure 10.13 Digital signal.

Compared with the standard analog pneumatic and electronic signals, digital signals vary immensely among manufacturers. There have been attempts over the years to standardize, but variation still exists. The International Society of Automation (ISA), a worldwide instrumentation organization, has provided a set of standards for fieldbuses.

Mechanical (Link or Linkage)

Mechanical link a way of mechanically transmitting the motion of a primary sensor to a controlling mechanism; conveys linear or rotary motion by using a pivoting crank.

A **mechanical link** transmits the motion of a primary sensor to a controlling mechanism. By providing a mechanical link, the need for a standard instrument signal such as a 4–20 mA or 3–15 PSIG is eliminated. The major limitation of this type of transmission device is that the controlling device needs to be very close to the process sensor. This type of link may convey linear motion or rotary motion by using a pivoting crank. A fulcrum is sometimes used to produce a mechanical advantage.

An example of a mechanical link is found in a displacer level device. A displacer is a device that can infer level by measuring the buoyant force of the process fluid. The displacer is attached to the controlling device by a mechanical link (Figure 10.14) called a *torque tube*. The torque tube transmits the measured buoyancy force to a controlling device.

Another very common device is a control valve positioner which has a mechanical link to the valve that senses the position of the valve and ensures that the position corresponds to the position signaled by the controller.

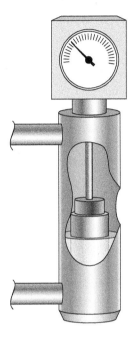

Figure 10.14 Mechanical link.

Loop Error

Each device in a control loop has a **loop error** (level of accuracy) associated with it. Device manufacturers usually list the accuracy of their own devices. For example, a manufacturer lists an old style temperature transmitter accuracy as E = +/− 0.5% of calibrated span. If the span is 500 degrees Fahrenheit (260 degrees Celsius), then E = (500 degrees Fahrenheit [260 degrees Celsius]) X (+/− 0.005) = +/− 2.5 degrees Fahrenheit (−16.39 degrees Celsius). The accumulated error for a control loop can be higher than expected and can be calculated as follows:

$$E_{loop} = +/- \sqrt{(E_{transmitter})^2 + (E_{valve})^2 + \ldots}$$

Loop error the accumulated error of each device in the loop; calculated as the square root of the sum of the squares of individual device accuracy.

Live Zero

A **live zero** means a standard bias has been added to the instrument signal (e.g., pneumatic 3 to 15 PSIG; electronic 4 to 20 mA). Instead of reading zero, the reading is 3 PSIG or 4 mA. This is done for two reasons:

1. Calibration is much easier to perform on an instrument around a positive output like 3 PSIG or 4 mA rather than on zero (0), since a negative pressure would imply the need for a vacuum source, or a negative current would require a polarity change. Either of these would be harder to do.

2. Process technicians, in essence, are given more information with a live zero condition. If the signal is truly indicating zero, then the tube or wire is broken or the transmitter or controller instrument has failed, whereas at 3 PSIG or 4 mA, functionality still exists but the instrument is at 0 percent of scale.

Live zero a standard bias added to the instrument signal (e.g., pneumatic 3 to 15 PSIG or electronic 4 to 20 mA); instead of reading zero, 0 percent of scale would have a reading of 3 PSIG (pneumatic) or 4 mA (electronic).

Summary

To maintain process control, one or more process variables must be sensed so that the desired product quality can be maintained and produced. Each process has a desired set point for comparison. The difference from the sensed process variable to the desired set point causes an output to move the measured value to approach the set point. For simple control situations, the process variable itself may be the sensed (controlled), measured to determine the condition of the variable, and also manipulated as a function of the actuating error signal.

Simple control loops consist of individual instruments necessary to control process variables in addition to any instruments required to convert signals or mathematically compute process variable measurements so that a signal can be manipulated. Control loops are basically considered either open or closed loop types, with the primary difference between the two based on whether or not the loop has feedback. Open loops do NOT have feedback and closed loops DO have feedback. Most control loops in process industries are closed loop, and this type is often called feedback control or feedback control loop.

All closed control loops must contain the following basic control components:

- Sensor
- Transmitter
- Controller
- Transducer
- Final control element

Sensing is the act of detecting, and a sensor or sensing element is the first device in a control loop. A sensor must be capable of detecting the process variable accurately, precisely, and quickly. This is quite possibly the most important function of a control system.

The next control loop component is the transmitter that converts the sensed process variable electrical response and produces a standard instrument signal output (3–15 PSI, 4–20 mA, or digital) from the transducer. Measuring transducers must also be accurate and precise. Next, the controller receives the signal from the transmitter and compares the signal to a desired value (set point) in the comparator part of the controller. If there is no difference, then there is no output. If there is a difference, then the output of the comparator causes a change in the calculating function of the controller. Preset algorithms (mechanical, analog, digital, or mathematical) respond to the magnitude and direction of the error (difference) signal received from the comparator. An appropriate response, called the corrective output, is produced to reduce or eliminate the error.

Converting devices may be used at any place in a control loop to receive information in one form of an instrument signal and change it into another. These devices may also be called converters, transducers, and/or relays. The most prominent converting device is the current-to-pneumatic converter (I/P), since the final control element is usually a pneumatically actuated control valve. Computing relays may also be used in control loops to compute values based on instrument signals. These are used anywhere that mathematical functionality is required within the control loop.

For most simple control loops, the manipulated process stream is the same as the controlled process stream. In other situations, one process variable may have a controlling effect on another. The final control element in a control system is the instrument that is manipulated to maintain the desired value of the process variable. In most cases, the final control element is a control valve.

Signals are the language of instruments. The following basic types of signals are used for transmitting information within processing facilities:

- *Pneumatic*: range of 3 to 15 PSIG; requires an instrument air supply of 20 to 25 PSIG to operate; are characteristically slow
- *Electronic*: either analog electronic or digital electronic
- *Analog electronic signals*: continuously represent a process variable and have a range of 4 to 20 mA
- *Digital electronic signals*: have coded information (binary numbers) that characterizes or simulates the analog PV and is transmitted to and from field transmitters using twisted pair wires, computer components on a data highway, or a fieldbus; transmission may be either serial (more common) or parallel (short distances only) data; digital signals vary greatly between manufacturers; fieldbuses have a standard set by the ISA
- *Mechanical*: motion by a primary sensor is transmitted to a controlling mechanism through a mechanical link (linkage); no standard signal is needed; controlling devices must be very close to the process sensor for this signal to operate properly; conveys linear or rotary motion when using a pivoting crank; may use a fulcrum to produce a mechanical advantage

Control loop devices have accuracy error associations. The respective manufacturer identifies these. Instruments transmitting signals are calibrated using a live zero that adds a small amount of positive output (e.g., 3 PSIG or 4 mA) to the signal scale zero point.

Checking Your Knowledge

1. Process control is the act of _____ one or more process variables.
   a. sensing
   b. measuring
   c. tweaking
   d. regulating

2. Which of the following is an example of a transduced variable? (Select all that apply.)
   a. Temperature
   b. Pressure
   c. Air humidity
   d. Sound pressure

3. (*True or False*) A closed loop is a signal path WITH feedback.

4. Write OPEN or CLOSED in the Loop column for the following devices.

Device	Loop
a. Light switch	
b. Toilet flush	
c. Water faucet	
d. Automobile cruise control	

5. Which of the following is the most prominent converting device in the process industries?
   a. I/P converter
   b. P/E converter
   c. E/P converter
   d. D/A converter

6. Match the type of signal with the description of how the signal is transmitted.

Signal	Description
I. Pneumatic	a. Transmits motion
II. Electronic analog	b. Transmits pounds per square inch
III. Electronic digital	c. Transmits binary code
IV. Mechanical link	d. Transmits milliamps

7. A control loop is composed of a transmitter, controller, I/P transducer, and a valve. Find the loop error if the individual accuracies (+ or −) are: transmitter accuracy = 0.5%; controller accuracy = 0.25%; I/P accuracy = 0.5%; and valve accuracy = 1.5%.
   a. 2.75
   b. 2.00
   c. 1.75
   d. 1.67

8. Why would a live zero be used? (Select all that apply.)
   a. To ensure a negative value is never achieved
   b. To provide a bias for instrumentation errors
   c. To provide easier calibration
   d. To make calculations easier for process technicians

9. Identify each instrument loop pictured below as either OPEN LOOP or CLOSED LOOP control.

   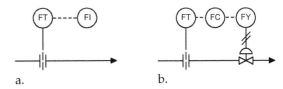

   a.   b.

10. For the instrument control loop illustrated below, fill in the appropriate information in the table below.

    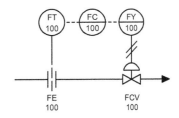

INSTRUMENT CONTROL LOOP			
Component	Element Type	Process Variable Being Controlled	Component Function

11. For the instrument control loop illustrated below, fill in the appropriate information in the table below.

    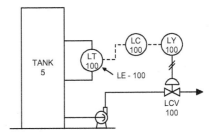

INSTRUMENT CONTROL LOOP			
Component	Element Type	Process Variable Being Controlled	Component Function

12. For the instrument control loop illustrated below, fill in the appropriate information in the table below.

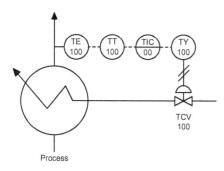

INSTRUMENT CONTROL LOOP			
Component	Element Type	Process Variable Being Controlled	Component Function

13. For the control loop illustrated below, fill in the appropriate information in the table below.

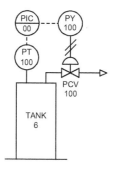

INSTRUMENT CONTROL LOOP			
Component	Element Type	Process Variable Being Controlled	Component Function

NOTE: Answers to Checking Your Knowledge questions are in the Appendix.

Student Activities

1. Photocopy individual control loop components from this textbook or other reputable sources (e.g., instructor classroom materials). Write the name of the component on the bottom front of each copy. On the back of each copy, write the types of signal combinations that may be used for input and output to the component. Using the copies as actual placement in a control loop, identify the process variable, and then select the combination of control loop components, in succession, that would form an appropriate control loop for the application. Be able to justify to another person (e.g., your instructor or another member of your small group) why you chose this particular arrangement of components for the specified process variable.

2. Identify open and closed control systems around your home and school. Be sure to explain how you derived your decisions on why the system is open or closed. Share your findings with other class members to see how your lists compare.

3. Identify control loops and/or control loop components in the lab.

4. Use pilot plant equipment and instrumentation and locate the control loop components.

Chapter 11
Control Loops: Primary Sensors, Transmitters, and Transducers

 ## Objectives

After completing this chapter, you will be able to:

11.1 Describe the function of measuring instruments for pressure, temperature, level, and flow and their role in the overall control loop process. (NAPTA Control Loops: Primary Sensors, Transmitters, and Transducers 1*) p. 194

11.2 Describe the purpose and operation of the elements in a control loop. (NAPTA Control Loops: Primary Sensors, Transmitters, and Transducers 2) p. 194

11.3 Discuss differential pressure in relation to the process input to the transmitter. (Control Loops: Primary Sensors, Transmitters, and Transducers 4) p. 197

11.4 Compare and contrast the transmitter input and output signals. (NAPTA Control Loops: Primary Sensors, Transmitters, and Transducers 3) p. 198

11.5 Describe the function of a current to pneumatic or pneumatic to current transducer (signal converter). (NAPTA Control Loops: Primary Sensors, Transmitters, and Transducers 5) p. 200

11.6 Describe the relationship between a 3 PSIG to 15 PSIG air signal and a 4 mA to 20 mA electric signal. (NAPTA Control Loops: Primary Sensors, Transmitters, and Transducers 6) p. 201

---

*North American Process Technology Alliance (NAPTA) developed curriculum to ensure that Process Technology courses will produce knowledgeable graduates to become entry-level employees in process technology. Objectives from that curriculum are named here in abbreviated form. For example, "(NAPTA Control Loops: Primary Sensors, Transmitters, and Transducers 1)" means that this chapter's objective 1 relates to objective 1 of the NAPTA curriculum about primary sensors, transmitters, and transducers.

11.7 Given a process control scheme, explain how a control loop functions. (NAPTA Control Loops: Primary Sensors, Transmitters, and Transducers 7; Instrumentation Sketching 3) p. 201

Key Terms

Discrete sensing element—a sensing element that can stand alone or is individually distinct; connected to the transmitter by sensor wires, **p. 196.**

Impulse tubing—a tube that is usually made of stainless steel and allows the process variable to be sensed by the sensor located in the transmitter, **p. 196.**

Instrument scale—a range of ordered marks that indicate the numerical values of the process variable, **p. 199.**

Integrally mounted sensing element—a sensing element that is a physical part of the transmitter, **p. 196.**

Linear scaling—a linear relationship between two scales (input versus output), **p. 199.**

Lower range value (LRV)—the number at the bottom of the scale, **p. 199.**

Scaling—the act of equating the numerical value of one scale to its mathematically proportional value on another scale, **p. 199.**

Sensor—(also known as a *primary element*) the control loop component that is in direct contact with the process variable to be measured; it senses or detects change and provides a signal to the transmitter, **p. 195.**

Signal converter (transducer)—part of a transmitter that converts the process variable into a standard instrument signal; a device that converts one energy form into another, **p. 201.**

Span—a number that expresses the difference between the upper and lower range values (URV and LRV) on a scale; sometimes referred to as an *operating range*, **p. 199.**

Standard signal—the language that instruments use to communicate with one another (4 to 20 mA, 3 to 15 PSIG, or digital), **p. 197.**

Upper range value (URV)—the number at the top of the scale, **p. 199.**

11.1 Introduction

In this chapter, the purpose and operation of control loop components' primary sensors, transmitters, and transducers are discussed in more detail. Most transmitters house both the sensing and measuring functions that ultimately produce a signal that is transmitted to the next control loop element. The next control loop element may be connected to a controller, recorder, indicator, programmable logic controller (PLC), or digital control system (DCS), or to a combination of these items.

The functions of sensing, measuring, and transmitting are also discussed. We include the purpose and operation of each portion of the process. We discuss how each component relates to others in the loop based on the process variable (PV).

Figure 11.1 shows a simple control loop in a liquid level measuring system. This illustration is used in this chapter as the basis for identifying the location of a transmitter and its functionality within a control loop.

11.2 Component Purpose and Operation

When control loops were introduced in the previous chapter, sensing and measuring were shown as separate functions. Some transmitter instruments (Figure 11.2) are capable of both sensing and measuring the process variable. Such transmitters produce an output signal that carries the measurement information to the next instrument in the loop.

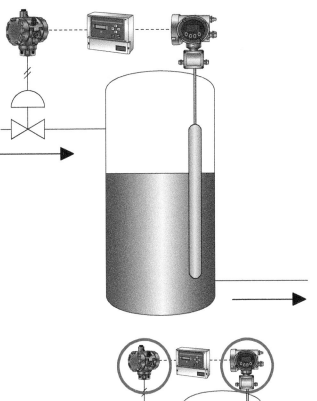

Figure 11.1 Simple control loop (liquid level measuring system).

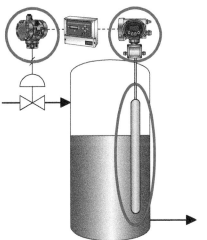

Figure 11.2 Primary sensors, transmitters, and transducers.

There are actually several elements that perform the sensing and measuring functions or act in tandem with the transmitter to perform these functions. These elements include the **sensor** and the *transducer*. The sensor detects the process variable, and the transducer converts one energy form into another.

In order for a control loop to function properly, the sensors and transmitters must provide the process variable measurement to the loop in a usable form. In a sensor driven loop, the *sensor* serves as a transducer and provides an input directly to the controller. In a transmitter driven loop, the *transducer* is part of the transmitter and converts the process variable into a standard instrument signal.

Sensor (also known as a *primary element*) the control loop component that is in direct contact with the process variable to be measured; it senses or detects change and provides a signal to the transmitter.

Sensors

Sensors (Figure 11.3) can be mechanical or electronic. A thermocouple is an example of an electronic sensor that can be connected directly to a controller. The controller in a sensor driven loop must be specifically designed to accept the *nonstandard input* rather than a standard instrument signal. With the thermocouple, the nonstandard input signal is in millivolts. These dedicated input controllers are only used when they clearly provide a significant advantage functionally or financially.

The role of a primary sensor is to detect a process variable. A sensor must also be capable of responding accurately, precisely, and quickly if good control is to be accomplished by the

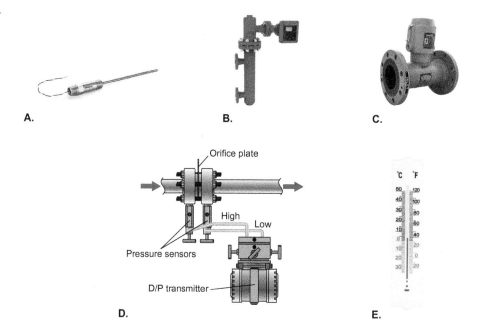

Figure 11.3 Sensors.
A. Threader thermowell sensor.
B. Displacer torque tube.
C. Turbine meter. **D.** Differential pressure transmitter.
E. Thermometer.

CREDIT: **A–C.** © Emerson 2019. **E.** Tomas Ragina/Shutterstock.

Discrete sensing element a sensing element that can stand alone or is individually distinct; connected to the transmitter by sensor wires.

Integrally mounted sensing element a sensing element that is a physical part of the transmitter.

Impulse tubing a tube that is usually made of stainless steel and allows the process variable to be sensed by the sensor located in the transmitter.

loop. **Discrete sensing elements** (stand-alone or individually distinct elements) are devices connected to the transmitter by sensor wires. **Integrally mounted sensing elements** are a physical part of the transmitter.

Thermocouples and resistance temperature devices (RTDs) are examples of discrete temperature sensors that are usually installed in a thermowell extending into the process. The thermocouple wires, or the lead wires extending out of the sheathing of the RTD, are used to connect these sensors to the transmitter. The transmitter may be physically connected to the open end of the thermowell or located a short distance away. In either case, the sensors are entities that are separate from the transmitter.

A differential pressure (D/P) transmitter (Figure 11.4), however, has an integrally mounted sensor. Since the D/P sensing cell (Figure 11.5) is part of the transmitter, the process variable has to be brought to it. This is accomplished by installing **impulse tubing** between the process and the transmitter. (Impulse indicates a wave transmitting through something.) The tubing is usually made of stainless steel and allows the process variable to be sensed by the D/P cell located in the transmitter. In remote seal applications, stainless steel capillary provides a connection between the pressure instrument and the diaphragm seal (see Figure 11.5). It protects the pressure instrument from high or low process temperatures and provides distant

Figure 11.4 Differential pressure (D/P) transmitter.

CREDIT: © Emerson 2019.

or remote readings. The impulse tubing, at a minimum, must meet the material specification of the process where it is connected.

Figure 11.5 A. Cross section of a differential pressure sensing cell. **B.** DP pressure product seal system.
CREDIT: B. © Emerson 2019.

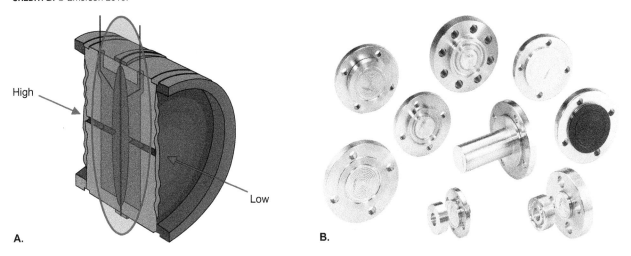

11.3 Transmitters

The most common method for providing a measured input to a control loop is through a transmitter. As stated previously, transmitters have a primary sensor either mounted within them or connected to them. After being sensed and measured, the process variable measurement is transduced (converted) by the transmitter circuit into a standard instrument signal. The standard instrument signals are 4 to 20 mA (electronic), 3 to 15 PSIG (pneumatic), or digital.

Unlike a directly connected sensor where the signal is nonstandard, transmitters provide a standardized signal that is accepted by all other instruments in the loop. For example, a differential pressure (D/P) transmitter contains a special type of pressure sensing and measuring component called a D/P cell. The D/P cell is designed to measure the difference between two pressures and then produce an output representing that difference. Inside the transmitter is a transducer that transforms the output of the D/P cell into a standard instrument signal, such as 4 to 20 mA or 3 to 15 PSIG.

One type of electronic transmitter uses a capacitance D/P cell. As pressure increases, the distance between the capacitance plates decreases, thus increasing the capacitance. The increased capacitance is transduced to an increased 4 to 20 mA output signal, making the output signal directly proportional to the differential pressure. Since the output of the transmitter is also the input to the next instrument in the loop, a meaningful and recognizable **standard signal** is an absolute necessity.

The D/P transmitter is a commonly used transmitter in the processing industry. D/P transmitters can measure pressure and differential pressure as well as determine level (Figure 11.6), flow rate, and even density. The application and variable being controlled determine how and where the impulse tubing is connected to the process. For example, to measure level using a closed pressure vessel when the contained medium is only in a liquid state, the high pressure side of the transmitter would be connected near the bottom of the tank (the zero level) and the low pressure side would be connected near the top of the tank. To measure flow, the high pressure side would be connected upstream of the orifice plate, and the low pressure side would be connected downstream. Connecting the high pressure side to the process and allowing the low pressure side to be vented to atmosphere provides a gauge pressure measurement. D/P transmitters are very flexible, but they are not simply used as a gauge or pressure transmitter since simple Bourdon tube type pressure gauges function well in such applications. D/P transmitters are used where control applications are required to maintain process conditions at optimum levels.

Standard signal the language that instruments use to communicate with one another (4 to 20 mA, 3 to 15 PSIG, or digital).

Figure 11.6 Differential pressure level transmitter.

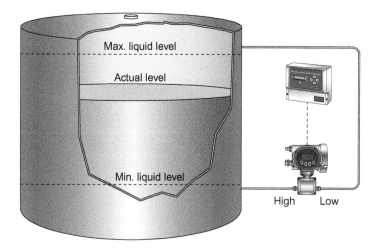

11.4 Transmitter Input and Output Signals and Scaling

Regardless of the type or magnitude of the process variable, the output of the transmitter is typically converted into one of the common standard instrument signals (Figure 11.7). In an analog electronic control loop, a thermocouple with a millivolt input to the transmitter has a 4 to 20 mA output. An RTD with its resistance input to the transmitter also has a 4 to 20 mA output signal. Pressure inputs, as well as analytical measurements applied to transmitters, also have a corresponding 4 to 20 mA output signal. All these process variables are converted into a standard instrument signal representing 0 to 100 percent of their respective measurements. That is why a standard instrument signal is so important. The process variables are different, but the output signals are the same.

Figure 11.7 Transmitter signals.

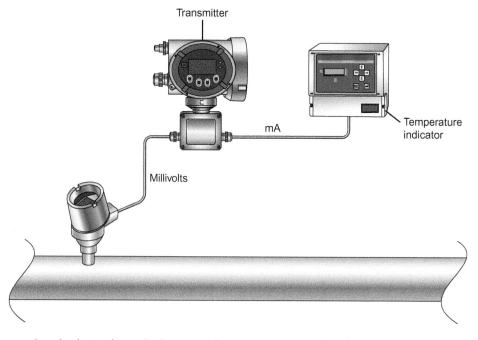

Standard signals are the language that instruments speak when communicating with one another. For every input value applied to an analog electronic transmitter, there is a unique and corresponding milliampere output value. For example, since the output of a transmitter represents 0 to 100 percent of the measured process variable, then 4.0 milliamperes of a 4 to 20 mA standard output signal (in a direct acting transmitter) would represent 0 percent of the measured process variable and 20.0 milliamperes would represent 100 percent. That means that 12.0 milliamperes represents 50 percent.

For the example given on the next page, what is the percent **span** for the 25 PSIG value?

$$\%\text{Span} = \frac{\text{Value} - \text{LRV}}{\text{Span}} \times 100\% = \frac{(25 - 0)\text{ PSIG}}{50 \text{ PSIG}} \times 100\% = 50\%$$

To understand better the relationship between the input of a transmitter and its output, calculating the input of a transmitter by observing its output is useful. The concept of instrument **scaling** must be addressed before calculations related to the input/output of a transmitter (Table 11.1).

Span a number that expresses the difference between the upper and lower range values (URV and LRV) on a scale; sometimes referred to as an *operating range*.

Scaling the act of equating the numerical value of one scale to its mathematically proportional value on another scale.

Table 11.1 Point Calibration on a Transmitter

Percent of Scale	Input	Output
0%	500 degrees F (260 degrees C)	4 mA
25%	625 degrees F (329 degrees C)	8 mA
50%	750 degrees F (399 degrees C)	12 mA
75%	875 degrees F (468 degrees C)	16 mA
100%	1000 degrees F (538 degrees C)	20 mA

An **instrument scale** is a range of ordered marks at fixed intervals inscribed on an indicator plate. Scales are used to indicate actual measurement values. When calibrating a transmitter, an instrument technician observes two scales: the input scale and the output scale. The act of calibrating a transmitter involves applying a known set of input values to the measuring component and then adjusting the transmitter output to correlate the measurement range to a standard signal range. Scaling converts process signals so that they are compatible with instrument and control systems. The conversion process (scaling) was discussed in greater detail in Chapter 9, *Signal Transmission and Conversion*, of this textbook.

Scaling is the act of equating the numerical value of one scale to its mathematically proportional value on another scale. For example, the measurement applied to a standard analog electronic pressure transmitter is represented on an appropriate pressure scale, while the output signal is represented on a milliampere scale.

Process technicians may use formulas (Figure 11.8) to convert the numerical value of one scale to another. To know how to plug in the appropriate numbers in the formula, several scaling terms must be understood:

Instrument scale a range of ordered marks that indicate the numerical values of the process variable.

- **Upper range value (URV)**
- **Lower range value (LRV)**
- **Operating range**
- **Span**

The formula for span follows:

$$\text{Span} = \text{HRV} - \text{LRV}$$

Where: HRV = High Range Value   and   LRV = Low Range Value

So, if URV = 150 PSIG and LRV = 50 PSIG, then span = 100 PSIG

The relationship between the input and output of a transmitter can be either linear or nonlinear. This discussion is limited to **linear scaling**. When there is a linear relationship between two scales (transmitter input versus output), the following formula applies:

$$\text{VALUE}_B = \frac{\text{VALUE}_A - \text{LRV}_A}{\text{SPAN}_A} \times \text{SPAN}_B + \text{LRV}_B$$

Upper range value (URV) the number at the top of the scale.

Lower range value (LRV) the number at the bottom of the scale.

Linear scaling a linear relationship between two scales (input versus output).

Where:

A = Original scale
B = New scale
Proportioning factor = Decimal representation of the original value on scale "A"

Figure 11.8 Calculating a transmitter's output by measuring its input.

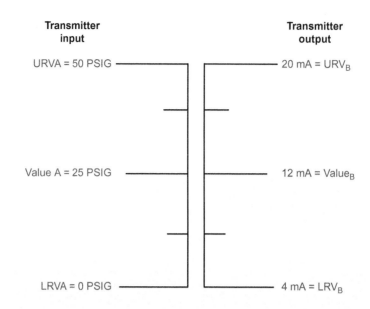

Example:
Refer to Figure 11.8. If a pressure transmitter is calibrated to measure 0 to 50 PSIG and is measuring 25 PSIG, what is the output of a standard 4 to 20 mA transmitter?

$$VALUE_B = \frac{VALUE_A - LRV_A}{SPAN_A} \times SPAN_B + LRV_B$$

$$VALUE_B = \frac{25 \text{ \sout{PSIG}} - 0}{50 \text{ \sout{PSIG}}} \times 16 \text{ mA} + 4 \text{ mA}$$

$$VALUE_B = 12 \text{ mA}$$

11.5 Transducers and Signals

Generally speaking, a transducer (Figure 11.9) is a device that converts one energy form into another. For example, transducers can convert quantities such as temperature and pressure into an electronic form. More specifically, a primary sensor such as a diaphragm produces motion in response to pressure, and then a transducer converts the motion into a measurable electrical quantity.

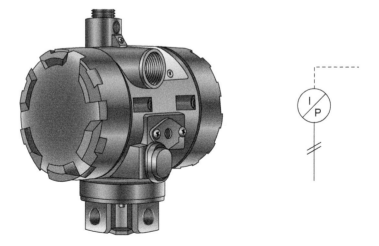

Figure 11.9 Transducers. *(Left)* Illustration of a transducer. *(Right)* Transducer symbol.

Sometimes the sensor and the transducer are one and the same. For example, a piezoelectric transducer is a crystalline wafer that responds to an applied pressure by producing an electrical output. Another example is a thermocouple that responds to a difference in temperature by generating a millivolt signal. Unlike their macro counterparts, such as current-to-pneumatic (I/P) transducers (converters), these sensor related transducers may produce any form or level of electrical impulse.

In the previous chapter, the discussion covering instrument converters was in fact a discussion about transducers. In the strictest sense, any device that converts one signal form into another is a transducer.

A common instrument signal transducer converts an analog electronic signal (4 to 20 mA) into a pneumatic signal (3 to 15 PSIG). This device is called an I/P (current-to-pneumatic) transducer. A **signal converter (transducer)** is necessary to convert the electronic signal into a pneumatic signal in any control loop that uses pneumatically actuated control valves.

Signal converter (transducer) part of a transmitter that converts the process variable into a standard instrument signal; a device that converts one energy form into another.

11.6 Pneumatic and Electrical Signals

Pneumatic and electronic signals provide a simple language that instruments use to communicate information between one another. As mentioned, commonly used instrument signals in the process industry are the 3 to 15 PSIG pneumatic signal and the 4 to 20 mA electronic signal. Most modern smart transmitters, which are microprocessor based, produce a 4 to 20 mA output signal as well as a digital signal.

Both of these analog signals have what is referred to as a live zero, which means that the zero value of the signal is greater than true zero (e.g., 3 PSIG instead of 0 PSIG). A live zero output makes corrections easier to detect and implement. With a live zero scale, if the output of a pneumatic transmitter is reading 3.2 PSIG when its measurement is 0 percent, it is easier to correct the zero value by adjusting the pressure downward. If the value of the signal is 2.8 PSIG on the live zero scale, the problem is easy to correct by adjusting the value upward. But, if the output of the transmitter were below true zero on a scale without a live zero, the output signal would go unnoticed and consequently continue to provide an incorrect measurement.

11.7 Control Schemes and Control Loops

When observing a control loop in the field or in a schematic (Figure 11.10), the best place to start is at the measurement point. The information (measurement) is carried and dealt with throughout the control loop. Each instrument in the control loop functions according to how the incoming signal is interpreted. The sensor or transmitter processes the signal received from the measurement transducer and produces a standard instrument signal representing that measurement.

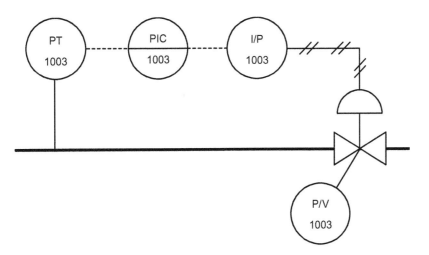

Figure 11.10 Pressure control loop diagram.

Summary

This chapter covered the first components of a control loop: the primary sensor, the transmitter, and the transducer. These components can be housed in the same device. Their definitions and operational purposes have specifics that make each of them unique and clearly distinct from each other.

Sensors and transmitters are capable of both sensing and measuring the process variable and then producing an output signal that carries the measurement information to the next instrument within the loop arrangement. The sensor performs the detection of the process variable such as flow rate, level, and/or pressure. Sensors come in various shapes and sizes depending on the particular application involved. Sensors may be connected directly to a controller, such as in a thermocouple arrangement. However, this type of direct connection is not always either financially or functionally possible. The primary sensor must perform accurately, precisely, and quickly for satisfactory control of the process. The sensor itself may be a stand-alone unit, or it may be an integral part of a transmitter.

A sensor or transmitter takes the sensed process measurement and transduces it within the transmitter circuitry, making the output signal a standard instrument signal. This signal is the language of instruments. The input value is applied to the transmitter, and the output is a unique and corresponding output value.

Scales on a transmitter are adjusted to fit both the input and output of an instrument, so there is a correlation to both. The transmitter output must correlate from the measurement range to a standard signal range. This means that one scale is equated numerically to its mathematically proportional value on another scale. While input to output relationships of a transmitter can be either linear or nonlinear, this chapter deals exclusively with a linear relationship.

An I/P (current-to-pneumatic) transducer converts an analog electronic signal (4 to 20 mA) into a pneumatic signal (3 to 15 PSIG). This type of transducer is the most common since almost all the final control elements, such as a control valve within a control loop, are pneumatically actuated. Both the 4 to 20 mA and 3 to 15 PSIG signals have a live zero attribute. The zero is greater than actual zero.

When observing an instrument control loop, the best place to start is at the measurement point or the primary sensor of process variable. Each component of the control loop functions according to this measurement. In most control loops, the process variable measurement must be changed to a standard instrument signal before being processed by the next control loop component.

Checking Your Knowledge

1. The primary role of a sensor in a control loop is to _____ the process variable.
   a. transmit
   b. transduce
   c. measure
   d. detect

2. The D/P cell transmitter converts a process variable measurement into a standard instrument _____.
   a. value
   b. signal
   c. pressure
   d. difference
   e. None of the above

3. A differential pressure transmitter can be used to measure differential pressure as well as _____. (Select all that apply.)
   a. temperature
   b. density
   c. rhythm
   d. flow rate
   e. level

4. The most common electronic instrument signal used in process industries is expressed in _____.
   a. millivolts
   b. milliamps
   c. pounds per square inch
   d. span
   e. range

5. What output would a transmitter produce at 75 percent of scale?
   a. 4 mA
   b. 8 mA
   c. 16 mA
   d. 20 mA

6. Calculate the span with an HRV of 225 PSIG and an LRV of 50 PSIG.

7. In order to increase production rate, a reactor set point was increased from 160 PSIG to 190 PSIG. The reactor pressure transmitter was recalibrated from 0–200 PSIG to 0–300 PSIG. What is the percent span for the new set points?
   a. 50 percent
   b. 63 percent
   c. 80 percent
   d. 90 percent

8. In a standard I/P transducer, an 8 mA input corresponds to _____ of output?
   a. 4 PSIG
   b. 6 PSIG
   c. 9 PSIG
   d. 7.2 PSIG

9. Convert the following to find the missing factors in the items below:

Process Range	Output Signal	Current Process Reading	Current Signal Reading
200–800° F	4–20 mA	_____ °F	14 mA
0–1000 GPM	4–20 mA	250 GPM	_____ mA
100–300 PSI	3–15 PSI	_____ PSI	6 PSI

10. I/P transducers are commonly used in the process industry with _____ actuated valves.
    a. electronically
    b. digitally
    c. electrically
    d. pneumatically
    e. None of the above

11. What is the term used to describe the condition where the zero value of a standard instrument signal is greater than true zero?
    a. Positive zero
    b. Live zero
    c. Integral zero
    d. Zero span

NOTE: Answers to Checking Your Knowledge questions are in the Appendix.

Student Activities

1. Using a cell phone, describe how it compares to a process primary sensor, transmitter, and transducer configuration in a control loop.

2. Describe how a home computer network, compares to a process primary sensor, transmitter, and transducer configuration in a control loop.

3. Identify or locate primary sensors, transmitters, and transducers in the lab.

4. Use pilot plant equipment and instrumentation to locate primary sensor, transmitter, and transducer components.

5. Using an I/P converter, explain the relationship between the PSIG air signal and the mA electric signal.

6. View a photograph of a process control loop from one of the local industries or make an actual visit to view a predetermined control loop at a particular site.

7. Perform a bench calibration on a transmitter and signal converter or transducer, or at minimum watch a calibration being performed or demonstrated.

8. In class activity–Process Control Loop Operation

   Instructions

   a. Review the following process control loop.

   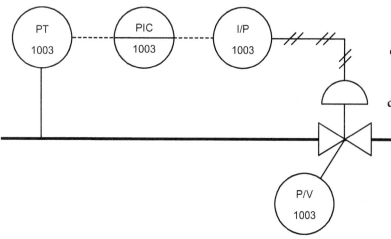

   Electronic Pressure Control Loop

   b. Write a description of how the control loop operates starting in the field at the measurement point. Explain how the information (measurement) is carried and dealt with throughout the control loop by addressing each instrument in the control loop according to the function of the instrument within the control loop and how it affects the incoming signal.

   c. Give examples and any reasoning for justification of where this type of control loop could be found in process operations.

   d. Follow your specific instructor guidelines for submitting responses to this activity.

Chapter 12
Control Loops: Controllers and Final Control Element Overview

 Objectives

After completing this chapter, you will be able to:

12.1 Explain terms associated with controllers:
automatic/manual switch
local/remote switch
set point
direct acting
reverse acting
tuning
proportional band or gain
integral and reset
derivative and rate (NAPTA Control Loops, Controllers 1, 8*) p. 207

12.2 Explain bumpless transfer of automatic to manual and manual to automatic control. (NAPTA Control Loops, Controllers 4, 5) p. 211

12.3 Describe the process for switching from automatic control to manual control and from manual to automatic on a local controller. (NAPTA Control Loops, Controllers 6, 7) p. 212

---

*North American Process Technology Alliance (NAPTA) developed curriculum to ensure that Process Technology courses will produce knowledgeable graduates to become entry-level employees in process technology. Objectives from that curriculum are named here in abbreviated form. For example, "(NAPTA Control Loops, Controllers 1, 8)" means that this chapter's objective 1 relates to objectives 1 and 8 of the NAPTA curriculum about control loop controllers.

12.4 Given a drawing or actual device, describe the application and operation of the following:

local controller

remote controller

split-range controller

cascade or remote set point (RSP) controller

ratio controller (NAPTA Control Loops, Controllers 2, 3) p. 212

12.5 Describe three types of final control elements and provide an application for each type. (NAPTA Control Valves and Final Control Elements 5, 6) p. 216

Key Terms

Automatic/manual switch—a switch that allows a process technician to select either automatic or manual control from the front of a controller, **p. 208.**

Automatic to manual and manual to automatic switching—the shift of controller action from automatic to manual or vice versa by adjusting the set point of the controller to the actual controlled point and then switching the mode; or switching from automatic to manual by simply repositioning the mode selector, **p. 211.**

Bump—a process upset occurring when a controller is switched from automatic to manual mode, **p. 211.**

Bumpless transfer—the act of changing the controller from manual to automatic (or vice versa) without a significant change in controller output, **p. 211.**

Cascade control loop—control loop characterized by the output of one controller becoming the set point of another, **p. 209.**

Control mode—the control action or the control algorithm response such as PID or a programmed function, **p. 210.**

Control valve—the final component of a control loop, operated either remotely or by hand, that receives its signal from the controller and manipulates the process variable in response to a system disturbance or change, **p. 207.**

Derivative action—a controller output response that is proportional to the rate at which the controlled variable deviates from the set point, **p. 211.**

Duty cycle—the period of time it takes for a signal to complete an on-and-off cycle, **p. 219.**

Fail last (fail in place)—the position of the valve seat upon loss of instrument air or power failure, **p. 217.**

Gain—change in output multiplied by the change in input, **p. 210.**

Integral action (reset)—a controller output response that is proportional to the length of time the controlled variable has been away from the set point, **p. 210.**

Local controller—when the controller is physically mounted in the processing area near the other instruments in the loop, **p. 212.**

Louvers (dampers)—devices used to control airflow, **p. 218.**

Primary controller—perceives the secondary loop as a separate entity; responds to one process variable (the controlled variable) while the secondary controller responds to another (the manipulated variable), **p. 214.**

Proportional band—the amount of deviation of the controlled variable from the set point required to move the output of the controller through its entire range (expressed as a percent of span), **p. 210.**

Rate action—the faster the rate of change of the process variable, the greater the output response (derivative action), **p. 211.**

Ratio controller—controllers that are designed to ratio (or proportion) flow rate between two separate flows entering a mixing point; designed so that its output represents the exact flow rate needed by the controlled flow loop to remain in alignment with the desired ratio to the uncontrolled flow; may also be capable of receiving two separate flow inputs and ratioing its output to the control valve located in the control line, **p. 215.**

Remote controller—any controller that is not located in the processing area with the transmitter and control valve, **p. 213.**

Remote set point (RSP)—a set point received from an external source, **p. 209.**

Secondary controller—a special type of controller, sometimes called a cascade controller or remote set point controller; a standard controller with the added capability of choosing to receive a remote set point from an external source or a local (internal) set point, **p. 214.**

Set point control—the mechanism by which a technician can manipulate the set point, **p. 209.**

Split-range controller—where the output signal is divided between two final control elements; if there are more than two valves, it is called a dual split-range controller, **p. 213.**

Tuning—adjusting the control action settings so that they produce an appropriate dynamic response to the process resulting in good control, **p. 209.**

Uncontrolled flow—flow in a process line where there is no control valve, **p. 215.**

12.1 Introduction

The controller (Figure 12.1) is the device in a control loop that operates automatically to regulate a process variable. In a simple feedback control loop, the controller first compares the value of the measurement signal to a predetermined or desired set point value. The result of this comparison produces another value, called the error. The error signal is then acted upon by one or more separate action components (control algorithms) in the controller to generate an appropriate output signal. The output signal of the controller then becomes the input to the final control element or to the transducer that in turn is connected to the final control element. The final control element (usually a **control valve**) manipulates the process, producing a change in the measurement sensed by the sensor or transmitter. The sensor or transmitter, once again, feeds the revised measurement information back to the controller and the cycle begins again.

Control valve the final component of a control loop, operated either remotely or by hand, that receives its signal from the controller and manipulates the process variable in response to a system disturbance or change.

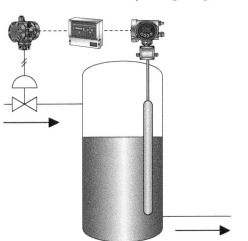

Figure 12.1 Controller.

The information in this chapter takes a closer look at controllers to include how they function, what type of responses that they may make to various signals, and what type of signal they provide as output. The last and final element of the control loop (discussed in depth in Chapter 13, *Control Loops: Control Valves and Regulators*) is also introduced here to include what types of final control elements are used, and how they relate to the signals produced by the controller.

The theory of control is easier to understand by studying analog systems first. Digital control is discussed in Chapter 17, *Introduction to Digital Control*.

Controller Characteristics

The purpose of a controller is to ensure a process variable remains at a desired value, or set point, or is returned to that value/set point if there is a process change. The following discussion has been divided based on the characteristics of controllers as found on either the front or side panels of controllers.

Front Panel Controllers

To perform as intended for the application, various controller configurations may be found on a typical front panel (Figure 12.2). They include the following:

- Automatic/manual switch
- Local and/or remote set point indicator
- Local/remote switch
- Set point adjustment controls
- Manual output controls
- Manual output indicator
- Process variable indicator
- Set point indicator
- Set point variable
- Tag number

Figure 12.2 Controller face plate.

CREDIT: © Emerson 2019.

Automatic/manual switch
a switch that allows a process technician to select either automatic or manual control from the front of a controller.

AUTOMATIC/MANUAL (A/M) SWITCH The **automatic/manual switch** enables the process technician to select either the automatic or manual mode from the front of a panel-mounted controller. When the process technician selects automatic mode, the controller automatically controls the process. When manual mode is selected, the process technician manually controls the process by adjusting the output of the controller accordingly. A manual output switch control is provided to send an output to the final element when the controller is in manual mode. During start-up, shutdown, and when certain process problems occur, it is necessary to use manual control. However, once the problem has been resolved or after the process has stabilized, the technician should switch the controller back to automatic.

SET POINT ADJUSTMENT CONTROLS Set point, also known as desired value, is the point at which the controller is adjusted to regulate the process. The **set point control**, usually found on the front plate of a panel-mounted controller, is the mechanism that a technician would adjust to change the local set point. A **remote set point (RSP)** is received from an external source (usually another controller) rather than from within the same controller.

Set point control the mechanism by which a technician can manipulate the set point.

Remote set point (RSP) a set point received from an external source.

LOCAL/REMOTE (L/R) SWITCH The local/remote switch on a controller allows the process technician to switch between a local set point and a remote (external) set point. A controller with a local/remote selector switch is usually the secondary controller in a **cascade control loop**. The cascade control loop will be discussed later in this chapter.

Cascade control loop control loop characterized by the output of one controller becoming the set point of another.

Side Panel

The components used for **tuning** a controller (Figure 12.3) can be found on the side panel of a typical pneumatic or electronic analog controller. These components for tuning include the various control actions (modes) and commonly used abbreviations for them, as follows:

- Direct acting/reverse acting (DIR and REV)
- Proportional action (GAIN)
- Integral action (RESET)
- Derivative action (RATE)

Tuning adjusting the control action settings so that they produce an appropriate dynamic response to the process resulting in good control.

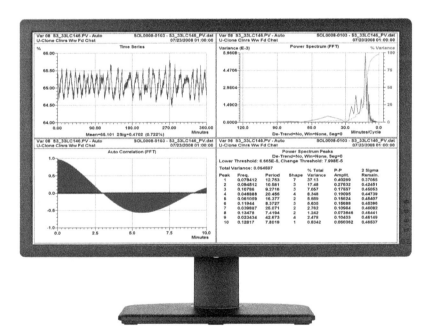

Figure 12.3 Controller—advanced analysis and tuning technology.

CREDIT: © Emerson 2019.

DIRECT ACTING AND REVERSE ACTING (DIR AND REV) A controller selected for direct action responds to an increasing input (measurement) by producing a corresponding increasing output. One convention denotes this selector switch option as increase/decrease. A controller selected for reverse action responds to an increasing input (measurement) by producing a corresponding decreasing output. Tuning a controller means adjusting the control action settings so they produce an appropriate dynamic response to the process resulting in *good* or *tuned control*. Once a physical process (the plant) has been designed and built, the controller is the only device in the system that is capable of counteracting the dynamics (process gain) of the process to the control system components.

Tuning

Proportional band the amount of deviation of the controlled variable from the set point required to move the output of the controller through its entire range (expressed as a percent of span).

PROPORTIONAL BAND **Proportional band** (PB) is the input change to the controller (band or part of the range) required to produce a full range change in output from a controller. A proportional band of 100 percent means that it takes a full range, or 100 percent, change in input to drive the output of the controller through its full range.

$$PB = \left(\frac{1}{\text{Gain}}\right)(100\%)$$

Gain change in output multiplied by the change in input.

Gain is another term used to describe this input to output relationship. Gain is a unitless number and is calculated as:

$$\text{Gain} = \frac{(\text{change in output})/(\text{output transmitter span})}{(\text{change in input})/(\text{input transmitter span})}$$

Note that (Process Gain) × (Controller Gain) = 1 = $G_P \times G_C$

Gain is a proportioning factor describing how the magnitude of the input relates to the magnitude of the output. For example, if a controller has a gain of 1, the controller will respond to an input change of 10 percent by producing an output change of 10 percent; if it had a gain of 2 instead, then the output change would be 20 percent. Regardless of how much change in input there is, the output of a controller can only produce a fixed output range (4 to 20 mA, which is 100 percent of the output range). Therefore, a gain of 1 is equal to a proportional band of 100 percent because it takes a full range change in input (100% percent) to produce a full range output. Proportional-only controllers do exist; however, proportional action is usually teamed up with integral action (PI).

Example (relationship between PB and gain):

$$\% \text{ PB} = \frac{1}{\text{gain}} \times 100$$

200% PB = 0.5 (gain)
100% PB = 1.0 (gain)
50% PB = 2.0 (gain)

Calculating Process Gain. Cooling water is fed into a reactor jacket for controlling reactor temperature. When the input cooling water flow is changed from 60 GPM to 55 GPM, the reactor temperature changes from 250 degrees Fahrenheit (121 degrees Celsius) to 275 degrees Fahrenheit (135 degrees Celsius). The cooling water flow transmitter is calibrated from 0 to 100 GPM, and the temperature transmitter is calibrated from 0 to 400 degrees Fahrenheit (−17.8 to 204.4 degrees Celsius)

In calculating the process gain for this process, one would find that it is reverse acting as shown below (in degrees Fahrenheit):

$$G_P = \frac{(275 - 250)°F/(400 - 0)°F}{(56 - 60) \text{ GPM}/(100 - 0) \text{ GPM}} = -1.25$$

Calculating Controller Gain. For this formula, the controller must be selected as direct acting to counter the process.

$$G_c = \frac{1}{1.25} = 0.8$$

Integral action (reset) a controller output response that is proportional to the length of time the controlled variable has been away from the set point.

Control mode the control action or the control algorithm response such as PID or a programmed function.

INTEGRAL ACTION (RESET) The **integral action (reset)** in a controller is designed to eliminate the offset inherent in feedback control loops. Reset describes the actual function of this **control mode**. Reset action repeats the proportional action in a given time period and is given as repeats per minute or repeats per second depending on the brand of the controller. Integral action is the inverse of reset and is given as minutes/repeat or seconds/repeat.

The controller senses a steady-state offset between the measurement and the set point and then produces a ramping response based on the proportional action. The output of the controller continues to change until the measurement is brought back to set point. Unlike proportional band, the integral action mode is never found alone and is combined with proportional action (PI) or in some cases with proportional and derivative action (PID).

DERIVATIVE ACTION (RATE) The **derivative action** of a controller responds to the rate of change of the controlled variable (i.e., measurement or PV). This is why derivative action is also known as **rate action**. The faster the rate of change in the error signal, the greater the derivative response. Once the PV measurement stops moving away from set point, the derivative action drops to zero. Some controllers may be programmed with only derivative action, but derivative is usually found in combination with proportional, integral, and derivative action (PID). Derivative action is added to a PI (proportional and integral) controller to provide a better response to slow loops, such as temperature loops.

Controller switching makes the following types of mode transfers:

- Bumpless
- Automatic to manual mode
- Manual to automatic control

All of the above described functions are accessible through the DCS console by clicking on the faceplate of the controller on the screen. Some of these functions will not be modifiable by technicians without certain permissions.

> **Derivative action** a controller output response that is proportional to the rate at which the controlled variable deviates from the set point.
>
> **Rate action** the faster the rate of change of the process variable, the greater the output response (derivative action).

12.2 Controller Switching—Bumpless Transfer

Bumpless transfer means to change the controller mode from manual to automatic, or vice versa, without the controller responding with a change in output. **Automatic to manual and manual to automatic switching** procedures vary according to the manufacturer equipment specifications. The most modern controllers have a set point tracking circuit that makes switching between these modes as simple as selecting the desired mode position. However, if a controller does not have an automatic tracking circuit, then a more involved procedure must be performed. For these controllers, set point tracking is selected when the controller is in the manual mode, making SP = PV. Because they are equal, the control error is zero and therefore no proportional or integral action is calculated. If SP = PV, then the integral or reset action would drive the internal output (called reset windup) to 0 percent or 100 percent output. Switching to automatic mode at this time would create a process bump. Remember that manual mode would bypass these internal calculations.

Generally, switching from manual to automatic requires adjusting the set point of the controller to the actual controlled point (PV) and then switching the mode (e.g., aligning the set point indicator to the measurement indicator PV). On the other hand, switching from automatic to manual may require only a simple repositioning of the mode selector.

In the old pneumatic systems and some of the older electronic controllers, a **bump** or mild upset could occur when switching the controller back from manual to automatic if the process variable has drifted from the set point. The controller causes the process value to move quickly back to set point when the switch is made, causing a bump. By setting the set point so that the process variable matches the PV, the switch becomes bumpless. If the technician had changed the controller to manual for a specific reason, there should not be a reason for returning the controller to a previous (old) set point at this time.

> **Bumpless transfer** the act of changing the controller from manual to automatic (or vice versa) without a significant change in controller output.
>
> **Automatic to manual and manual to automatic switching** the shift of controller action from automatic to manual or vice versa by adjusting the set point of the controller to the actual controlled point and then switching the mode; or switching from automatic to manual by simply repositioning the mode selector.
>
> **Bump** a process upset occurring when a controller is switched from automatic to manual mode.

12.3 Controller Switching—Mode Changes

Automatic to Manual Mode

The process for switching from automatic to manual mode includes the following steps:

1. With controller in automatic, the PV matches the set point.
2. Note the output signal to the control valve.
3. Switch automatic to manual.
4. Adjust manual output to achieve previous reading when it was in automatic.
5. Check PV and tweak the output as needed if the PV is not at desired value. The output needs to be adjusted as often as necessary to maintain the desired process value.

Manual to Automatic Mode

The process for switching from manual mode to automatic mode without bumping the process includes the following steps:

1. Adjust the set point to match the PV.
2. Switch manual to automatic.
3. Monitor the signal to see if the valve changes or if the PV drifts.
4. Once in automatic, adjust the set point if it is not at the desired value.

12.4 Controller Applications

Local Controllers

Local controller when the controller is physically mounted in the processing area near the other instruments in the loop.

A controller is considered to be a **local controller** (Figure 12.4) if it is physically mounted in the processing area near the other instruments in the loop.

Local control systems are still found on remotely located equipment such as knockout pots (or drums) located at the base of flares. The knockout drum is the last point where liquids can be separated from burn off gases before they rise up to the ignition point at the top of a flare stack. These vessels are generally remotely located and their levels are isolated and unrelated to other processes, making them ideal candidates for local control.

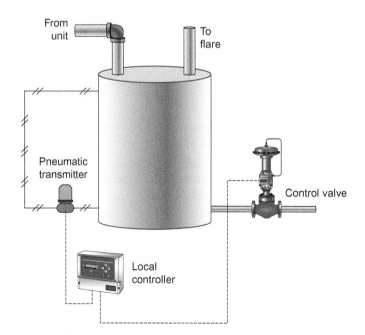

Figure 12.4 Local controller.

Remote Controllers

Any controller that is not located in the processing area is considered to be a **remote controller** (Figure 12.5). Most remote controllers are found in panels located in centralized control rooms. Remote controllers receive a signal from a transmitter located in the processing area, generate a corrective signal, and then transmit an appropriate signal out to the final control element, which is also in the processing area.

With few exceptions, modern process control schemes place the controller away from the processing area in centralized control rooms. Controllers began moving out of the field and into centralized control rooms shortly after the development of the pneumatic transmitters, around 1939. By the early 1950s, most controllers were located in control rooms, where they have remained ever since. Figure 12.5 depicts a pressure loop with the transmitter and control valve located in the processing area and the controller in a centralized control room. A large multitube bundle connects each point (bulkhead fitting) in the junction box (for example, JB-100) to a corresponding point (bulkhead fitting in JB-200).

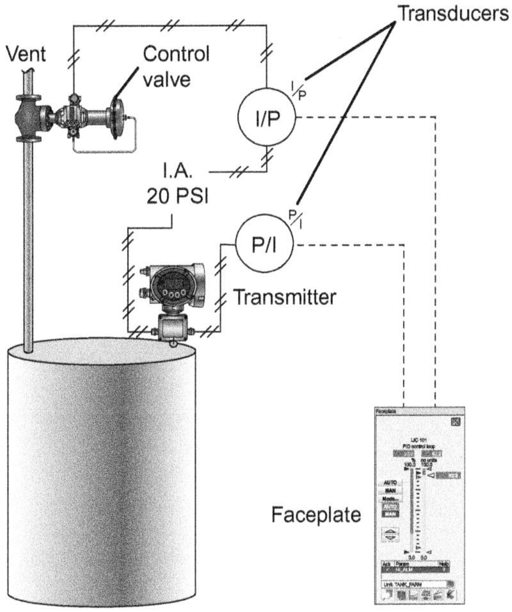

Remote controller any controller that is not located in the processing area with the transmitter and control valve.

Figure 12.5 Remote controller.

Controllers with Split-Range Valves

The output of a controller may be divided between two final control elements. The most common scenario is in which one final control element (valve A) responds to one portion of the output signal and the other portion of the output signal (valve B) responds to the upper half. Modern digital controllers can have more than one output; therefore, they can be **split-range controllers** (Figure 12.6). For example, the signal from output 1 could be 4 to 11.4 mA (3 to 8.5 PSIG) to valve A, while output 2 could be 12.6 to 20 mA (9.5 to 15 PSIG) to valve B. Both valves receive the full range output of the controller.

For example, a processing tank requires a constant vapor space pressure (space above the liquid) of 10 PSIG. During the filling of a vessel, the pressure in the vapor space increases, requiring the vent valve A to open. During removal, the pressure may drop, requiring valve B to open. This type of situation requires control output to be applied to two separate control valves.

The most common way split-range valves work together is to have one valve, valve B, respond by opening to an increasing signal from 9.5 PSI to 15 PSI and the other valve, valve A, respond by opening in an air to close arrangement from 8.5 PSI to 3 PSI. When the valves are receiving a signal of 8.5 PSI to 9.5 PSI, they are both closed. As the pressure drops in the tank, valve B opens, allowing pressure to enter the tank while the other remains closed. Conversely, when pressure rises in the tank, the valve A opens allowing pressure to be relieved from the tank while valve B remains closed.

Split-range controller where the output signal is divided between two final control elements; if there are more than two valves, it is called a dual split-range controller.

Figure 12.6 Split-range controller.

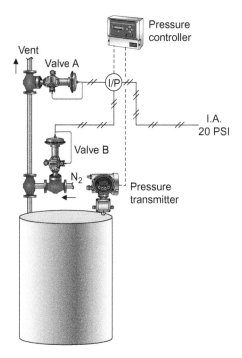

Cascade/Remote Set Point (RSP) Controllers

A cascade control loop is characterized by the output of one controller becoming the remote set point of another. As odd as this type of control may seem, cascade control is very common. In Figure 12.7, the primary (master) controller is responding to the temperature of the product while the secondary controller is operating to control steam flow. The temperature could be controlled without the secondary loop, but it would not be as fast or as sure.

There are two important reasons for using a cascade loop:

- Better control
- Reduced lag times

The **primary controller** on the shell side is a temperature controller. The **secondary controller** is a controller with a remote set point option. The cascade controller is the same as a standard controller, plus a cascade controller has the capability of choosing to receive a remote set point from an external source or a local (internal) set point.

In Figure 12.7, the primary (master) control loop is designed to control the temperature of the product leaving the shell side of the heat exchanger. This is done by controlling the pressure of the fuel gas into the furnace. The secondary, or intercontrol, loop is specifically controlling hot gas flow into the tube side of the heat exchanger. Since the secondary controller is a cascaded controller, it can be selected to control the hot oil flow locally (local position) or selected (remote position) to follow the needs of the primary controller. The secondary controller works to control the outlet temperature of the shell side of the heat exchanger.

Notice that the output of the primary controller is the remote set point of the secondary controller. This is just one of many different applications where a cascade control is found. In all cases, however, the secondary loop must be tuned to be faster than the primary loop to improve the control of a primary loop.

NOTE: When the secondary controller is not in the cascade mode, the primary controller's output continues to wind up (increase or decrease output) and causes a process bump when the secondary controller is switched back to the cascade mode. In modern controllers, output tracking can be selected for the primary controller. When the secondary controller is not in the cascade mode, the output of the primary controller tracks the set point of the secondary controller (Output$_{pri}$ = SP$_{sec}$), and the integral action of the primary is turned off. Also, the secondary controller should always have set point tracking selected.

Primary controller perceives the secondary loop as a separate entity; responds to one process variable (the controlled variable) while the secondary controller responds to another (the manipulated variable).

Secondary controller a special type of controller, sometimes called a cascade controller or remote set point controller; a standard controller with the added capability of choosing to receive a remote set point from an external source or a local (internal) set point.

Figure 12.7 Cascade or remote set point (RSP) controller.
CREDIT: © Simulation Solutions.

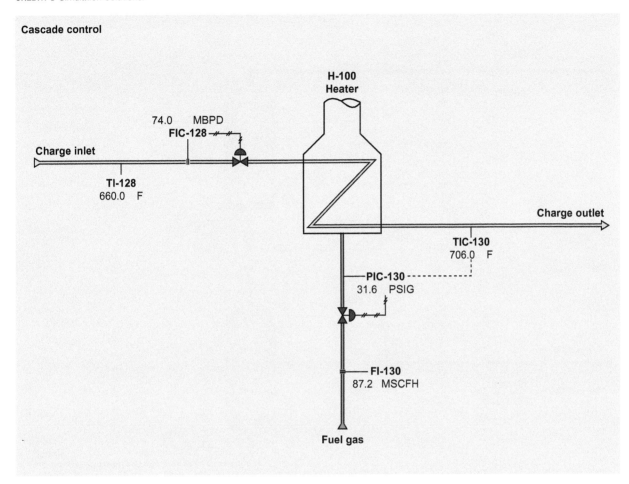

Ratio Controllers

Ratio control loops are designed to create a specific ratio (or proportion) between the rates of two separate flows entering a mixing point. Notice the similarities between the ratio control loop and a standard cascade loop. The **ratio controller** (Figure 12.8) is designed so that its output represents the exact flow rate needed by the controlled flow loop to remain in alignment with the desired ratio to the **uncontrolled flow**. A ratio controller is capable of receiving two separate flow inputs (the uncontrolled and the controlled) and ratioing the controller output to the control valve located in the controlled flow line. By doing so, a ratio between the uncontrolled and controlled flows is maintained.

If this example represents a feedstock line (uncontrolled flow) and a catalyst feed line (controlled flow) entering a reactor, then the catalyst is proportioned according to how much feed is entering the reactor. Too much catalyst is dangerous and expensive, and too little catalyst impacts the quality and quantity of the product. A proper ratio is imperative.

In the example, the uncontrolled flow transmitter produces an output that is received by the ratio controller where it produces an output representing the exact flow rate needed by the controlled flow loop. If the uncontrolled flow rate in the pipe increases, then the flow rate in the controlled pipe needs to increase proportionally as well. Conversely, if the uncontrolled flow decreases, then the controlled flow needs to decrease accordingly. The ratio controller usually has a set point range to allow the process technician to make minor adjustments to the ratio.

Ratio controller controllers that are designed to ratio (or proportion) flow rate between two separate flows entering a mixing point; designed so that its output represents the exact flow rate needed by the controlled flow loop to remain in alignment with the desired ratio to the uncontrolled flow; may also be capable of receiving two separate flow inputs and ratioing its output to the control valve located in the control line.

Uncontrolled flow flow in a process line where there is no control valve.

NOTE: Cascade control and ratio control are also discussed in Chapter 16, *Advanced Control Schemes*.

Figure 12.8 Ratio controller.

CREDIT: © Simulation Solutions.

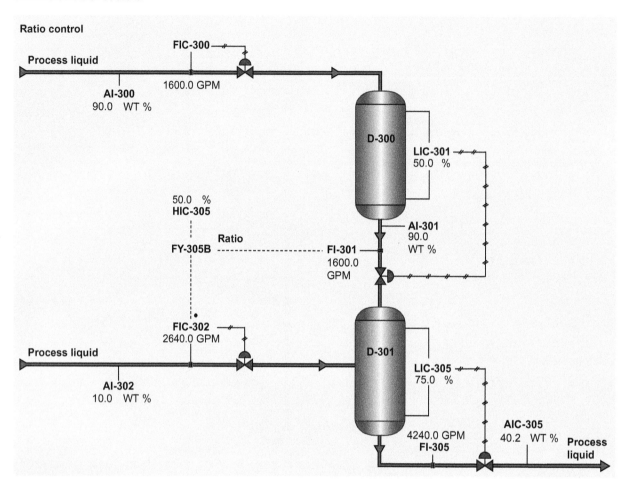

12.5 Final Control Element Overview

A final control element (Figure 12.9) is the last active device in the instrument control loop. The final control element is the device that directly controls the manipulated variable. This final control element performs a manipulating operation on the process that brings about a change in the controlled variable. While there are many different types of final control elements, the most common are control valves, louvers, and variable-speed drives.

Figure 12.9 Final control elements. **A.** Valve. **B.** Louver.

CREDIT: A. © Emerson 2019.

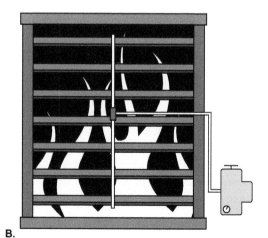

Control Valves

Control valves are the most common final control element in the processing industry. The function of a control valve is to manipulate the flow rate of some component in a process in order to affect the measured value of the controlled variable. Of course, this manipulated stream may also be the controlled variable stream. An actuating device is mounted to the control valve. The actuator changes an instrument signal into a linear or rotary motion. This motion drives the flow controlling mechanism (the plug or disc) in a valve.

Recall that control valves are designed to go to a fail-safe position upon air failure. For instance, a control valve controlling cooling water to a condenser may be designed to go to the open position upon air failure to allow full flow of cooling water. This would be referred to as a fail open valve. If a control valve were controlling steam to a jacket of a reactor, the valve might be designed to go to the closed position upon air failure to shut down the flow of steam to the reactor. This valve would be referred to as a fail closed valve. There are other terms used for fail open and fail closed. Fail open might also be referred to as air-to-close or reverse acting. Fail closed might also be referred to as air-to-open or direct acting. There is also a **fail last (fail in place)** feature for control valves that is not as common. Fail last means that when there is an instrument air failure, the valve stays in the same position that it was in at the time of the failure.

Fail last (fail in place) the position of the valve seat upon loss of instrument air or power failure.

Example 1:
A simple feedback control loop is controlling flow rate where the controlled variable is the same stream as the manipulated variable. In this example, the two streams are the same (Figures 12.10A and B). The flow rate measured by the flow transmitter is the same flowing material that is being manipulated by the control valve. In this example, the controller adjusts the signal to the control valve, which manipulates the flow rate of the material moving through the process. The transmitter then measures the resulting flow rate and feeds the new measurement value back to the controller where the control cycle continues.

Figure 12.10 A. Manipulated stream = controlled stream (actual). **B.** Diagram showing manipulated stream = controlled stream.

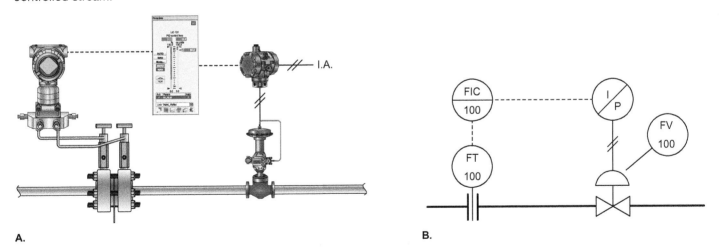

Example 2:
There is a heat exchanger where the temperature transmitter (product side) is located on the outlet of the tube side, and the control valve is located on the inlet to the shell side. The outlet of the shell side is connected to a steam trap. In this example, the manipulated stream is not the same as the controlled stream (Figures 12.11A and B), but the control valve reacts in the same manner. It responds to the controller output by reducing the flow rate and then waits for new instructions. The manipulated stream is the steam flowing into the shell side of the heat exchanger where it heats the process stream flowing through the tube side. The feedback signal originates in the transmitter located on the outlet line on the tube side

Figure 12.11 A. Heat exchanger showing manipulated stream ≠ controlled stream (actual). **B.** Diagram showing manipulated stream ≠ controlled stream.

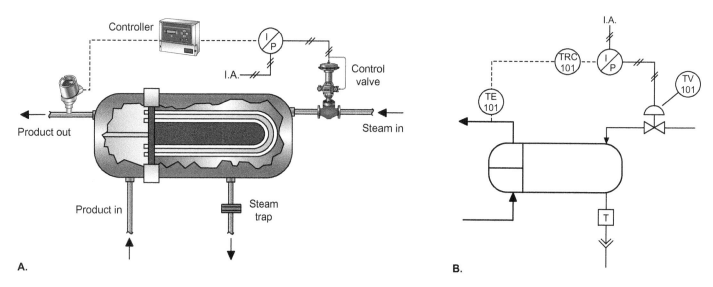

of the exchanger. As the temperature of the process cools below set point, the steam valve opens, allowing more heat energy to enter the exchanger. Conversely, as the temperature of the process exceeds set point, the steam control valve reduces the flow of steam into the exchanger causing the temperature of the process material to move closer to set point.

Additionally, in Chapter 13, *Control Loops: Control Valves and Regulators*, control valve positioners are discussed that are attached to the control valve stem. They are used in conjunction with the controller to match the controller output signal to the desired valve position (open or closed).

Louver and Damper

Many power plants, excluding nuclear plants, burn a hydrocarbon fuel to produce steam. If the boiler is to run efficiently, the burner in the firebox of the boiler must mix the right ratio of air to fuel. The most common way to do this is to ratio the airflow to the fuel flowing into the burner. A large amount of air is needed to burn the fuel completely, yet too much air flowing through the system reduces efficiency.

Louvers (dampers) (see Figure 12.9B) are devices similar in design to shutters or mini-blinds used to control airflow. These louvers are parallel to one another and can be fully opened to allow more airflow into the firebox or closed to diminish (dampen) the airflow. Each louver in an industrial airflow system usually pivots on a shaft that runs parallel across its midsection and connects to the frame on both sides. Another rod links the louvers together so that they can be manipulated simultaneously. An actuator then drives the connecting rod, controlling the airflow through the louvers.

When the louvers are positioned parallel to the airflow, there is very little resistance. When positioned at an angle to the airflow, they can effectively regulate the draft through a firebox in a furnace or steam boiler. This is an important factor in adjusting the air–fuel ratio needed for good combustion. Many louvers and dampers have blowers to induce a force to provide airflow. There is usually an air to mechanical transducer (positioner) to adjust the louvers.

Louvers (dampers) devices used to control airflow.

Electric Motors

If an electric motor is automatically turned on and off by a controller responding to a process variable, then the motor can be considered as a final control element or at least the actuator of the final control element. An example of this situation is a tank (Figure 12.12) in which the level is controlled between a high-level point and a low-level point. When the level reaches the high mark, an electric motor turns on, driving a pump that lowers the level in the tank. When it reaches a predetermined low point, the motor and pump are turned off.

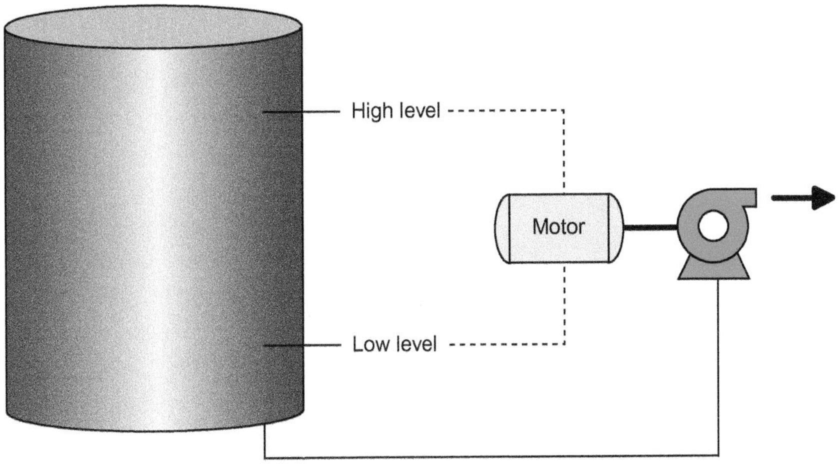

Figure 12.12 Electric motor.

Variable-speed electric motors can also be considered as the final control element. For example, if an electric motor drives a pump that is energized when the level reaches a certain height and is turned off at a designated low-level point, then it is acting as an on–off final control element. Variable-speed electric motors are capable of operating within a selectable revolutions per minute (rpm) range. Variable-speed motor control may be achieved by controlling the level of direct current (DC) applied to the motor or by controlling the **duty cycle** of the applied alternating current (AC). Usually, an mA signal from a speed controller is sent to a variable frequency drive (controller) that varies the frequency (0 to 70 Hz) supplied to the motor, which in turn adjusts the rpm of the electric motor. An example (4 to 20 mA = 0 to 70 Hz = 0 to 2200 rpm) might be values seen with a combustion air blower on an incinerator.

Most people are familiar with a variable-speed electric drill, which is driven by a variable-speed motor. The concept is the same in larger industrial variable-speed motors. Variable-speed motors are used to control fluid pumping rates, conveyor belt speeds, and other final controlling devices. If a variable-speed motor is driving a pump, then the flow rate can be modulated just like a pneumatically driven control valve.

A reversible electric motor with a gearbox is capable of actuating block valves or control valves just like a pneumatic actuator. Unlike the spring and diaphragm actuator with a loss of signal, the electric motor actuator typically fails in place.

Duty cycle the period of time it takes for a signal to complete an on-and-off cycle.

Summary

The controller is the device in a control loop that operates automatically to regulate a process variable by first comparing the value of the measurement signal to a predetermined set point value. The amount of error is acted upon by one or more separate action components within the controller so that an appropriate output signal is generated. The output signal of the controller becomes the input signal to the final control element (control valve, louvers and dampers, or a motor) or to an I/P transducer that in turn is connected to the final control element. The final control element manipulates the process that produces a change in the measurement sensed by the transmitter. This cycle repeats over and over again within the control loop.

Controllers have various types of controls located on either the front or side panel. The front panel has various switches that include the following:

- *Automatic/manual (A/M) switch* When in automatic control, the controller is in control of the process; when in manual control, the process technician is in control of the process by adjusting the output of the controller, especially in situations such as start-up or shutdown.
- *Set point adjustment controls* The set point is the desired value to which the controller is adjusted so that it regulates the process. If the set point indicator is local, it adjusts the readout from a nearby source. If the set point

indicator is remote, then it receives information from an external source outside of the controller.

- *Local/remote (L/R) switch* The L/R switch allows technicians to switch between two different set point information sources (local and remote).

In addition, the side panel has tuning capability, to include the various mode settings as follows:

- *Direct acting/reverse acting* Direct acting controllers respond to an increasing input measurement by producing a corresponding increasing output. Reverse action controllers respond to an increasing input by producing a corresponding decreasing output.
- *Proportional band (gain)* The gain is where the input change to the controller is required to produce a full range change in output from the controller; how the magnitude of the output relates to the magnitude of the error; proportional action controllers are usually teamed up with integral action (PI).
- *Integral action (reset)* Integral action in a controller is designed to eliminate the offset inherent in feedback control loops; the controller continues to reset the offset between the measurement and the set point until the measurement is brought back to set point.
- *Derivative action (rate)* Derivative action is also known as rate action. The controller responds to the rate of change of the controlled variable; the faster the rate of change, the greater the derivative response. Derivative action stops when the rate of change stops; derivative action is most often found in combination with proportional and integral action (PID).

Controller modality changes or transfers include the following:

- *Bumpless*: When the controller is changed from automatic to manual control or vice versa and the controller does not have to respond with a change in output; usually performed by setting the output so that the process variable matches the set point before making the switch.
- *Automatic to manual control:* see Procedure below.
- *Manual to automatic control:* see Procedure below.

Controllers are identified by their type, to include the following:

- *Local* The controller is physically mounted in the processing area near the other instruments in the loop; usually in a pneumatic control loop; most often used where the process is isolated and unrelated to other processes such as knockout drums.

- *Remote* Where the controller is not located in the processing area with the transmitter and control valve; pneumatic remote controllers are usually located no more than 300 feet (91.4 meters) away, and electronic remote controllers may be significantly farther away from the other loop components.
- *Split-range* Where the output of the controller is divided between two final control elements as in the case where one valve responds to the lower half of an output signal and another valve responds to the upper half of the output signal; feed or product tankage is a common place for this type of controller so that pressure is controlled on incoming and outgoing operations.
- *Cascade* Where the output of one controller becomes the set point of another as in where the primary (master) controller responds to temperature and the secondary controller operates steam flow; used where better control is needed or lag times need to be reduced.
- *Ratio* Where the rates of two separate flows enter a mixing point need to be proportionally introduced as in one part of stream A to four parts of stream B or a ratio between one uncontrolled and one controlled flow is maintained.

A final control element is the last active device in the instrument control loop and directly controls the manipulated variable so that when one process is manipulated, then the other is changed. These are the three most common final control elements:

- *Control valves* The most common final control element; manipulate the flow rate of some component in a process so that the measured value of the controlled variable is affected; may have an actuating device that changes the instrument signal into a linear or rotary motion that drives the flow controlling mechanism such as a plug or disc in a valve.
- *Louvers or dampers* Commonly used to increase or decrease airflow to burners; when the slats are parallel and open, airflow is increased and has little resistance; when the slats are closed, then no airflow is available and the burners may not function.
- *Electric motors* Where an electric motor drives a pump that is energized when the level reaches a certain height and is turned off at a designated low-level point; a reversible electric motor has a gearbox capable of operating block valves or control valves; variable-speed motors are used to control how many rpms are allowed by controlling the level of DC applied to the motor or the duty cycle of an applied AC.

Checking Your Knowledge

1. Match the following comments to the type of mode found in controllers.

I. Controller responds to the rate at which the variable changes.		a. Direct
II. Controller responds to an increasing input with a decreasing output.		b. Reverse
III. Controller responds to reset the measurement back to the set point.		c. Proportional
IV. Controller responds to an increasing input with an increasing output		d. Integral
V. Controller takes the full range of input to drive the output.		e. Derivative

2. If a controller has a gain of 3, the controller will respond to an input change of 30 percent by producing an output change of _____ percent.
   a. 10
   b. 20
   c. 30
   d. 90

3. Calculate the proportional band for a gain of 4.0.
   a. 12%
   b. 25%
   c. 35%
   d. 40%

4. How can a switch between manual and automatic be made bumpless?
   a. Setting the set point so that the process variable matches the PV
   b. Setting the set point so that the process variable is greater than the PV
   c. Setting the set point so that the process variable is less than the PV
   d. A switch can never be bumpless.

5. Match the following comments to the type of mode found in controllers.

I. Two of the same type of process variables must be controlled proportionally.		a. Local
II. One controller is needed to manipulate two final control elements.		b. Remote
III. The controller and the final control element are remotely located and their actions are isolated and unrelated to other processes.		c. Split-range
IV. The output of one controller serves as the set point for another controller.		d. Cascade
V. The controller is removed from the process area.		e. Ratio

6. Which of the following is considered the device that directly controls the manipulated variable?
   a. The measuring instrument
   b. The transmitting element
   c. The comparing and controlling element
   d. The final control element
   e. None of the above

7. Which of the following is the most common final control element in the processing industry?
   a. Control valve
   b. Electric motor
   c. Louver and dampers
   d. Compressor
   e. None of the above

8. Which of the following do variable-speed motors control? (Select all that apply.)
   a. Fluid pumping Rates
   b. Conveyor belt speeds
   c. Internal timers
   d. Louvers and dampers

9. Properly identify each type of controller by selecting the appropriate letter for the image of the controller:
   I. Split-range
   II. Ratio
   III. Remote
   IV. Local
   V. Cascade

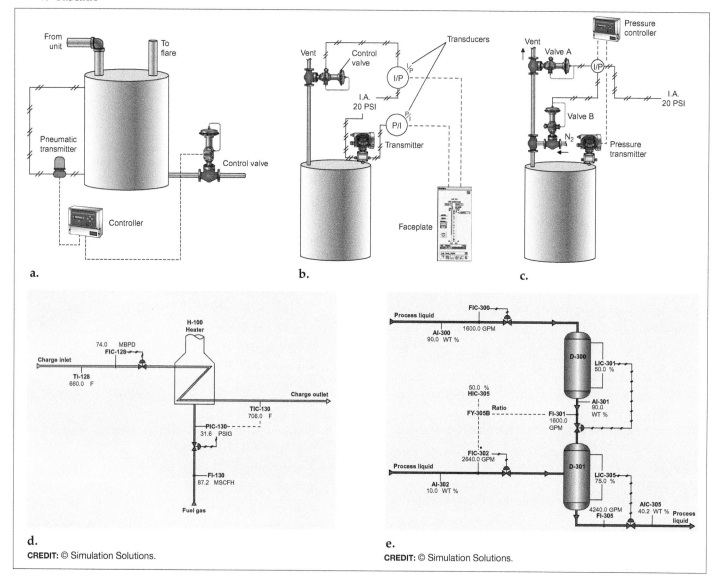

NOTE: Answers to Checking Your Knowledge questions are in the Appendix.

Student Activities

1. Obtain control loop graphics without labels. Write descriptors on each of the control loop elements. Check your responses against the textbook images. Repeat this process to improve your visual recognition of each element and recall how it works within the control loop, paying special attention to the controller functionality.

2. Identify or locate various types of controllers in the lab.

3. Given a drawing or actual local pneumatic controller, read the chart and state the high and low range values.

4. Perform the following control skills depending on available resources:
   - Make set point adjustments on a local controller.
   - Operate a local controller in manual mode.
   - Make set point adjustments on a remote pneumatic controller.
   - Switch from manual to automatic control remotely without bumping the process.

5. **CONTROLLER SWITCHING–A**
   Using available instrumentation within your classroom facilities, perform the following in the presence of your instructor:
   - Make set point adjustments on a local controller.
   - Operate a local controller in manual mode.
   - Make set point adjustments on a remote pneumatic controller.
   - Switch from manual to automatic control on a remote pneumatic controller without bumping the process.

6. **CONTROLLER SWITCHING–B**
   Obtain the *Pneumatic Controller High and Low Range Values* handout from your instructor, and explain how to read the chart.

7. **CONTROLLER SWITCHING–C**
   Obtain the *Indicator Illustrations* handout from your instructor and determine the following:
   - Does the illustration indicate linear or nonlinear?
   - What are the upper and lower range values?
   - What is the range?
   - What is the span?

8. **CONTROLLER SWITCHING–D**
   Determine if a control loop is in or out of control, and explain why it is in or out of control when the following indicators are observed:
   - The control loop is in control if all elements of the loop are yielding the desired process result (automatic and closed loop).
   - If the process result is off from what is desired, what indicator shows an out of control condition? Possible places in a loop where the malfunction could exist include: the primary element, transmitter, controller, and/or final control element.
   - The process technician must troubleshoot the cause of the problem.
   - Placing the problem element on manual control can help regain control of the process.

9. **CONTROLLER LABORATORY**

Background Information

The term *bumpless transfer* describes the manner in which the output of the controller responds to a change in set point value brought on by switching from manual control. In a cascade control system, the act of switching from a local set point to an external set point can cause a bump as well. Most modern controllers have a set point tracking circuit (or subroutine in computer systems) that allows switching between these modes without a bump in output. If a controller does not have an automatic tracking circuit, then a more involved procedure must be performed. Generally, switching from manual to automatic requires adjusting the set point of the controller to the actual controlled point and then switching the mode, that is, aligning the set point indicator to the measurement indicator. On the other hand, switching from automatic to manual may only require a simple repositioning of the mode selector. It should be noted that the automatic to manual and manual to automatic switching procedure varies according to the manufacturer's equipment specifications.

When switching between manual and automatic control, a technician must first match the process variable to the set point and then switch to automatic or the controller will react aggressively to match the set point. If the PV and set point are significantly different, the controller will produce an output change to the final control element that will cause a spike in the process. In some cases, this is not a problem but in others, it could be problematic.

Materials Needed

A control loop or process simulator
Instruction manual for the automatic controller

Safety Requirements

Safety glasses should always be worn while in the lab.

Procedure

1. Start up the process plant in the lab and identify a specific control loop to manipulate.
2. Switch the controller from automatic to manual.
   Line out the process so that the process variable is in a steady-state condition.
   With the controller in automatic, the PV indicator pointer should match the set point indicator pointer. Note the output pressure to the control valve.
   Switch automatic to manual.
3. Operate the controller in manual.
   Under dynamic conditions, the controlled variable may have changed because the control valve is now frozen in position. Therefore, adjust the manual output control to achieve the same desired value that you had while in automatic control.
   Periodically check the process PV and tweak the output controls as needed if the PV is not at desired value.
   Adjust the output as often as needed to maintain the desired process value.
4. Switch between manual and automatic.
   Adjust the set point controls to place the set point indicator pointer to match the process PV.
   Switch from manual to automatic.
   Monitor the pressure to see if the valve changes or if the PV drifts.
   Once in automatic, adjust set point if it is not at the desired value.
5. Make a remote set point adjustment.
   With the cascade controller selected to remote set point (control), make a set point adjustment.

Change the set point by moving it above the measurement indicator and record what happens to the output.

Change the set point by moving it below the measurement indicator and record what happens to the output.

Additional Information

All controllers have instruction manuals containing specific operating instructions. If you have a cascade control system, transferring control from local (secondary controller's set point) to remote (primary controller's set point) requires the same logical approach as transferring from manual to automatic in a single automatic controller. With the primary controller in manual and the secondary controller in local automatic, line out the primary controller's PV. Make sure the output from the primary controller (its output) is equal to the secondary controller's set point. Do this by adjusting the primary controller's output manually. Then switch the secondary controller from local set point to remote set point. Now the primary controller should be directing the entire cascade loop in manual. Finally, switch the primary controller from manual to automatic in the same manner mentioned in the lab procedure.

Findings

Compile your observations and readings into a report that includes your recorded data, an explanation of the results of the experiment, and a conclusion. Include a paragraph explaining the differences between a local set point and a remote set point. Be sure to discuss the importance of bumpless transfer when switching an analog controller from automatic to manual control and from manual to automatic. Follow your instructor's guidelines on how to submit these findings.

Chapter 13
Control Loops: Control Valves and Regulators

 Objectives

After completing this chapter, you will be able to:

13.1 Given a drawing or actual device, identify the following components of a control valve system:

body

bonnet

disc

actuator

stem

seat

spring

valve positioner

handwheel

hand jack

I/P transducer

current-to-pneumatic (I/P) transducer (NAPTA Control Valves and Final Control Elements 2, 4, 10*) p. 227

13.2 Define terms associated with valves and other final control elements and operating scenarios in which they are desirable:

air to close (fail open)

air to open (fail closed)

fail last/in place/as is

---

*North American Process Technology Alliance (NAPTA) developed curriculum to ensure that Process Technology courses will produce knowledgeable graduates to become entry-level employees in process technology. Objectives from that curriculum are named here in abbreviated form. For example, "(NAPTA Control Loops: Control Valves and Final Control Elements 2, 4, 10)" means that this chapter's objective 1 relates to objectives 2, 4, and 10 of the NAPTA curriculum about control valves and final control elements used in control loops.

225

double-acting diaphragm

double-active piston

solenoid

variable speed motor (NAPTA Control Valves and Final Control Elements 3, 7, 9, 13) p. 229

13.3 Discuss the purpose and operation of actuators and valve positioners and explain why the action of a valve actuator may not correspond with the action of the valve. (NAPTA Control Valves and Final Control Elements 3, 6, 8, 10–13) p. 230

13.4 Explain how a controller's output signal can be reversed at the valve, a control scheme that uses such a signal, and effects on the valve's fail-safe position upon loss of instrument air (NAPTA Control Valves and Final Control Elements 12, 13) p. 234

13.5 Explain the purpose and operation of the following:

globe valves

three-way valves

butterfly valves

ball valves (NAPTA Control Valves and Final Control Elements 1) p. 235

13.6 Identify the location and use of pressure regulators used in process control, including:

back-pressure regulator (self-actuated)

pressure-reducing regulator. (NAPTA Control Valves and Final Control Elements 6–9) p. 238

Key Terms

Actuator—an electrical, pneumatic, or hydraulic device that manipulates a control device, usually a valve, **p. 228.**

Back-pressure regulator—a device used to regulate and/or control the pressure of a process fluid upstream of the device location, **p. 239.**

Body—the housing component of a valve, **p. 227.**

Bonnet—a bell-shaped dome mounted on the body of a valve where the valve disc is held when the valve is opened, **p. 227.**

Handwheel—the mechanism that raises and lowers a valve stem to control the flow of fluid through a valve, **p. 228.**

I/P transducer (current-to-pneumatic transducer)—a device that converts a milliampere signal into a pneumatic pressure signal, **p. 228.**

Plug—the only movable component in the valve; it is actuated to open or close the flow path through the valve. May also be called a *disc* or *ball*, **p. 227.**

Pressure-reducing regulator—a device used to regulate and/or control the pressure of a process fluid downstream of the device location, **p. 239.**

Regulator—a self-contained and self-actuating controlling device used to manipulate variables such as pressure, flow, level, and temperature in a process, **p. 228.**

Seat—the stationary part of the valve trim, connected to the body, that comes in contact with the valve plug; when the valve plug is fully seated, the flow through the valve ceases, **p. 227.**

Spring—the device that provides the energy to move a valve in the opposite direction of the diaphragm loading motion so that the valve can be opened and closed proportionate to the instrument signal; also provides energy to return the valve back to its fail-safe condition, **p. 228.**

Stem—the pushing rod that transfers the motion of the actuator to the valve plug, **p. 228.**

Valve positioner—a device that ensures a valve is positioned properly according to the signal received from the controller, **p. 228.**

13.1 Introduction

A final control element is the last active device in the instrument control loop. The final control element directly controls the manipulated variable by performing an operation on the process to bring about change in the controlled variable. Of the many different types of final control elements, the most common final control element is a control valve.

This chapter describes the various components of control valves as well as their purpose and operation. It discusses what happens during valve failure and provides some simple troubleshooting tips. It also covers actuators and valve positioners. Additionally, input and output signal differences, fail-safe positions, and special valve types are discussed to show how various configurations can be made to enhance the safety and operability of the process.

Instrument air is commonly used to actuate the valve actuator. This air may need to be regulated, clean, and dry. Dust can plug small orifices found in all pneumatic instruments. Moisture or water in instrument air lines can interfere with control and can freeze in cold weather. For these reasons, this chapter includes a discussion about instrument air regulators, and how they are used in the operation of control valves. Most examples and discussions are about sliding stem valves, but other types, such as butterfly valves, have important differences that are reviewed.

The Control Valve

Control valves are the most common final control element in the processing industry. Control valves regulate the flow of materials in the process by producing a differential pressure (D/P) drop across the valve. For example, material may be added to or taken away from a vessel during the opening and closing operation of a control valve. For good control, a throttling valve must drop at least 10 percent of the process pressure in the line.

A control valve has an actuating device, or actuator. The actuator changes an instrument signal into either a linear or a rotary motion. This motion drives the flow-controlling mechanism (e.g., **plug** or disc) in a valve. Key components of a sliding stem control valve are shown in Figure 13.1 and described below:

- The **body** of a valve is usually joined to the process by a threaded or flanged connection. In a critical service, however, the connection might be welded. Welding may complicate valve replacement.
- The valve **bonnet** connects the valve body to the actuator. The bonnet can be removed to allow entry into the valve body cavity. It usually contains the packing box, which provides a seal around the sliding stem mechanism that connects the actuator to the valve plug.
- The valve plug assembly that includes the valve stem moves to open or close the flow path through the valve. As the plug moves away from the **seat**, the flow through the valve increases. As the plug nears the seat, flow decreases. Once the plug reaches the seat, a reasonably tight seal is formed that stops the flow through the valve. In other valves, such as the butterfly and ball valve, the plug component is referred to as a *disc* or *ball*.

Plug the only movable component in the valve; it is actuated to open or close the flow path through the valve. May also be called a *disc* or *ball*.

Body the housing component of a valve.

Bonnet a bell-shaped dome mounted on the body of a valve where the valve disc is held when the valve is opened.

Seat the stationary part of the valve trim, connected to the body, that comes in contact with the valve plug; when the valve plug is fully seated, the flow through the valve ceases.

Figure 13.1 Control valve key components. A. Valve illustration. B. Photo of a valve. C. Pneumatic control valve symbol.
CREDIT: B. © Emerson 2019.

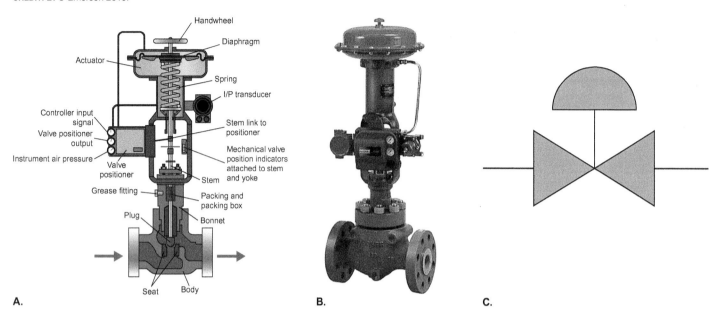

Actuator an electrical, pneumatic, or hydraulic device that manipulates a control device, usually a valve.

Regulator a self-contained and self-actuating controlling device used to manipulate variables such as pressure, flow, level, and temperature in a process.

Stem the pushing rod that transfers the motion of the actuator to the valve plug.

Spring the device that provides the energy to move a valve in the opposite direction of the diaphragm loading motion so that the valve can be opened and closed proportionate to the instrument signal; also provides energy to return the valve back to its fail-safe condition.

Valve positioner a device that ensures a valve is positioned properly according to the signal received from the controller.

Handwheel the mechanism that raises and lowers a valve stem to control the flow of fluid through a valve.

I/P transducer (current-to-pneumatic transducer) a device that converts a milliampere signal into a pneumatic pressure signal.

- The **actuator** is the device that provides motion to the valve using a spring diaphragm, spring piston, or double-acting piston.
- A **regulator** is a device used to control the pressure of a process fluid upstream of the device location.
- The **stem** is the pushing and pulling rod that transfers the motion of the actuator to the valve plug.
- The seat in a valve is the stationary part of the valve trim connected to the body that comes in contact with the valve plug. When the plug is fully seated, the flow through the valve stops.
- The **spring** provides the energy to move the valve in the opposite direction of the diaphragm loading motion. This provision is made so that the valve can be opened and closed in proportion to the instrument signal. The spring provides the energy to return the valve back to its fail-safe condition.
- The diaphragm is the flexible membrane that creates a force to move the stem.
- A **valve positioner** is actually a controller that is proportional only. A mechanical link that is directly connected to the positioner senses the position of the valve stem. The position of the valve stem is then compared to the value of the instrument signal, and a response is produced to make the position of the valve and the signal equal.
- A **handwheel** is an actuator accessory that is used to override the actuator manually or to limit its motion. The handwheel may be located on the top of the actuator or on its side. Process technicians may manually limit or close a problem valve if it is equipped with the handwheel. A hand jack is a wheel attached to a control valve that can manually move the valve, but few control valves have a hand jack. Many control valves have associated handwheel inlet, outlet, and bypass valves to permit maintenance activities, since these reduce installation costs.
- An **I/P transducer (current-to-pneumatic transducer)** is a device that converts a milliampere signal into a pneumatic pressure signal. The most common use for an I/P transducer is to provide the source of energy needed to drive a diaphragm or piston actuator. A current-to-pneumatic transducer typically receives a 4 to 20 mA current signal and converts it to a 3 to 15 PSIG pneumatic signal.

13.2 Control Valve Failure Conditions

Control valves are ultimately responsible for regulating the movement of fluids in a process. If there is a power or air failure, they should move to a safe position. If there is an I/P transducer before the valve, loop (current) failure should also be considered when determining valve fail position. Design engineers address these possibilities with each control valve placed into the process. Control valves are generally designed to fail in an open, closed, or last position. Actuators without return mechanisms (such as springs) typically fail in last position. External devices, such as solenoids and external air tanks, may be placed into the system to drive the control valve actuator to a predetermined intermediate position in the event of power failure or reduction in air pressure.

During the design phase of a plant, the design engineers must recognize the required fail-safe conditions needed if a plant loses power or instrument air and must choose each control valve response accordingly. Every process technician should be aware of how control valves fail under power or instrument air loss. Since most control valves are pneumatically actuated the following discussion assumes a pneumatic diaphragm actuator with a spring return device. When using P&IDs, the control valve failure position should be indicated under the valve symbol. These symbols were discussed in Chapter 12, *Control Loops: Controllers and Final Control Element Overview*.

Fail Open

When an air to close control valve (Figure 13.2) loses its instrument air signal or supply, the valve fails open because the return spring provides more opposing force than the diminishing instrument air applied to the diaphragm and the force applied by the process (if any). For example, during a power failure, you would want a pressure-relieving valve on a reactor to fail in the open position. This position would prevent a pressure buildup that could rupture the vessel.

Fail Closed

When an air to open control valve (Figure 13.3) loses its instrument air signal or supply, the valve fails closed because a return spring provides more opposing force than the diminishing instrument air applied to the diaphragm and the force applied by the process pressure. For example, during a power failure, a valve in the controlling position that is allowing material to flow into a tank should close. If the valve stayed open and power is not restored, the tank could overfill. This scenario requires the control valve to fail closed when there is loss of instrument air.

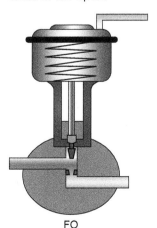

Figure 13.2 Fail conditions—air to close or fail open.

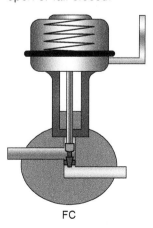

Figure 13.3 Fail conditions—air to open or fail closed.

Fail in Place or Fail Last

Pneumatic actuators with opposing springs naturally fail in the direction of their spring tension, unless they have a lockup relay attached. A *lockup relay* seals in the existing signal applied to the actuator at the point of power loss. Actuators without a spring or other return mechanism usually fail in their last position (Figure 13.4) just prior to loss of power unless the process pressure is high enough to change the valve position. Electric motor actuators, for example, naturally fail in place. Electronic fail-safe actuators use super capacitors that discharge stored energy to the motor, and the actuator is driven open or closed on a loss of power.

Figure 13.4 Fail conditions—fail in place or fail last.

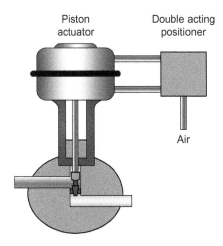

13.3 Control Valve Actuators

An actuator is a device that responds to an applied instrument signal by creating a linear or a rotational motion. An actuator is what makes a valve a control valve. By responding to an instrument signal, an actuator provides the motion necessary to throttle the valve.

In most control situations, a valve and an actuator may need to be selected for accommodation of the needs of the control loop as well as for safety and shutdown considerations. For this and other reasons, process technicians should know what combinations can be expected in the field. The following sections cover the possible combinations of actuator and valve actions used in the processing industry.

Actuator and Valve Actions for Sliding Stem Valves

To begin with, actuators and valves are either direct acting or reverse acting (Figure 13.5). An actuator is considered to be direct acting when an increased application of pressure

Figure 13.5 **A.** Actuator operation. **B.** Actuator symbols for P&ID.

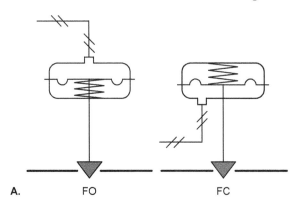

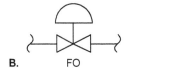 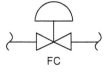

causes the stem to extend. Conversely, an actuator is considered to be reverse acting when an increased application of pressure causes the stem to retract. A valve is considered to be direct acting when the valve stem is pushed down to close and reverse acting when the valve stem is pulled up to close. There are four possible combinations of actuator and valve actions:

- Direct/direct
- Direct/reverse
- Reverse/direct
- Reverse/reverse

The easiest and probably the best way to determine what the actuator–valve combination is capable of doing is to look at the identification (ID) plate attached to the actuator. The ID plate provides information about the combined actuator and valve action (air to close [ATC/atc] or air to open [ATO/ato]) as well as other information.

Pneumatically Driven Actuators

There are two major subcategories of pneumatically driven actuators: (1) spring and diaphragm actuators, and (2) piston-type actuators.

The most common spring and diaphragm actuator has a single diaphragm supported by a diaphragm plate connected to a steel rod, called a stem. A spring is placed on the opposite side of the plate to create an opposing return force. In a direct acting actuator, the diaphragm responds to the pneumatic signal by extending the stem component while compressing the return spring. As the pressure decreases, the spring tension pressure exceeds that of the diaphragm and pushes back on the diaphragm, retracting the stem.

Spring and diaphragm actuators (Figure 13.6) are very popular because of their low cost and high mechanical advantage. The spring also provides a mechanical fail-safe condition upon loss of signal. This type of actuator also provides excellent throttling control with or without a positioner.

Piston-type actuators fall into two major categories: single acting and double acting.

- In the single-acting category, there are two subcategories: spring-opposed actuators and air-cushion actuators. In single acting, spring-opposed actuators, like the diaphragm type, the piston is opposed by a spring. In the air cushion type, the piston actuator has pressure trapped under the piston, and the air compresses as the piston is pushed down. Then, as the instrument signal is reduced, the trapped compressed air pushes the piston back up.

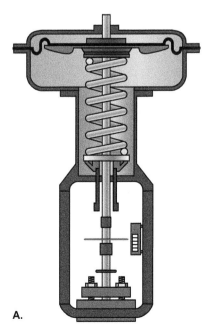
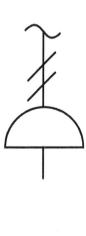

Figure 13.6 **A.** Spring and diaphragm actuator. **B.** Spring and diaphragm actuator symbol.

- In the double-acting type, instrument air pressure is routed to both sides of the piston, driving the stem to a required position by balancing the pressures on either side of the piston. Many piston-type actuators are double-acting type, which means that they need to be controlled with a positioner.

Piston actuators (Figure 13.7) tend to have longer strokes and can accept much higher input pressures. Higher pressure produces more force and provides more power, which can be important in some applications where there is a high pressure drop across the valve. Piston-type actuators may also be of the spring-opposed or air-cushion control types that provide a fail-safe condition.

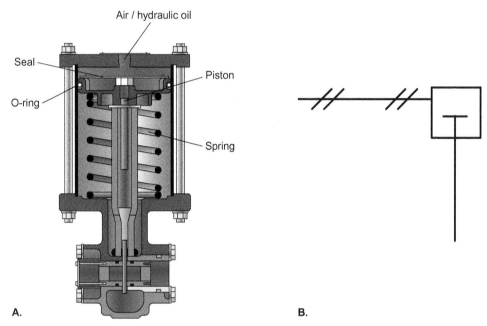

Figure 13.7 **A.** Piston-type actuator. **B.** Piston actuator symbol.

Valve Positioners

The function of a positioner is to make the valve position match the controller output signal. The valve positioner locates the moving parts of a valve in accordance with a predetermined relationship with the instrument signal received from the loop controller. The positioner can be used to modify the relationship between the input and output instrument air signal.

The uses of a valve positioner are to:

- Position the valve
- Reverse the action
- Mimic a valve trim type
- Provide split-range control.

A valve positioner may be used to adjust the position of the valve (Figure 13.8) according to specific needs or to change the amount of pneumatic signal needed to fully stroke the valve. A split-range application would be an example.

A valve positioner may be used to reverse the action (Figure 13.9) of the valve. For example, an air to open/fail closed valve can be made to operate as if it were an air to close/fail open valve.

Older positioners have trim characteristics which are the design of the control valve trim so that the valve opens up gradually during the initial stem travel from the closed position and then opens up more aggressively during the final stages of stem travel as it nears the full-open position. These characteristics can be changed by replacing the cam to quick opening, linear, or equal percentage. See Figure 13.10 to see the flow/percent open characteristic of these trim types. In the newer digital positioners, the trim characteristics can be programmed online and downloaded as needed.

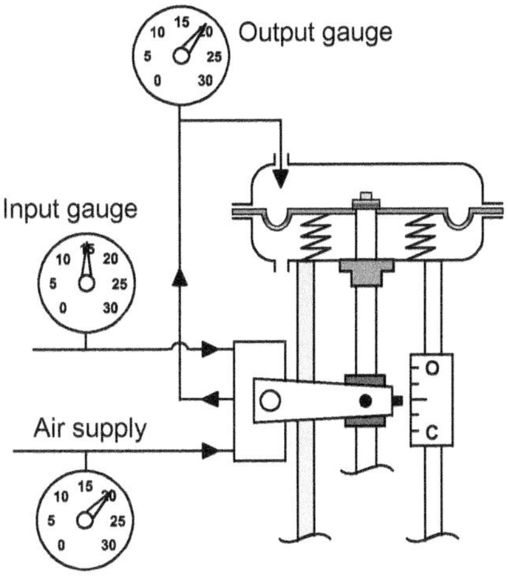

Figure 13.8 Valve positioners—position the valve.

Figure 13.9 Valve positioners—reverse the action.

Figure 13.10 Valve positioners—trim characteristics curves.

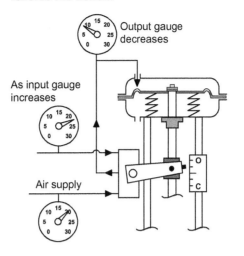

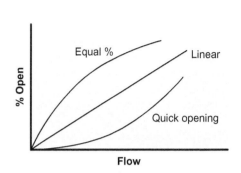

A valve positioner may also be selected to mimic (Figure 13.11) the characteristics of different valve trim types such as quick opening, linear, and equal percentage.

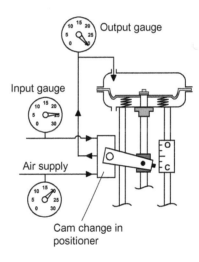

Figure 13.11 Valve positioners—mimic a valve trim type.

Valve Positioner Operation

Pneumatic valve positioners (Figure 13.12) on spring and diaphragm actuators may come equipped with three gauges. The movement of the valve may be predicted by observing the three associated pressure gauges.

- *Instrument input pressure gauge:* This gauge indicates the signal pressure from the controller or I/P transducer and is especially important while troubleshooting the loop for problems. The process technician can look at this gauge to determine the output signal from the controller.
- *Output pressure gauge:* This gauge indicates the output pressure applied to the actuator that may or may not be equal to the instrument signal pressure. This gauge also reads the pressure required for the valve position.
- *Air supply pressure gauge:* This gauge indicates instrument air supply pressure. All pneumatic instruments capable of producing a pneumatic output have an instrument air supply connected to them.

Valve positioners are very versatile. The operating parameters of the positioner can be as simple as direct acting and linear. The positioner can also be set up to have nonlinear characteristics. For example, if a valve with linear trim (arranged in or extending along a

Figure 13.12 **A.** Valve positioner element. **B.** Valve positioner in operation.
CREDITS: **A.** © Emerson 2019. **B.** Oil and Gas Photographer/Shutterstock.

A.

B.

straight or nearly straight line) were set up by the positioner to mimic quick opening trim, then the relationship between the instrument signal to the actuator output signal would be defined by this quick opening property.

A simple and fairly common application would have a direct-acting actuator combined with a direct-acting valve with linear trim. A valve with linear trim has a linear relationship between stem travel and flow rate. If the positioner were set up to respond in a direct manner to the full stem travel of the valve, then the following scenario would hold true.

Example:
If a direct linear relationship existed under ideal conditions and the signal gauge indicated 9 PSIG (50 percent signal), the output gauge would probably also indicate 9 PSIG. In reality, the output gauge may read more or less than 9 PSIG because the positioner drives the output to the actuator until the valve stem reaches the position equal to the instrument signal.

13.4 Output Signals

In some process control applications, the output of the controller may need to be reversed at the control valve. This can be accomplished by configuring the current-to-pneumatic (I/P) transducer or the valve positioner to respond in reverse to the signal from the controller. For example, the I/P transducer or the valve positioner would reverse an increasing signal from the controller to a decreasing signal to the control valve. In older controllers, this may make the controller output indication opposite to the valve closed or open position. Also, note that in modern programmable controllers the output signal can be matched to the valve so that the output indicator shows 0 percent as fully closed and 100 percent as fully opened. This is done in addition to choosing reverse or direct acting.

One example of where this reversing action is necessary is in a split-range control system. Both valves may need to be closed when the signal from the controller is at midrange, with one open with an increasing signal above midscale and the other open with a decreasing signal from midscale. In this situation, one valve has to respond in a reverse manner to the signal from the controller or choose the first valve to fail closed (calibrated to 9 to 15 PSIG; close-to-open) and the second valve to fail open (calibrated 9 to 3 PSIG; close-to-open).
NOTE: It is also common to have a dead-band zone in the middle, say 3–8.5, low range, 8.5 to 9.5 both closed, and 9.5 to 15 upper range.

Since spring and diaphragm actuators are designed with a particular fail-safe position, this type of actuator is forced into the designed fail-safe position if instrument air is lost. Therefore, whether or not the output of the controller is reversed does not matter, since the actuator fails into the position demanded by the opposing spring.

13.5 Valves for Control Configurations

The following are the main types of valves used in control configurations:

- Globe
- Three-way
- Butterfly
- Ball or segmented ball

Globe Control Valves

The globe-style valve body is the most common type of valve used in the processing industry. The plug and seat, often called *valve trim*, are located within the inner cavity, or body, of the valve and provide a connection between the inlet and outlet. The globe valve gets its name from its globular shape. Although a true globe shape is like a sphere, any reasonably or somewhat rounded valve body style usually falls into this category. Flow through a globe valve body changes direction (Figure 13.13).

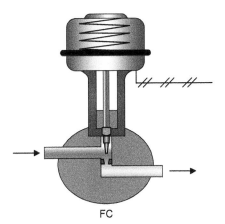

Figure 13.13 Globe control valve—single port.

The globe valve, like all valves, controls the flow of material through its inner cavity with its plug and seat components. Control is accomplished with one or more ports. A valve port is the restrictive orifice in a valve defined by the diameter of the seat. The single-port valve is the most common of the globe body valves, and a variation of this valve, a cage-style valve, is the single most popular control valve in the processing industry. Double-ported valves were developed to balance the high differential pressure (D/P) acting on single-port valves. The double-seated valve generally has a higher flow capacity than a single-ported valve of the same size and requires less actuator force to drive the stem. The problem with double-ported valves is that they tend to leak because there are two sets of plugs and seats that must shut at the same time. Unfortunately, over time with the inevitable uneven wearing across the two ports, this becomes almost impossible to manage.

Three-Way Control Valves

A three-way valve is a special type of globe body valve that has three connecting ports instead of two. They are designed to either *mix* (Figure 13.14) two flowing input streams together with one outlet port or *divert* one flowing input stream between two output ports.

The diverting valve (Figure 13.15) could be used as a switching valve diverting a flowing stream from one vessel to another, or as a temperature control valve diverting part of or the entire process stream into a heat exchanger. The mixing three-way valve could be used for blending two separate streams into one, producing a proportioned mixture of the two.

As shown in Figure 13.15, if the reactor inlet temperature drops, the diverting three-way control valve closes port A and opens port B to put hotter reactor effluent into the exchanger. This increases reactor inlet temperature.

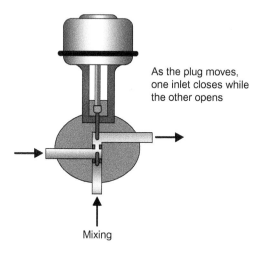

Figure 13.14 Mixing three-way control valve.

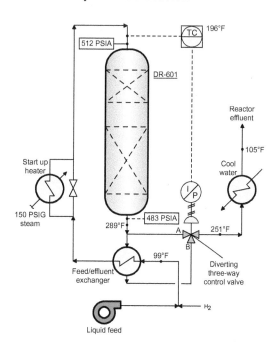

Figure 13.15 Diverting three-way control valve for a hydrotreater reactor.

Butterfly Control Valves

Butterfly control valves (Figures 13.16A and B) have lower manufacturing costs and higher flow capacities than globe valves. Actuators used with butterfly valves use a rotary motor. The rotary stem produces less wear on the packing than a sliding stem valve does.

Figure 13.16 **A.** Butterfly valve—spring diaphragm. **B.** Butterfly valve—piston-type actuator.

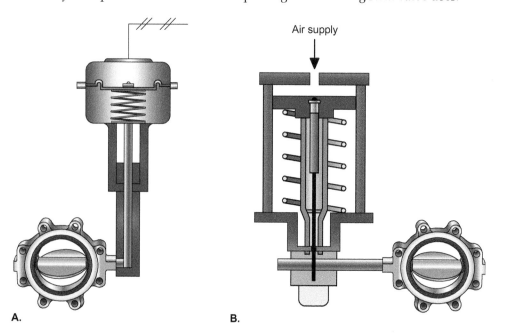

Butterfly valves, also called rotary valves, are equipped with spring and diaphragm actuators (Figure 13.16A) or piston-type actuators (Figure 13.16B). The butterfly valve has flow characteristics somewhere between linear and quick opening, so this type of valve exhibits a nonlinear relationship between the percent opening and the rate of flow through the opening. The butterfly valve is used to control all types of fluids, including both liquids and gases.

Ball or Segmented Ball Control Valves

The full ball control valve (Figure 13.17) is a rotary valve that contains a spherical plug with a hole that runs through it. The control valve actuator uses a stem that is attached to the top of the ball to rotate it in order to control the flow of fluid through the valve body. By comparison, a segmented ball valve has one edge either contoured or with a V-shaped edge to yield a desired flow characteristic (e.g., linear or quick opening).

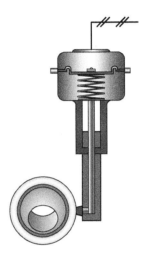

Figure 13.17 Ball control valve—spring diaphragm.

Ball control valves may be used as a tight shutoff or as a modulating valve offering high flow capacity because there are no internal obstructions when the valve is fully open. Ball valves are also often used in applications where the process fluid is slurry in order to minimize the settling and straining of these materials.

The spherical plug of the ball control valve is adaptable to function in three-way service (Figure 13.18). The control valve in the effluent from the reactor has a single, hot inlet with two outlets. The inlet admits hot effluent to heat up colder reactor feed while one outlet bypasses unneeded hot material directly to cooling. Rotation of the ball allows more or less

Figure 13.18 Three-way ball control valve.

CREDIT: © Emerson 2019.

hot effluent to heat the feed (depending on the inlet temperature controller) while bypassing the remainder to cooling.

13.6 Instrument Air Regulators

A regulator is a self-contained and self-actuating controlling device used to regulate variables such as pressure, flow, level, and temperature in a process. Regulators generally receive the energy necessary to move the internal valve mechanism from the process stream itself. Although pressure and flow regulators are the most common types used in industry, level and temperature regulators are also found. The discussion in this section is limited to the air pressure regulator since it is the most common type within industry.

An instrument air regulator can either reduce the supply pressure (upstream side) or relieve pressure from the downstream side by releasing excess downstream pressure into the atmosphere.

Among the important parts of an air pressure regulator (Figure 13. 19) are the following:

- Inlet: supplies pressure from a source
- Outlet: regulates pressure
- Diaphragm: senses the pressure on the outlet side of the regulator
- Pilot valve assembly: working part of the regulator; entire mechanism works to either supply pressure or relieve pressure from outlet side

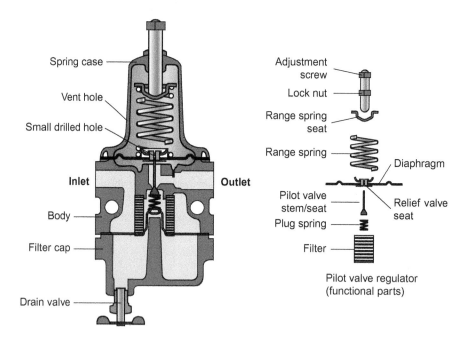

Figure 13.19 Regulator cutaway.

Instrument air regulators (Figures 13.20 and 13.21) drop the pressure received from the instrument air distribution system (usually 80 to 150 PSIG) down to a supply pressure range that most pneumatic instrumentation is capable of tolerating. The supply pressure applied to pneumatic instruments such as transmitters, controllers, and control valves cannot exceed approximately 25 to 30 PSIG, and many of them recommend a supply pressure of around 20 PSI. In an electronic control loop, the I/P transducer also needs a regulated supply pressure in this range.

An instrument air regulator procedure follows this sequence:

1. The handwheel is turned.
2. The spring compresses.
3. The diaphragm is pushed down.

Figure 13.20 **A.** Back-pressure regulator. **B.** Back-pressure regulator symbology.
CREDIT: **A.** © Emerson 2019.

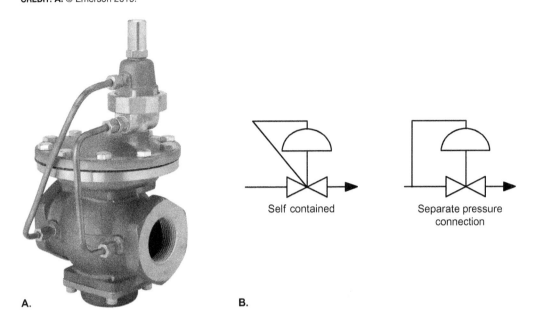

4. The pen and plug move from their seat.
5. Air rushes in, increasing downstream pressure (downstream of the regulator).
6. The output gauge responds accordingly (moves in the expected direction).
7. Equilibrium is established. (The valve closes, and the system is in balance.)

A pressure control system would include a sensing device, a controlling device, and a final controlling element. Since the regulator is a self-contained controlling device, it has these elements integrally located inside. The two major types of pressure regulators are the **back-pressure regulator** and the **pressure-reducing regulator**.

Back-Pressure Regulators

A back-pressure regulator (Figures 13.20A and B) is a device used to regulate and/or control the pressure of a process fluid *upstream* of the location of the regulator. For example, this type of regulator is used to maintain the pressure in the vapor space of a vessel. Most sealed tanks in industry have a pressure control system associated with them.

Pressure-Reducing Regulators

The other major type of pressure regulator is the pressure-reducing regulator (Figures 13.21A and B). This device is used to control the pressure of a process fluid *downstream* of the location of the regulator.

An application where a pressure-reducing regulator might be used is where steam leaves a boiler and could potentially have hundreds of pounds of pressure that need to be reduced to specific unit requirements that may only be 150 PSI. This type of regulator is called a steam pressure-reducing regulator. Notice that in both the back-pressure and pressure-reducing regulator symbols, there are two distinct sensor points. One is internally sensed and the other is externally sensed. In both regulator examples, the pressure reduction occurs at the regulator. Even though the sensor point may be external, this does not negate the fact that a regulator is self-contained because it still gets its manipulating power from within the regulating system.

Back-pressure regulator a device used to regulate and/or control the pressure of a process fluid upstream of the device location.

Pressure-reducing regulator a device used to regulate and/or control the pressure of a process fluid downstream of the device location.

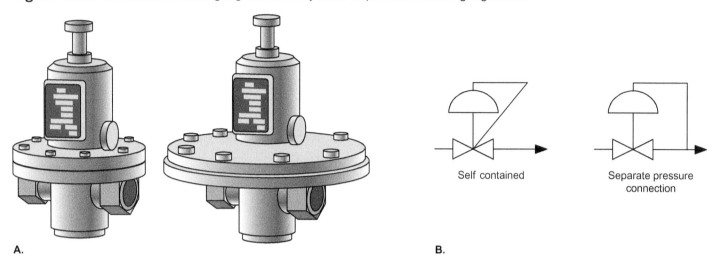

Figure 13.21 **A.** Pressure-reducing regulators. **B.** Symbols of pressure-reducing regulators.

Summary

A valve itself is not controlled unless it has some type of actuator and positioning arrangement. Actuators begin the movement of the valve either to an open or a closed position, while the positioner determines the percent it is open. The movement of the valve may be predicted by observing the three associated pressure gauges.

Key components of a sliding stem control valve include the body of the valve that is connected to the process and houses the actual place where the valve seat is located and the disc or plug mechanism that allows the valve to move. Above the valve body is the bonnet that connects the valve body to the actuator and provides a seal around the sliding stem mechanism that connects the actuator to the valve plug located inside the valve body. The actuator provides pushing or pulling motion to the valve stem so that the valve plug opens or closes appropriately. The actuator may be pneumatically controlled via an instrument air signal to a movable diaphragm. The pressure on the diaphragm in the actuator is converted by mechanical linkage to a positioner into a proportional value. This value produces a counter response that positions the valve accordingly. In some cases, a handwheel may be used to override an actuator or to limit its motion. I/P transducers may be used to convert signals going to actuators or positioners into their more commonly needed air pressure requirements.

During unusual conditions, such as power failures, control valves should move to a position that promotes safe operating conditions. These are called fail open, fail closed, or fail in place. Devices such as return springs, lockup relays, or solenoids may be used in addition to instrument air to aid the appropriate fail-safe position.

Actuators respond to instrument air signals by creating either a linear or a rotational motion on the stem that makes a valve either open or close. Actuators may be direct or reverse, depending on which way the stem moves when pressure is applied. Since there are four potential types of actuator–valve combinations, the best way to determine their capability is to look at the valve ID plate.

The two major subcategories of pneumatically driven actuators are (1) spring and diaphragm actuators and (2) piston-type actuators. Spring and diaphragm actuators are popular due to their lower cost and higher mechanical advantage, which allow them to provide a mechanical fail-safe condition upon loss of signal. Piston-type actuators have longer strokes so that they can accept higher input process pressures. Piston-type actuators may be either of the spring-opposed or air-cushion subcategories. Piston-type actuators may be either single- or double-acting types. Double-acting types allow instrument air pressure to both sides of the piston driving the stem.

Valve positioners arrange the moving parts of a valve in relation to the control loop controller signal. Valve positioners are used to position the valve, reverse the action, mimic a valve trim type, or provide split-range control.

Pneumatic valve positioners may come equipped with three gauges: (1) an instrument pressure gauge that indicates the signal pressure from the controller, (2) an output pressure gauge that may indicate a measurement equal to the instrument pressure, and (3) a supply pressure gauge that indicates the instrument air supply pressure.

In some conditions, the output of a controller may need to be reversed at the control valve. The valve positioner can be configured to respond in reverse to the signal from the controller, as in a split-range control system. The fail-safe condition that would be triggered by failure of the instrument air is controlled by the opposing spring in the actuator.

The main types of valves used in controlling operations are globe, three-way, butterfly, and ball or segmented ball. Globe valves may have one or two ports depending on the particular design and are usually shaped like a globe. Three-way valves are also globe valves, but they have three ports. This type is used for mixing (two inlets–one outlet) or diverting (one inlet–two outlets). Butterfly valves have lower manufacturing costs and allow higher flow capacities for both liquids and gases. Ball or segmented ball may be used where a tight shutoff or modulation is needed.

While there are many types of regulators used within processing operations, this chapter has focused on instrument air regulators, which have the ability to regulate both the amount of supply pressure on the upstream side of the regulator and the discharge pressure on the downstream side. The two main types of instrument air regulators are the back-pressure regulator and the pressure-reducing regulator. Both types are self-contained mechanisms.

Checking Your Knowledge

1. In a globe valve, as the plug moves away from the set, the flow through the valve _____.
   a. increases
   b. decreases
   c. remains constant
   d. forms a tight seal

2. A control valve configured with a spring and diaphragm actuator is said to be required air to close; if there is an air or power loss, the valve will:
   a. be forced completely closed by the spring.
   b. be forced completely open by the spring.
   c. remain in its last position.
   d. hunt for the best option.

3. During a power failure, you want a valve that is allowing the addition of material to a tank to stop the flow. Therefore, this valve should be a:
   a. fail closed valve.
   b. fail open valve.
   c. fail in place valve.
   d. It can be any type because the power is off.

4. A(n) _____ is a device that reacts to an instrument signal by creating linear or rotational motion.
   a. modulator
   b. agitator
   c. controller
   d. actuator

5. If a rising stem control valve and its spring and diaphragm actuator are both direct acting, then the control valve will _____ if air is lost to it.
   a. fail open
   b. fail closed
   c. fail intermediate
   d. None of the above

6. Which of the following can a valve positioner do?
   a. Position the valve stem in reference to the instrument signal
   b. Reverse the direction of flow through the valve
   c. Reverse the action of the signal received from the controller
   d. None of the above

7. What are the three pressure gauges on a pneumatic valve positioner?
   a. Instrument, output, and supply
   b. Instrument, input, and supply
   c. Input, output, and valve
   d. Input, output, and position

8. The output signal from a controller can be reversed by _____.
   a. a regulator
   b. reversing the lead wires
   c. the valve positioner
   d. the valve pressure

9. In the processing industry, the most common control valve body style is the _____.
   a. butterfly
   b. three-way
   c. globe
   d. double-port

10. A butterfly valve is opened with a _____.
    a. linear motion
    b. rotary motion
    c. sliding-stem
    d. flipper modulator

242 Chapter 13

11. A three-way valve can be used to _____.
    (Select all that apply)
    a. divert a flowing stream into two separate pipes
    b. mix two separate flowing streams together into one pipe
    c. control the amount of air flowing into a furnace
    d. replace two valves

12. What is the recommended supply pressure for many air supply to control valves?
    a. 10
    b. 20
    c. 50
    d. 100

13. On the following diagram, identify the following:
    I. Body
    II. Actuator
    III. Diaphragm
    IV. Seat
    V. I/P transducer
    VI. Stem
    VII. Bonnet
    VIII. Plug or disc
    IX. Handwheel
    X. Spring
    XI. Valve positioner
    XII. Packing and packing box

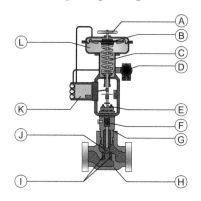

14. Properly identify each valve by putting the correct letter of the appropriate name with its image.
    I. Globe body valve
    II. Butterfly valve
    III. Three-way valve

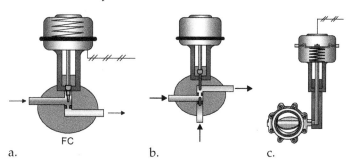

a.   b.   c.

NOTE: Answers to Checking Your Knowledge questions are in the Appendix.

Student Activities

1. In small groups, make answer keys for the photographs used in the item above. Test each other's recognition of each of the instruments. Compare and contrast various manufacturers and configurations of control valves, actuators, positioners, and their instrument air regulators.

2. Identify/locate various types of control valves and regulators in the lab.

3. Stroke an assembled control valve with instrument air to check P versus % open.

4. Given several pressure indications for a valve positioner, work in small groups to determine how the valve will react.

5. Describe operating scenarios in which fail open, fail closed, and fail last positions are desirable.

6. Using a simple control loop system within a pilot plant, change a variable (e.g., tank level). Record the mA meter signal, then the controller output, then the air signal from the transducer, and record how the valve moves. Record the readings at several different milliamp and air points between 0 and 100 percent, 4 to 20 milliamps, and 3 to 15 PSIG on the air signal. If pilot equipment is not available, use graphics for each of the changes. Calculate the milliamp and air pressure values based on the relationship between the two and determine the expected valve movements and positions.

Chapter 14
Controllers

Objectives

After completing this chapter, you will be able to:

14.1 Explain the following states and conditions associated with process dynamics, process control, and controllers:

  process equilibrium
  load change
  lag time
  dead time
  over range
  spanning
  feedback control
  feedforward control
  direct or reverse acting controller
  on/off control
  proportional band/gain
  proportional action
  integral action (reset)
  derivative action (rate, preact)
  windup (NAPTA Control Loops: Controllers 1*) p. 245

14.2 Define the following terms associated with process recording:

  offset
  trend
  hunting
  steady state (NAPTA Instrumentation Troubleshooting 5) p. 252

---

*North American Process Technology Alliance (NAPTA) developed curriculum to ensure that Process Technology courses will produce knowledgeable graduates to become entry-level employees in process technology. Objectives from that curriculum are named here in abbreviated form. For example, "(NAPTA Control Loops: Controllers 1)" means that this chapter's objective 1 relates to objective 1 of NAPTA's course content about controllers in control loops.

14.3 Identify tuning modes and troubleshooting for selected controller tuning errors and process problems. (NAPTA Instrumentation Troubleshooting 3) p. 255

Key Terms

Chart recorders—devices that physically or electronically record data that process and instrument technicians can use to troubleshoot process trends and instrumentation problems, **p. 245**.

Cycling—the moving or shifting of a process variable above and/or below the set point, **p. 262**.

Dead time—the elapsed time between the initiation of an input (measured, controlled, or manipulated variable), change, or other stimulus and the point at which the resulting response can be measured or observed, **p. 246**.

Deviation—the difference between the set point and the controlled variable under dynamic (changing) conditions, **p. 253**.

Direct acting controller—a controller whose output signal value increases as the controller input signal value increases or decreases as the controller input value decreases, **p. 249**.

Feedback control—a closed loop control strategy where the difference in the set point and measurement drives the output of the controller to the final control element, **p. 248**.

Feedforward control—an open loop control strategy in which a disturbance in the process is anticipated, and the controller output is adjusted to compensate for its effect, **p. 249**.

Hunting—a condition that exists when a control loop is improperly designed, installed, or calibrated, or when the controller itself is not properly aligned; overcorrection above and/or below the set point, **p. 254**.

Lag time—a measure of the elapsed time between two events, states, or processes, **p. 246**.

Load change—a change in any variable in the process that affects the value or state of the controlled variable, **p. 245**.

Offset—the difference between the set point and the controlled variable under steady state (not changing) conditions, **p. 253**.

On/off control—a controller that simply drives the manipulated variable from fully closed to fully open, depending on the position of the controlled variable relative to the set point, **p. 250**.

Over range—the condition that exists when the signal value of a device or system exceeds the maximum allowable value, **p. 246**.

Process equilibrium—the condition that exists when there is a balance of material and energy within a given system or process; also referred to as *steady state*, **p. 245**.

Proportional action—a controller output response that is proportional to the amount of deviation of the controlled variable from the set point, **p. 251**.

Reverse acting controller—a controller whose output signal value decreases as the controller input signal value increases, and vice versa, **p. 249**.

Spanning—the process of setting or calibrating the lower and upper range values of a device; also the process of testing a device or system by traversing up and down the scale through the entire calibrated or operational range, **p. 247**.

Steady state—a trend condition in which the process is in equilibrium and running smoothly, producing a relatively straight line on a graph or chart, **p. 254**.

Trend—the plotting of a process variable over time, **p. 253**.

Windup—the saturation of the controller output signal to its minimum or maximum value if the process variable does not return to the set point, **p. 251**.

14.1 Introduction

When studying controllers and process dynamics, it is important to understand that processes have certain common and predictable behaviors. For example, process can be in either a state of equilibrium (steady state) or in a state of change. During states of change, a process may experience load change, changes in dead time, or respanning, or it may be over range.

When tuning process controllers, you may experience direct or indirect controller action. The control schemes may include on/off control, feedback control, or feedforward control. Other key terms you may encounter in connection with process control and controllers are gain, reset, rate, windup, offset, and tuning.

As processes run, **chart recorders** record and document operations and maintenance data, which can be a very important troubleshooting tool. Through the use of historical graphs and charts, important observations can be made about trends. From these trends, a process technician can determine whether there are problems in the control loop (such as hunting) or if the process is running smoothly in a state of equilibrium (flatline).

Chart recorders devices that physically or electronically record data that process and instrument technicians can use to troubleshoot process trends and instrumentation problems.

Process Dynamics and Control

Processes behave in recognizable ways. Process instruments are used to maintain a desired condition or to alter a process in order to attain a desired condition. There are several terms associated with process dynamics. These include process equilibrium, load change, dead time or lag time, over range, and spanning.

Process Equilibrium

Process equilibrium (Figure 14.1) describes a condition in which a balance of material and energy exists within a given system or process. Process equilibrium is sometimes also referred to as *steady state*. A system that is in a steady state remains constant over time, but that constant state requires continual work. This condition is also referred to as a system in *dynamic equilibrium*. If the process control system stops working, the process variable would drift away from the equilibrium condition.

Process equilibrium the condition that exists when there is a balance of material and energy within a given system or process; also referred to as *steady state*.

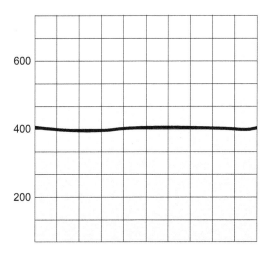

Figure 14.1 Process equilibrium.

Load Change

The term **load change** (Figure 14.2A) describes a change in any variable in a process that alters the value or state of the controlled variable. Examples of changes that might occur to affect variables are more cooling, more heat, or higher flow rate. When a load change occurs, the process will move to a new steady state value. This process may occur several times, as shown in Figure 14.2B.

Load change a change in any variable in the process that affects the value or state of the controlled variable.

Figure 14.2 A. Load change. **B.** Repeated load changes.

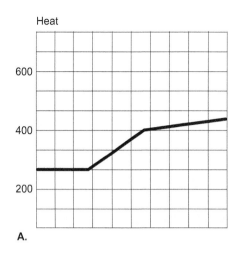
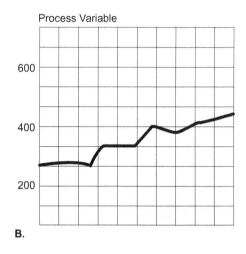

Lag Time

Lag time a measure of the elapsed time between two events, states, or processes.

Lag time (Figure 14.3) is the elapsed time between two events, states, or processes. For example, a process lag time is the time from when the process change can be measured or observed to the time when it reaches a new equilibrium value. There are various types of lag time in process industries, such as measurement lag time, transmission lag time, and controller lag time.

Figure 14.3 Dead time. **A.** Dead time reading on a terminal display. **B.** Graph of dead time (orange) and lag time (blue).
CREDIT: A. © Emerson 2019.

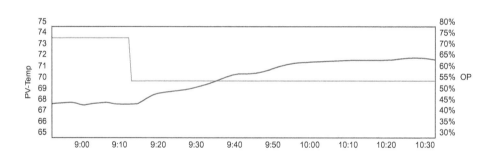

Dead Time

Dead time the elapsed time between the initiation of an input (measured, controlled, or manipulated variable) change, or other stimulus and the point at which the resulting response can be measured or observed.

With respect to a process reaction curve, **dead time** (see Figure 14.3B) is the difference between the time the manipulated variable (MV) starts to change and the time when initiation of the process variable (PV) change is observable and measurable.

Over Range

Over range the condition that exists when the signal value of a device or system exceeds the maximum allowable value.

Over range (Figure 14.4) refers to measurements beyond the maximum under range allowable values. For example, most transmitters can output less than 4 mA and more than 20 mA, even though they have been calibrated to specific process variable between 4 and 20 mA. Over or under range values can occur with controller input and output signals. These values may or may not be displayed on the controller:

- Device range limits (e.g., transmitter ABC may have a range of 0 to 62.5 feet), but the actual tank level is more than 62.5 ft
- Device span limits (e.g., transmitter ABC could be calibrated [spanned] 0 to 30 feet for a 33-foot high storage tank) and the tank level is at 32 ft, 1 ft from the top

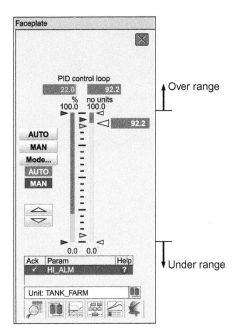

Figure 14.4 Over range and under range.

CREDIT: © Emerson 2019.

Spanning

Spanning (Figure 14.5) describes the process in which the lower range values and range of a device are set or calibrated. Devices or systems (such as transmitters or controllers) can be tested by moving up and down the scale through the entire calibrated range.

Spanning the process of setting or calibrating the lower and upper range values of a device; also the process of testing a device or system by traversing up and down the scale through the entire calibrated or operational range.

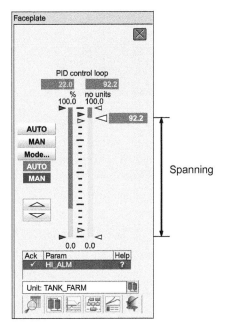

Figure 14.5 Spanning.

CREDIT: © Emerson 2019.

Controllers and Control Actions or Modes

Controllers receive measurements of process variables from instrumentation and send output signals to final control elements (Figures 14.6 and 14.7). Once a physical process (the plant) has been designed and built, the controller is the only device in the system that is capable of matching the dynamics (loop gain) of the process to the control system components. Different brands of controllers may be in use (Emerson, Siemens, Foxboro, Taylor, etc.). However, the principles of operation will be the same.

A controller is an analog device or computer program that can operate automatically in order to control a process variable (PV) based on predetermined conditions.

Figure 14.6 Panel controller.
CREDIT: © Emerson 2019.

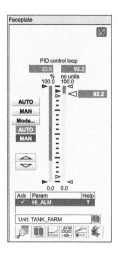

Figure 14.7 Local controller. **A.** Illustration. **B.** Photo.
CREDIT: **B.** Courtesy of Alisha Nash/The Dow Corporation.

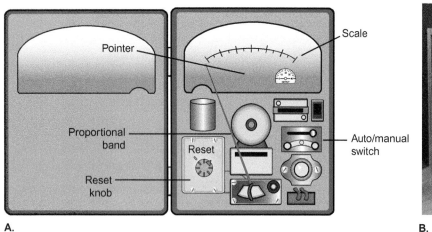

A.

B.

All controllers compare a process variable to a set point (SP), perform some type of calculation, and then produce an output. This output drives a control valve (or other final control element) that manipulates the process in a way that continually returns the measurement to the set point.

Feedback Control

Feedback control a closed loop control strategy where the difference in the set point and measurement drives the output of the controller to the final control element.

Feedback control (Figure 14.8) is a closed loop control strategy where the controller's output to the final control element manipulates a change to the process that is measured and is sent back as input (feedback) to the controller. This control action should continue until (PV) equals (SP). Effectively, feedback control operates to eliminate an existing error between SP and PV.

Figure 14.8 Feedback.

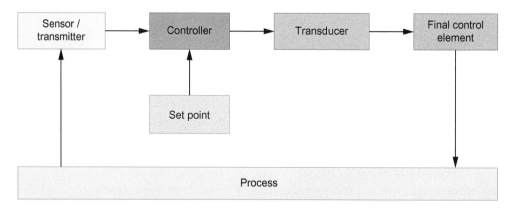

Feedforward Control

Feedforward control anticipates the effect of an uncontrolled disturbance (such as change in steam pressure or change in ambient temperature) and adjusts the output from a controller to a valve based on the magnitude of the change.

For example, an air-cooled reflux condenser may adjust the pitch of the fan blades to control reflux temperature in a distillation column. A feedforward controller monitoring air temperature would adjust the pitch of the blades, based on ambient temperature, to avoid overcooling as the sun sets.

Feedforward control an open loop control strategy in which a disturbance in the process is anticipated, and the controller output is adjusted to compensate for its effect.

Direct or Reverse Acting Controllers

A controller may be selected as *direct acting* or *reverse acting* to counteract the process dynamics. This can be accomplished through programming on newer controllers or by a selector switch on older controllers. All controllers must be set to act in one of these two modes (Figure 14.9). Whether a controller needs to be direct or reverse acting is determined by the configuration of the process.

Figure 14.9 Direct and reverse acting controllers.
CREDIT: © Emerson 2019 (for faceplate).

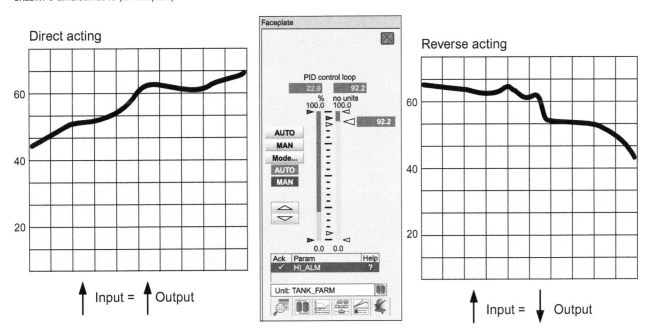

DIRECT ACTING CONTROLLER In a **direct acting controller**, the controller's output signal value increases as its input signal value increases (increased input = increased output). Direct acting also means that decreased input will result in decreased output.

Figure 14.10 illustrates the concept of direct acting controllers. For example, a controller receives an increasing signal as a tank fills above set point. The controller responds by sending an increasing output signal to the drain valve. The drain valve opens incrementally in response.

Direct acting controller a controller whose output signal value increases as the controller input signal value increases or decreases as the controller input value decreases.

REVERSE ACTING CONTROLLER In a **reverse acting controller**, the output signal decreases when the controller's input signal value increases, and the output signal increases when the controller's input signal value decreases. As mentioned, if a positive error increases the control output, the controller is said to be direct acting. In the opposite case, when a positive error *decreases* control output, the controller is said to be a reverse acting controller.

Reverse acting controller a controller whose output signal value decreases as the controller input signal value increases, and vice versa.

Figure 14.10 Direct action.

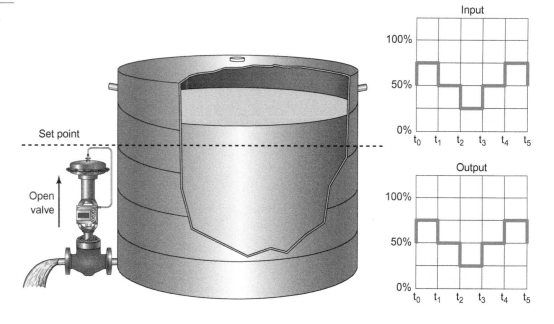

Figure 14.11 illustrates how reverse acting controllers decrease the output signal when the input signal increases. Note that the controller receives an increasing signal as the tank level rises above set point, and the controller responds by sending a decreasing output signal to the fill valve. The fill valve closes incrementally in response.

Figure 14.11 Reverse action.

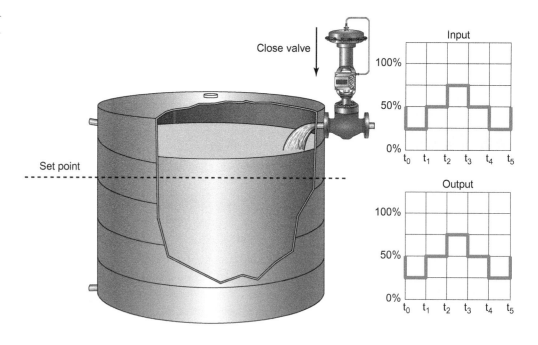

On/Off Control

On/off control may be used when controlled variable cycling above and below the set point is considered acceptable. It may also be used when the process quickly responds to an energy change that does not alter too quickly (Figure 14.12). A common example of on/off control is the temperature control in a domestic heating system.

On/off control a controller that simply drives the manipulated variable from fully closed to fully open, depending on the position of the controlled variable relative to the set point.

Figure 14.12 On/off control.

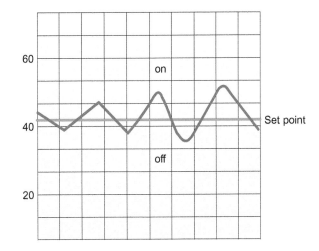

Other Control Terms

Tuning a controller means adjusting the control action settings (tuning parameters) so that they produce an appropriate dynamic response to the process, resulting in good control. Typical controller action settings or tuning parameters are proportional, integral, and derivative (PID). Some programmable controllers have special control algorithms with different tuning parameters.

PROPORTIONAL GAIN Proportional gain is the amount of deviation of the controlled variable from the set point that is required to move the controller's output through its deviated range. The controller output for a standard feedback control scheme will be the signal value that is being transmitted to the final control element.

PROPORTIONAL ACTION Proportional action balances the amount of deviation of the controlled variable from the set point. The proportional action's contribution to the controller's output responds once to the deviation of PV from the set point (SP). It acts in the present time, usually only once.

In other words, the question with proportional gain is: "How big is the error?" There is a reciprocal relationship between proportional band (PB) and gain:

$$100\%/PB = \text{Gain} \qquad \text{Gain } 100\%/\text{Gain} = PB$$

where PB is given in percent (%), and gain is a decimal number without units.

In other words, the wider the proportional band, the lower the gain.

Proportional action a controller output response that is proportional to the amount of deviation of the controlled variable from the set point.

INTEGRAL ACTION (RESET) Integral action (reset) is a controller output response that is proportional to the length of time the controlled variable has been away from the set point and that repeats the proportional action a given number of times, acting on the past error. The integral action's contribution to the controller's output is the controller's gain divided by the integral time multiplied by the sum of the (PV − SP) error. Integral action asks: "How long has the error existed?" and responds to the answer.

WINDUP Windup is the condition that occurs when the controller output signal reaches its minimum or maximum value and the process variable does not return to the set point. The elapsed time that it takes for this to occur will be based on the process characteristics and the gain and integral or reset modes of the controller. In many controllers, anti-reset windup parameters can be entered to prevent the output from going to a maximum or minimum.

Windup the saturation of the controller output signal to its minimum or maximum value if the process variable does not return to the set point.

DERIVATIVE ACTION (RATE, PREACT) Derivative action (rate, preact) is a controller output response that is proportional to the rate at which the controlled variable deviates from the set point. In other words, derivative action asks: "How fast is the process variable changing?" (Figures 14.13 and 14.14.)

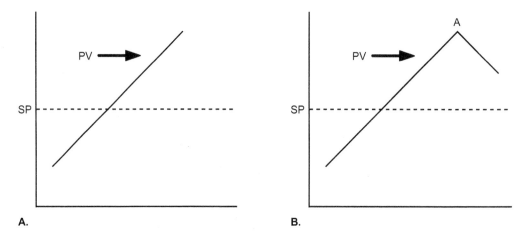

Figure 14.13 A. Derivative action (PV constant). **B.** Derivative action (PV changed).

Figure 14.14 Derivative action.
CREDIT: © Emerson 2019 (for faceplate).

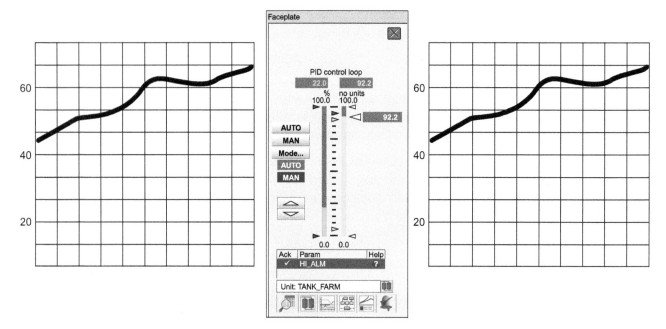

Derivative action can be compared to cruise control, with the selected speed as the set point. Cruise control adjusts the RPM to maintain the designated setting. When you go uphill in cruise control, the RPMs increase as the speed gets further from set point. The RPMs are reduced as speed gets closer to the set point so that it does not overshoot.

This control action acts strictly on the rate of change that the PV deviates from SP. Note that derivative does not act if the process technician makes a set point change. That would be a rate of change to the set point. This is shown on the graph in Figure 14.13A.

14.2 Process Recording

Chart recorders or historical files in a computer system provide both operations and maintenance with an important troubleshooting tool (Figure 14.15). Process technicians use chart-recorded data to spot both good and bad process trends. Instrument technicians use these

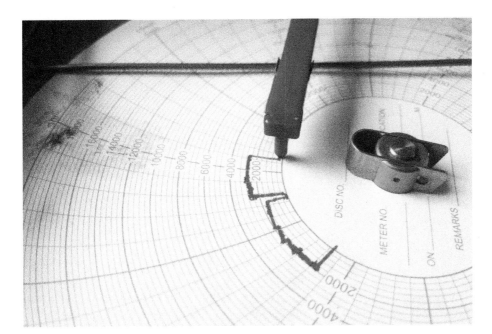

Figure 14.15 Chart recorder.
CREDIT: ABUN5M/Shutterstock.

trends to help troubleshoot problems in the instrumentation control loop and to calculate controller tuning parameters. The following are a few of the trends that can be observed through chart recordings.

Offset

When the set point and the controlled variable are in steady state (nonchanging) conditions, the difference between them is called the **offset**. Offset of the control variable under dynamic (changing) conditions is sometimes referred to as **deviation**. Figure 14.16 shows an example of offset.

Offset the difference between the set point and the controlled variable under steady state (not changing) conditions.

Deviation the difference between the set point and the controlled variable under dynamic (changing) conditions.

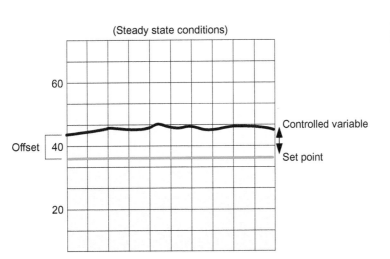

Figure 14.16 Offset.

Trend

Trend is the direction in which a process variable can be plotted over time. In a distributed control system (DCS), any output or input, an alarm, or a calculated value can be *trended* (watched and plotted over time to determine its state or direction). Trends can be recorded on media such as a hard drive, a tape drive, or an optical platter and can be displayed in different increments (e.g., hours, minutes, days). Figure 14.17 shows an example of a trend.

Trend the plotting of a process variable over time.

Figure 14.17 Trend.

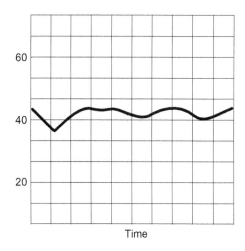

Hunting

Hunting a condition that exists when a control loop is improperly designed, installed, or calibrated, or when the controller itself is not properly aligned; overcorrection above and/or below the set point.

If a signal causes a system to overcorrect itself repeatedly, a phenomenon called **hunting** may occur. In hunting, the system first overcorrects itself in one direction and then overcorrects itself in the opposite direction. Because hunting is undesirable, measures are usually taken to correct it. The expression *hunting* came into use in the 19th century and describes how a system "hunts" for equilibrium. Figure 14.18 shows an example of hunting.

Figure 14.18 Hunting.

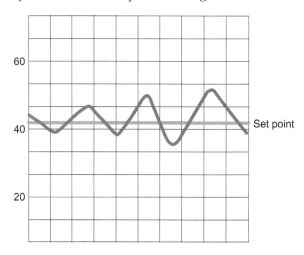

Steady State

Steady state a trend condition in which the process is in equilibrium and running smoothly, producing a relatively straight line on a graph or chart.

Steady state is sometimes used to describe the trend condition in which a process is in equilibrium and running smoothly. When a process is in steady state, chart recordings of the process show a relatively straight line. Figure 14.19 shows an example of steady state.

Figure 14.19 Steady state.

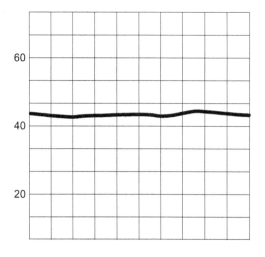

14.3 Tuning Modes (Gain, Reset, and Rate)

Control may be exercised using one mode or a combination of two or three modes: gain, reset, and rate. (*Note*: These three modes may also be referred to as proportional, integral, and derivative [PID] as noted earlier in the chapter). These modes must be selected and set correctly for effective process control. This is called *tuning a control loop*.

Tuning seeks to optimize process control in automatic mode much the same way a process technician would seek to optimize process control in manual mode. The following sections on gain, reset, rate, and PID illustrate this point.

Gain (Proportional Band) Mode

The required response to correct offset or deviation from set point may not be the same percentage as the percentage of change. The proper amount of correction may require some factor of the amount of deviation (e.g., a deviation may require a multiplier). This may be accomplished by introducing gain into the controller.

In order to change input or output to correct a deviation using the proportional band setting, consider the example shown in Figure 14.20. If the level in a particular tank drops 10 percent below set point and remains there, it may be necessary for the outlet valve to close 20 percent to compensate. Experience dictates that a 20 percent output change will correct a 10 percent offset.

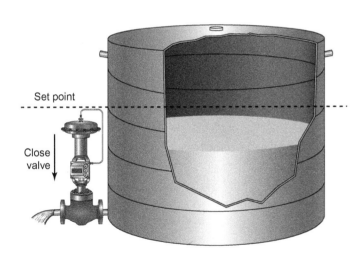
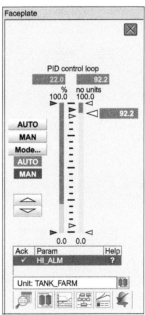

Figure 14.20 Proportional correction.

Using the same example, if a controller is placed in automatic mode, the same response ratio can be obtained by having a gain setting of 2. In this case, a 10 percent drop in tank level will result in a 20 percent closing of the drain valve (Figure 14.21). Thus, gain consists of an established ratio between the process deviation and the controller response.

If the percentage response required is exactly equal to the percentage change in the measured variable, a gain of 1 would be established (Figure 14.22). If the drain valve only needs to close 5 percent in response to a 10 percent drop in tank level, a gain set at 0.5 would accomplish this (Figure 14.23).

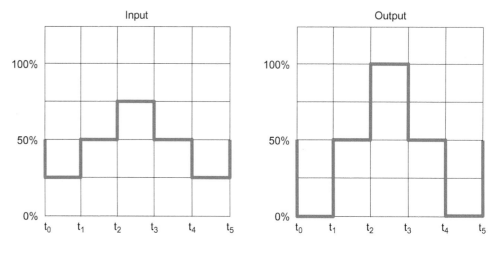

Figure 14.21 Gain of 2.

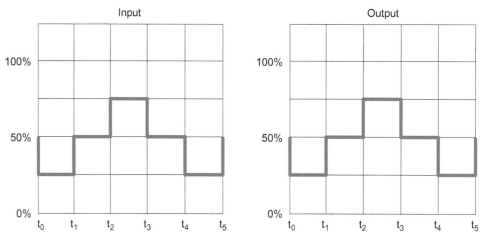

Figure 14.22 Gain of 1.

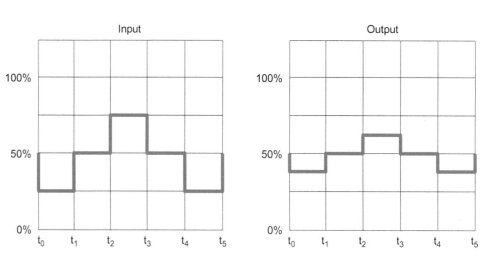

Figure 14.23 Gain of 0.5.

The optimum gain setting may depend upon several varying factors. If the gain is set too high, the controller will overcompensate for each deviation detected, and the process will begin to cycle (Figure 14.24).

If the gain is set too low, the controller will not provide enough correction to reach set point, and a constant offset will result (Figure 14.25). However, even a very careful gain setting is unlikely to maintain a process that is at a steady state at set point.

Suppose that gain has been set so accurately for a given process under a particular set of conditions that set point is maintained. What would happen when the process conditions change? Let us consider what happens when the level in a tank is maintained at set point by a controller with a gain of 1 (Figure 14.26).

Figure 14.24 Cycling.

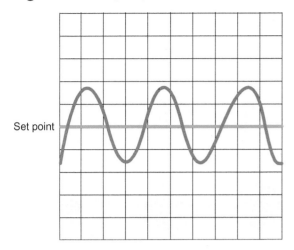

Figure 14.25 Offset.

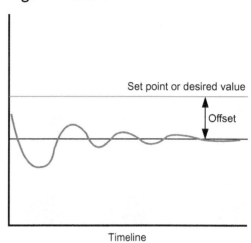

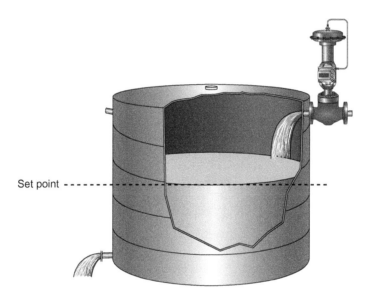

Figure 14.26 Process stable at set point.

Then, let us suppose that the downstream demand increases 20 percent, causing the tank level to begin to fall (Figure 14.27). In response to the dropping tank level, the inlet valve begins to open at the rate set by gain.

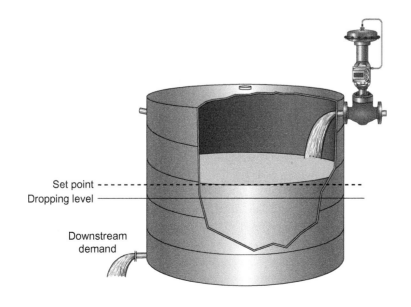

Figure 14.27 Twenty percent increase in downstream demand.

At some point, the inlet valve will have opened enough to compensate for the increased downstream demand, and the tank level will stop falling.

Suppose the inlet flow equals the outlet flow when the tank level has dropped 10 percent. Only the continued offset of tank level at 10 percent below set point will keep the inlet valve open the additional 10 percent necessary to compensate for the increased downstream demand (Figure 14.28).

Figure 14.28 Process stabilizes with an offset.

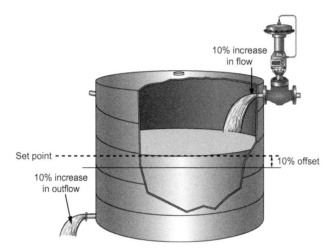

As this example illustrates, any change in product flow will render gain inadequate. In the real world, product flow is always fluctuating. This is a major reason why gain is rarely used alone. Gain only provides a ratio response to a deviation. Nothing in the gain mode will induce the controller to seek set point. So, reset mode is applied along with gain to solve this problem.

Reset (Integral) Mode

Gain is seldom used alone as a control mode since this mode is only capable of providing a ratio response to a deviation. It has no capacity to adjust an overcorrection or an undercorrection automatically. For this reason, most controllers utilize gain and reset together. Gain puts the correction closer to where it should be, while reset fine tunes the correction until the measured variable returns to set point.

Before discussing how reset mode works, consider how a process technician would react to continual offset using manual mode.

Suppose a process technician is monitoring a tank and notices the product level is continually below an acceptable level. In manual mode, the process technician gently bumps the inlet valve open until the product stabilizes near the desired level.

The controller in automatic mode uses the reset function to accomplish this same result. Reset opens the valve at a steady rate until set point is reached. But how rapidly should the valve be opened? Throwing the valve open quickly to 100 percent would likely induce cycling.

In order to establish the correction necessary to regain set point quickly without overshooting and cycling, the reset mode calculates a slope to draw the process variable up to set point. This slope is defined by gain across time. Typically, reset is defined in repeats per minute and refers to the gain correction command.

Consider a tank with a direct action level controller set with a gain of 1 and a reset of 1 minute. The level in the tank rises 10 percent above set point, resulting in a 10 percent increase in signal to the controller. The controller establishes a correction slope of 10 percent per minute. This means the controller signals the inlet valve to close at 10 percent per minute. The controller continues to decrease the valve opening at the rate of actual deviation multiplied by gain until the process has regained set point. The chart in Figure 14.29 illustrates how reset corrects an offset if the deviation remains constant.

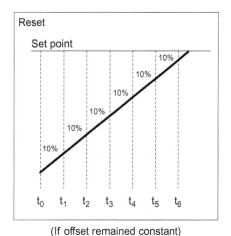

Figure 14.29 Reset (correcting a level deviation using reset).

Rate (Derivative or Preact) Mode

The third tuning mode, rate, is used far less frequently than gain or reset. When it used, it is most commonly used for temperature related applications.

To understand rate, examine the response necessary to correct a dramatic temperature drop in a vessel. How would a process technician respond to this deviation using manual mode?

The process technician would have to draw the measurement back to set point as quickly as possible without dramatically overshooting it. How much should the process technician increase the heat input to achieve this result? To estimate the optimum valve opening, the process technician will take into account how fast the temperature is dropping, as well as how far it has dropped. If the temperature is falling quickly, the process technician will most likely open the valve more than if the temperature is falling slowly.

The controller utilizes rate to perform the same operation in automatic mode. With the rate setting for the controller set at 5 minutes, the controller calculates the deviation from set point that will occur in 5 minutes if the process temperature is allowed to fall at the existing rate. The controller then makes a corrective response based on this calculation.

Consider this example. The temperature in a vessel begins to drop at a rate of 10 percent per minute. In response, the controller calculates how far the temperature could be expected to drop in 5 minutes at the present rate. A 10 percent drop for 5 minutes would result in a 50 percent total drop (Figure 14.30).

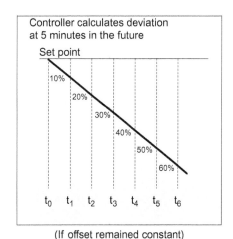

Figure 14.30 Rate calculation.

Based on this calculation, the controller immediately signals a 50 percent increase in heat input. This results in a quick, hard reversal back toward set point (Figure 14.31).

Rate calculates the correction based on an estimated future deviation. (In this example, it is how far from the set point the measurement will have fallen 5 minutes in the future.) Because of this, rate is often referred to as *preact* (*pre = before*).

Figure 14.31 Rate correction.

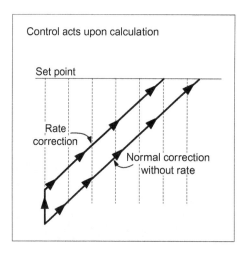

At what value should rate be set? This depends on a number of factors. A rate that is set too low will not optimize the time necessary to return to set point. An excessive rate setting could force the temperature up too hard, resulting in a large overshoot of set point and consequent cycling.

PID

To repeat, the three controller modes discussed above are sometimes referred to as proportional, integral, and derivative (PID):

- Proportional is the same as gain.
- Integral is the same as reset.
- Derivative is the same as rate.

Proportional control refers to *proportional band*, a term used to explain the ratio between the input and the output of a controller. Proportional band is a process technician's way of explaining how much offset is necessary between a process measurement and set point to move a controller across its entire range of measurement.

Imagine the two-headed arrow shown in Figure 14.32 is a solid bar supported in the middle by a fulcrum (like a seesaw). The left head of the arrow represents the input signal into a controller. The right arrowhead represents the output signal from a controller.

Figure 14.32 Proportional band showing a gain of 1.

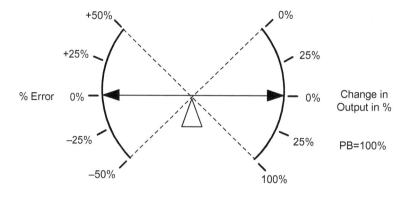

A 100% change in output is caused by a 100% change in error. The proportional band is therefore 100% or the gain is 1.

In this example, moving the right head of the arrow from 0 percent to 100 percent requires movement of the left head of the arrow from −50 percent to +50 percent. This movement (from −50 percent to +50 percent) equates to travel of 100 percent. In this case, the proportional band is said to be equal to 100 percent. Also, because an increase in the input signal results in an equal (one to one) increase in the output signal, the gain is said to be 1.

In the example in Figure 14.33, the fulcrum has been moved to the left. In this instance, the left arrow only needs to move half the range of the previous example to accomplish the same result. In essence, it only needs to travel from −25 percent to +25 percent to move the right arrow the full range of 0 percent to 100 percent. In this case, the proportional band is said to be only 50 percent. Also, because movement of the left arrow results in a movement of twice the distance by the right arrow (one to two), the gain is said to be 2.

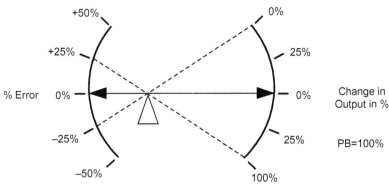

Figure 14.33 Proportional gain of 2.

A 100% change in output is caused by a 50% change in error. The proportional band is therefore 50% or the gain is 2.

The relationship between proportional band and gain can also be illustrated with the following chart (Figure 14.34).

Figure 14.34 Relationship between proportional band and gain.

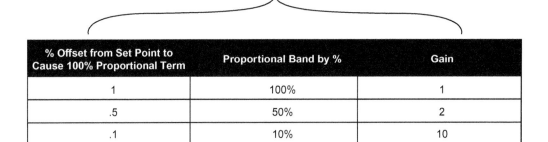

% Offset from Set Point to Cause 100% Proportional Term	Proportional Band by %	Gain
1	100%	1
.5	50%	2
.1	10%	10

Combined Control Action

It has already been shown that offset is inevitable with proportional-only control. However, some controls are available that reduce offset. A proportional plus integral (P+I) control reacts to the size of the change in the process variable. Proportional plus derivative (P+D) control detects the speed with which the deviation occurs and speeds up or slows down the valve reaction accordingly.

When used, derivative action is usually combined with a proportional plus integral control to give P+I+D, the most sophisticated control of all. Integral removes the offset while derivative action speeds the response. Adjustment of both the integral and derivative is provided on most combined controls. Figure 14.35 illustrates each of the various types of control (P, P+I, P+D, and P+I+D).

Figure 14.35 Examples of deviation from set point.

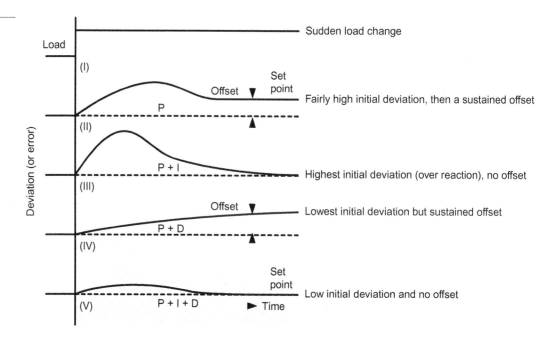

Troubleshooting

Tuning errors can create process problems. However, a process technician should never assume that cycling or some other problem is caused by a poorly tuned loop. A number of symptoms that appear to be tuning related can also be caused by other components of the control loop. Cycling and offset are two of those examples.

Cycling

Cycling the moving or shifting of a process variable above and/or below the set point.

Cycling Tuning Problem. Earlier in this chapter, it was stated that **cycling** can be caused by tuning errors, such as the following:

- If the gain is set too high, the controller will overreact to measurement deviation.
- If the time increment for reset is too small, the controller will also overreact, establishing a correction slope that is too steep. Overshoot and cycling will result.
- If the time increment for rate is too large, the controller may also exaggerate the initial reversal toward set point, also resulting in overshoot and cycling.

Cycling Process Problem. Cycling can also be caused by a process problem such as the following:

- A sticking valve (the valve may fail to react until the process has already deviated dramatically)
- Pump cavitations
- Faulty wiring (can result in an intermittent signal transmission, which leads to cycling)

Offset

Continual offset can be caused by a gain that is set too low. It can also be caused by a malfunctioning final control element (e.g., a valve that is unable to open fully and that prevents the process from reaching set point).

If a controller exhibits cycling or a continual offset, do not assume a tuning problem. Try correcting the problem in manual mode. If the problem persists in manual mode, tuning is not the culprit. In this case, the process technician needs to try to isolate which element in the control loop could be at fault. The process technician should only conclude that a loop needs to be tuned if it is determined that control can be reestablished in manual mode.

Summary

Processes have certain behaviors that are common and predictable. During states of change, a process may experience load change, dead time, spanning, or may be over range.

During processes, controllers receive measurements of process variables from instrumentation and produce output signals to final control elements. In any given process, there may be several types of control including feedback, feedforward, and on/off control.

In order to produce appropriate dynamic responses to a process and maintain good control, controllers must be tuned. When tuning a controller, it is important to take into consideration proportional gain, proportional action, integral action, windup, and derivative action (rate, preact).

When running a process, it is important to use process recording instrumentation (chart recorders) to document process trends. By recording process trends, process technicians and instrumentation technicians are better able to spot both good and bad process trends and troubleshoot problems.

When examining recorded process data, there are several things a technician might monitor. These include offset, hunting, cycling, and steady states. In addition, technicians might also examine gain, and reset rate (derivative or preact). All this information helps technicians determine if the instruments are correctly tuned and if everything is working properly.

Checking Your Knowledge

1. Match the following process dynamics terms with their correct definitions.

Term	Definition
I. Dead time	a. The condition that exists when the signal value of a device or system exceeds the maximum measurable value.
II. Load change	b. The process of setting or calibrating the lower and upper range values of a device, or testing a device or system by traversing up and down the scale through the entire calibrated or operational range.
III. Over range	c. A change in any variable in the process that affects the value or state of the controlled variable.
IV. Process equilibrium	d. The elapsed time between the initiation of an input and the point at which the resulting response can be measured or observed.
V. Spanning	e. The condition that exists when there is a balance of material and energy within a given system or process. This condition is also referred to as *steady state*.

2. Feedback control operates to eliminate an existing error between SP and _____.

   a. D/C
   b. PV
   c. span
   d. the controller

3. Which type of controller will supply a decreased output when the input is increased?

   a. Direct acting
   b. Reverse acting
   c. On/off controller
   d. Feedback controller

4. Match the following process control terms with their correct definitions.

Term	Definition
I. Proportional gain	a. A condition that exists when a control loop is improperly designed, installed, or calibrated, or when the controller itself is not properly aligned. This condition will be observed by the process technician as a random behavior or cycling above and/or below the set point (desired control point).
II. Windup	b. A trend condition in which the process is in equilibrium and running smoothly producing a relatively straight line.
III. Offset	c. The amount of deviation of the controlled variable from the set point required to move the controller's output through its entire range (expressed as 10 percent of span).
IV. Tuning	d. A plot of a process variable over time.
V. Trend	e. The difference between the set point and the controlled variable under steady state (not changing) conditions.
VI. Hunting	f. Setting the controller's parameters of gain, reset, and rate (PID) to optimize the control of the process variable.
VII. Steady state	g. The saturation of the controller output signal to its minimum or maximum value if the process variable does not return to the set point.

5. Consider a tank with a direct action level controller set with a gain of 1 and a reset of 1 minute. The level in the tank rises 20 percent above setpoint, resulting in a 20 percent increase in signal to the controller. The controller establishes a correction slope of _____ percent per

   a. 5
   b. 10
   c. 20
   d. 30

6. Which of the following situations will cause a controller to overreact? (Select all that apply.)

   a. The gain is set too low.
   b. The gain is set too high.
   c. The time increment for rate is too small.
   d. The time increment for reset is too small.

NOTE: Answers to Checking Your Knowledge questions are in the Appendix.

Student Activities

1. Given a set of trend charts, break into small groups and analyze the trend charts with respect to deviation and offset. Determine instances deviation and offset are clearly visible. Report your observations and conclusions to the class.

2. Given a computer program or process variable simulator, demonstrate the different behaviors of pressure, temperature, level, or flow control systems. Focus the discussion and lab activity on the resistance to change and the capacitive element of each system that you describe.

3. Given an analog pneumatic controller, perform a controller calibration and alignment procedure.

4. Given an analog electronic controller, perform a controller calibration and alignment procedure.

5. Given a single loop digital controller, perform a controller calibration and configuration procedure.

6. Given a control loop or a simulation of a control loop, conduct tuning operations and tune the loop.

7. Given a computer program or a process simulator, investigate the concepts of proportional band, proportional action, and proportional gain.

8. Given a computer program or a process simulator, investigate the concepts of integral action or reset.

9. Given a computer program or a process simulator, investigate the concepts of derivative or rate action. Note: For clarity and understanding, it is recommended that only gain, gain and integral, and gain and derivative be used to reinforce this session.

10. Given a drawing, picture, or actual trend, distinguish between the effects of proportional, integral, and derivative control. To do this: (a) study a chart with a measured variable plotted against time; (b) work in small groups to analyze the trend charts with respect to controller mode influences; (c) discuss with the group what you think is causing the trend to be the way it is; and (d) report your group's conclusions to the class.

Chapter 15
Control Schemes

Objectives

After completing this chapter, you will be able to:

15.1 Identify and describe types of control schemes:
on/off
lead/lag
feedback
feedforward (NAPTA Control Loops: Controllers 1*) p. 266

15.2 Describe a control scheme that allows the following modes:
local manual control
local automatic control
remote manual control
remote automatic control
cascading of the remote automatic control (NAPTA Control Loops: Controllers 2, 3) p. 268

Key Terms

Cascade control—employing two controllers so one process variable is controlled by controlling another; the output of one controller is the remote set point of another, **p. 270**.
Lead/lag control—the manner in which a process reacts to a disturbance or other manipulating conditions, **p. 266**.
Local automatic control—the act of controlling an instrument loop by automated means within the processing area, **p. 269**.

---

*North American Process Technology Alliance (NAPTA) developed curriculum to ensure that Process Technology courses will produce knowledgeable graduates to become entry-level employees in process technology. Objectives from that curriculum are named here in abbreviated form. For example, "(NAPTA Control Loops: Controllers 2)" means that this chapter's objective 2 relates to objective 2 of NAPTA's course content on controllers in control loops.

Local manual control—the act of controlling a process variable by hand within the processing area, **p. 268.**

Remote automatic control—the act of controlling an instrument loop remotely from a control room, **p. 270.**

Remote manual control—the act of controlling a valve manually from a remote location, such as a control room, **p. 270.**

15.1 Introduction

This chapter focuses on several different types of controls and their applications. These different control types fall into two categories: control schemes and control modes. *Scheme* refers to the approach used to control a particular process variable (PV). Different process problems require different approaches for optimal control.

Mode refers to the agent or location of control (e.g., manual versus automatic or local versus remote). The term *mode* can be confusing. You may see *control modes* or *control actions* such as proportional (P), integral (I), and derivative (D) described as a three-mode control or a PID controller. On modern programmable controllers, controller modes may be described as "Man" (manual), "Auto" (automatic), "RSP" (remote set point), or "Cas" (cascading).

Types of Control Schemes

Process control involves many different types of control schemes. Some of the most common types of control schemes include on/off, lead/lag, feedback, and feedforward.

On/Off Control

On/off control may be used when a controlled variable cycling above and below the set point value (SV) is considered acceptable. It occurs when a final control element is moved from one of these two fixed positions to the other with a small change of the controlled variable above or below the set point.

On/off control is the most common type of control used in our homes. Ovens, central air and heat, and water heaters all use on/off control schemes to operate. In industry, on/off control is used in places where a fairly large control range is not a problem. For example, a condensate collection vessel or other knockout pot typically uses two points for control: a high-level point and low-level point. At the high-level point, a pump is turned on to move the liquid out of the tank into another tank. At the low-level point, the pump is turned off.

Lead/Lag Control

Lead/lag is a broadly used term in the processing industry that describes how a process reacts to a disturbance (e.g., a process reactor temperature change leads to a corresponding pressure change) or other manipulating conditions (e.g., a measurement change lags a controller output change).

Lead/lag is also used to describe how a control mode action reacts to an error (rate = lead; reset = lag), or which measured variable will be used to control a loop. It is even used to describe equipment utilization strategies.

Lead/lag control the manner in which a process reacts to a disturbance or other manipulating conditions.

EXAMPLES OF LEAD/LAG CONTROL In a boiler system, **lead/lag control** specifically means to utilize both high and low selector relays to force the fuel to follow the airflow on a rising steam demand and to force the airflow to follow the fuel on a falling steam demand (Figure 15.1).

Figure 15.1 Boiler with lead/lag control.

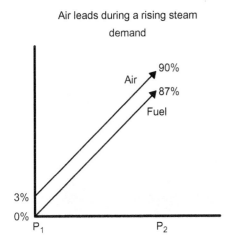
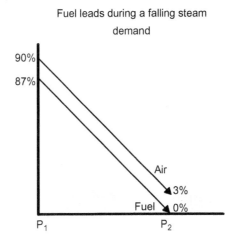

Another lead/lag controller circuit is used to swap the lead role between two equal parallel components, such as evaporators in a refrigerator system like the one shown in Figure 15.2. This is done to provide a mechanism that will ensure equal wear to the two components. It allows the lagging component to be used in a supplementary capacity in the event of an unusually high demand cycle.

Figure 15.2 Cooling installation units (chillers and RH [relative humidity] unit).

CREDIT: Lightpoet/Shutterstock.

Feedback Control

Feedback control is a closed loop control strategy where the difference in the set point and measurement (called error) drives the controller output to change the position of the final control element to make PV = SP. There must be a difference between set point and measurement for this control action to operate.

Feedback control is designed to eliminate or minimize error. The fundamental concept that truly defines this control strategy is the feeding back of the measured variable to the controller, which adjusts the process to meet the set point.

Feedforward Control

Feedforward control is an open loop control strategy where the controller output is based on knowledge (usually mathematical modeling) of the relationship between the output of the controller and the input received from the point in the process where the disturbance is measured (Figure 15.3). In feedforward control, a disturbance in the process is anticipated, and the controller output is adjusted to compensate for its effect.

Figure 15.3 Feedforward control example.

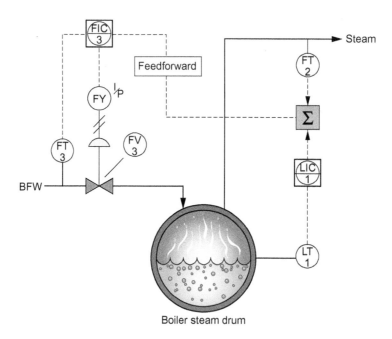

Feedforward control is designed to prevent an error from occurring in the first place by changing the position of the final control element before the measured variable senses the change.

The fundamental concept that defines this control strategy is the open loop aspect of the measured variable in relationship to the controller. The controller output is changed by the magnitude of the disturbance rather than just the process variable of interest.

In many industrial applications, feedforward and feedback control are used together. In high-pressure steam production (boiler), the steam demand signal is the feedforward input that is added to the boiler level controller output, which is the remote set point to the boiler feed water controller.

15.2 Controller Modes

Controller mode refers to the agent and location of control. There is a difference between controller mode and control mode. The term *controller mode* refers to the status of a controller and whether it is located adjacent to (local) or removed from (remote) the process equipment area. As discussed in Chapter 14, *Controllers*, the *control mode*, or action, refers to the proportional, integral, and derivative components or any other control algorithms of a controller.

Local Control (Field-Mounted)

In local control, the controllers are physically located in a process unit near a vessel, control valve, or transmitter. Figure 15.4 shows an example of a local controller.

Local Manual

Local manual control the act of controlling a process variable by hand within the processing area.

Local manual control involves a local controller in manual or the act of controlling a process variable by manually operating a valve within the processing area. For example, a cooling tower may require the addition of makeup water only one time per shift or less. If a process

Control Schemes **269**

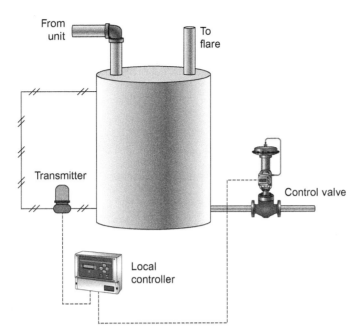

Figure 15.4 Local controller.

technician performs this duty by observing the basin water level and then regulating the makeup water valve by hand, this constitutes local manual control.

Local Automatic

With **local automatic control**, the automatic controller is also located in the processing area. An example is a steam letdown station consisting of a control valve and a pressure controller (integral mount) used to drop the main header steam pressure to a lesser pressure, making it appropriate for the steam requirements of a processing area.

Another example of local automatic control is large cooling towers that have a local control loop which regulates the level in the water basin automatically. If the controller is located in the process area, then it is a local automatic controller.

Local automatic control the act of controlling an instrument loop by automated means within the processing area.

Remote Control

In remote control, controllers are physically located outside of the process unit or in another remote location such as a control room, analyzer shack, distributed control system (DCS) monitor, or programmable logic controller (PLC) monitor. Figure 15.5 shows an example of a remote controller.

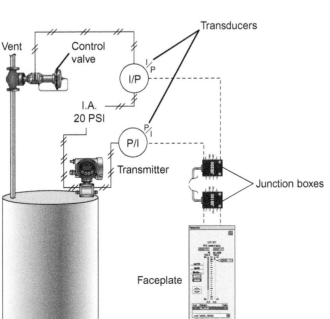

Figure 15.5 Remote controller with junction boxes.

Remote Manual

Remote manual control is typically done with the manual control component located on the front of an automatic controller. This component allows the process technician to have manual control over the final control element from the automatic controller in the control room (Figure 15.6).

Remote manual control the act of controlling a valve manually from a remote location, such as a control room.

Figure 15.6 A. Local setting. **B.** Remote setting.

CREDIT: © Emerson 2019.

A. B.

Sometimes, a device called a manual loading station is found as a stand-alone unit with an indicator. A manual loading station is capable of actuating a control valve from a remote location but does not have the ability to be switched into automatic control.

Remote Automatic Control

Controlling an instrument loop remotely from a control room is called **remote automatic control**. Most controllers in the processing industry are remotely located. A remote automatic controller is defined as a controller that is located away from the processing area, usually in a centralized control room, and that responds without direct intervention by the technician. Soon after the development of pneumatic transmitters, local controllers were moved to centralized control rooms where they could be monitored more efficiently.

Remote automatic control the act of controlling an instrument loop remotely from a control room.

Cascading of the Remote Automatic Control

In processes that exhibit excessive lag times or where the process variable has particularly tight constraints, it becomes necessary to control one process variable by controlling another. The control scheme that employs two controllers, where the output of one controller is the remote set point of another (e.g., a temperature controller output is the set point for a steam flow controller or slave) is called **cascade control**. An example of cascade control is shown in Figure 15.7. Chapter 16, *Advanced Control Schemes*, has more detail about cascading control.

As with most modern controllers, most cascaded controllers are remotely located away from the processing area.

The two cascaded controllers have controller modes as follows:

- Primary = AUTO/MAN
- Secondary = CAS/AUTO/MAN

Cascade control employing two controllers so one process variable is controlled by controlling another; the output of one controller is the remote set point of another.

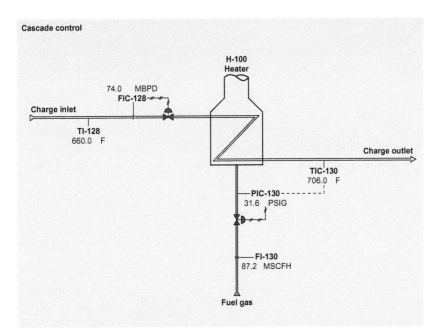

Figure 15.7 Cascading controller.
CREDIT: Simulation Solutions.

Summary

The control schemes selected to regulate the plant processes take into account many requirements such as quality, safety, economics, precision, production rates, and other factors. All plant processes are affected by many variables such as load changes, pressure swings, feedstock stream composition, and temperature variation. Selection of an appropriate control scheme is based on the acceptable level of process variability and the cost associated with reducing process fluctuations. Typically, precision control requirements mandate a more complex control scheme with more measurements and higher associated costs.

The impact of variability on the process should be the primary consideration when selecting a control scheme, sensors, and final control elements for each area of the process. If the impact is extremely minimal, then a simple on/off control may suffice. If the desired variability is somewhat less, then a feedback controller may be adequate. If precise control of a variable is mandated, then extra control hardware and combinations of feedback, feedforward, and/or lead/lag control may be implemented to minimize the effects of process disturbances.

The mode of a control scheme refers to the agent (manual versus automatic or cascade) and location of control (local versus remote). Local or field-mounted controllers are located within, or in close proximity to, the processing area. Remote controllers are installed in a location some distance away from the processing area, typically in a control room.

Most automatic controllers, whether local or remote, may be operated in multiple modes, such as manual, automatic, or cascade. The manual mode is usually used when process conditions are abnormal and a fixed position for the final control element is highly desirable. The automatic mode allows a controller set point to be entered and achieved without further manual intervention. The cascade mode requires at least two controllers and two measured variables. The primary controller input is the slower changing variable and the primary controller's output is the set point for the secondary controller. The secondary controller controls the faster changing variable, ultimately keeping the primary controller input variable on set point.

Checking Your Knowledge

1. A reactor's catalyst feed is manipulated due to a predicted load change. This is an example of:
   a. On/off control
   b. Feedback control
   c. Feedforward control
   d. Lead/lag control

2. A valve opens when the level in a tank exceeds the set point, and stays open until the level drops below that set point. This is an example of:
   a. On/off control
   b. Feedback control
   c. Feedforward control
   d. Lead/lag control

3. Hot oil flow to a heat exchanger increases when the furnace temperature drops. This is an example of:
   a. On/off control
   b. Feedback control
   c. Feedforward control
   d. Lead/lag control

4. A tank in a processing unit (see image below) needs to have an automatic control system installed to control its level. Manipulating a control valve on the outlet flow of the vessel controls the level. The level must be controlled at 93 percent with an acceptable error of plus or minus 2 percent.

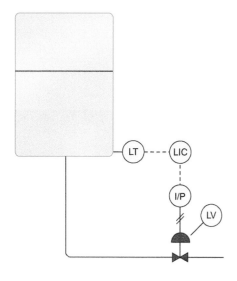

The image for this question is an example of:
   a. A feedback control loop
   b. A feedforward control loop

5. Which of the following is an example of a local automatic control? (Select all that apply.)
   a. Large cooling towers with a local control loop
   b. A process facility with a manual loading station
   c. A facility with a controller away from the processing area
   d. A steam letdown station with a control valve and a pressure controller

6. How is the primary mode of a cascading controller identified?
   a. AUTO/MAN
   b. CAS/AUTO/MAN
   c. CAS/MAN/PR
   d. PR/CAS/AUTO/MAN

NOTE: Answers to Checking Your Knowledge questions are in the Appendix.

Student Activities

1. Given a description of a process, draw a graphic that represents the control scheme used in the process.

2. Given a control scheme scenario and illustrations, identify the appropriated control scheme. Discuss your reasons for selecting this scheme.

3. Given proper equipment:
   a. Manipulate a process variable using automatic mode and then using manual mode.
   b. Change the set point in automatic mode and observe the variable measurement change to match the new value.
   c. Change the control mode to manual and return the variable measurement to the original value.
   d. Draw a P&ID describing the loop.

4. A tank in a processing unit needs to have an automatic control system installed to control its level. Manipulating a control valve on the outlet flow of the vessel controls the level. The level cannot tolerate even a 1 percent change. A single pipe emptying into the tank causes the only variable that can affect its level. Explain how feedforward can be used to accomplish this task?

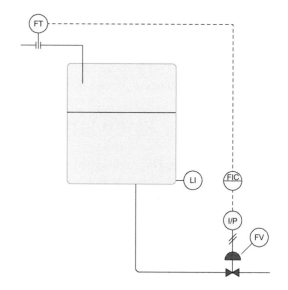

Chapter 16
Advanced Control Schemes

Objectives

After completing this chapter, you will be able to:

16.1 Contrast the basic rationale for advanced control schemes versus simple control loops. (NAPTA Control Loops: Controllers 1*) p. 274

16.2 Explain the purpose of a cascaded control scheme. (NAPTA Control Loops: Controllers 2, 3) p. 274

16.3 Explain the function of a ratio (fractional) control scheme. (NAPTA Control Loops: Controllers 2, 3) p. 278

16.4 Explain the purpose and function of a split-range control scheme. (NAPTA Control Loops: Controllers 2, 3) p. 282

16.5 Explain the purpose and function of a multivariable control scheme and steps required to change instrument controllers without bumping the process. (NAPTA Control Loops: Controllers 4–8) p. 283

Key Terms

Multivariable input—industrial processes that involve an interaction between two or more process variables, **p. 283.**

Override controller—a device or program that remains inactive until a specific constraint (highest or lowest permitted extreme) on the measured variable is about to be reached, **p. 285.**

Ratio—the proportion of two separate elements, such as the proportion of flow between two separate streams entering a mixing point, **p. 278.**

---

*North American Process Technology Alliance (NAPTA) developed curriculum to ensure that Process Technology courses will produce knowledgeable graduates to become entry-level employees in process technology. Objectives from that curriculum are named here in abbreviated form. For example, "(NAPTA Control Loops: Controllers 2, 3)" means that this chapter's objective 2 relates to objective 2 and 3 of NAPTA's course content on controllers in control loops.

Ratio control—control used in blending raw materials to make a final product and in regulating the correct proportion of flow rates of reactants feeding into a reactor; also a term used to describe maintaining the proper fuel-air ratio feeding into a furnace, **p. 278**.

Ratio control loop—control loop designed to mix two or more flowing streams together while maintaining a quantitative ratio between them, **p. 278**.

16.1 Introduction

Although single-loop, single-variable feedback controls are common, many applications require more complicated systems. A single-loop controller operates on the equilibrium theory that a control device will be configured to manipulate a variable to bring the controlled variable to a predetermined set point and maintain that value.

An advanced control scheme can be defined as a control system that goes beyond the single-loop controller. Advanced control refers to those systems that use the three basic control modes: proportional, integral, and derivative (PID), to yield control benefits not possible with simple loops. Such benefits include the following:

- Reduced lag times
- Faster response to process changes
- Better control
- Disturbance minimization
- High and/or low constant constraint controls
- Ratio and mixing control

There are many types of advanced control schemes. Their configurations vary, as do the processes they control.

16.2 Cascade Control

Cascade control is a control scheme in which the output of one controller becomes the set point for another. The two kinds of controllers found in cascade control are initiating controllers (referred to as primary controllers) and cascade (secondary) controllers.

Primary versus Secondary Controller

The initiating, or primary controller is a normal automatic controller with a set point adjustment located on the faceplate. This controller typically has controller modes of automatic and manual.

The secondary controller, which always controls the final control element, can be operated as a controller with a local set point (cascade, or CAS) or a set point received from an external source (remote set point, or RSP).

Figure 16.1 shows an example of primary and secondary controllers.

The drawing in Figure 16.2 is a schematic showing how the output of the primary controller becomes the RSP input to the secondary controller.

Regardless of whether the controller is locally or remotely operated, the controller will fundamentally operate the same as all other controllers. That is, it will operate to eliminate the difference between set point and measurement (error).

Purpose of Cascade Loops

Some important reasons for using a cascade loop include better control and reduced lag time.

BETTER CONTROL The secondary controller should always set point track its process variable in manual mode. (See description below in Cascade Control Tips.) Cascaded control schemes will use as many controllers as needed. For example, an upper controller

Advanced Control Schemes 275

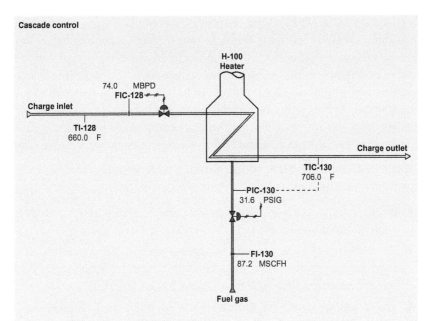

Figure 16.1 Cascade control loop.

CREDIT: © Simulation Solutions.

Figure 16.2 Cascading primary and secondary controllers.

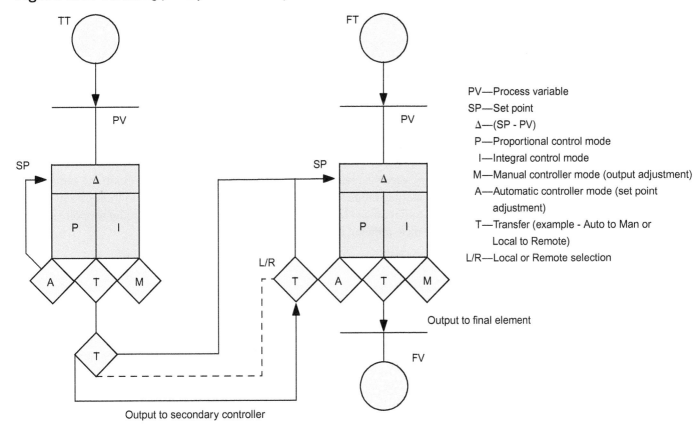

(% composition) should be cascaded to a middle controller (temperature), which is cascaded to a lower controller (flow). The significance of this cascade may not be apparent until you take a closer look at the system. Take as an example the valve that is used to supply heat to the process flowing through a heat exchanger. If the pressure in the line feeding this valve changes, the BTUs of heat flowing through the valve will change as well. Controlling the flow through the valve can eliminate this problem, producing better control of the temperature.

REDUCED LAG TIMES Using the previous example, let us consider how upstream pressure can increase or decrease the flow rate through a control valve. For example, if the flow rate through the control valve diminishes because of a pressure drop, the temperature controller will not be able to react to the reduced flow rate until the change in temperature measurement is sensed. This problem can be solved if a flow loop is added to the temperature loop so that the primary controller can send an RSP to the flow controller. With this modification, the flow loop will immediately compensate for pressure changes in the steam line feeding the control valve and effectively eliminate the process lag time.

CASCADE CONTROL TIPS When working with cascade control schemes, there are a few tips that should be considered. They are as follows:

- In order for a cascade system to function correctly, the secondary controller must respond more rapidly than the primary controller.
- The output from the primary controller should always follow the set point of the secondary controller when the secondary controller is not in RSP/CAS mode. This is sometimes called *set point track* or *output track*.
- The secondary controller should always set point track its process variable in manual mode. Some cascaded control schemes will use as many controllers as needed (e.g., an upper controller [% composition] should be cascaded to a middle controller [temperature], which is cascaded to a lower controller [flow]).

The general procedure for a bumpless transfer in cascade control is described here. You must check all the valves, controllers, and so on for their correct position before activating a process, This check is called *lining out*. Line out the primary variable with the secondary controller in local manual mode as follows:

1. With the secondary controller still in local (local set point), switch it from manual to automatic.
2. With the primary controller in manual, adjust the output so the remote set point indicator on the secondary controller lines up with the local set point indicator.
3. Toggle the secondary controller set point switch from local to remote.
4. If the primary variable and the set point on the primary controller are the same, then switch from manual mode to automatic. If the primary variable and the set point on the primary controller are *not* the same, then adjust the primary controller set point so that it equals the measurement and then switch it to automatic.

Examples of Cascade Control

EXAMPLE 1: HEAT EXCHANGER In Figure 16.3, the temperature transmitter located on the shell side outlet pipe of the heat exchanger measures product temperature and sends it to the primary controller TIC-115. The output signal from primary controller TIC-115 becomes the external set point of the secondary controller, FIC-110. Recall that the secondary controller can receive an external set point. The secondary controller then controls the flow rate of the steam entering the tube side of the heat exchanger. FIC-110 controls steam flow to the shell of E-100. If steam flow drops off due to another steam user coming online or a reduction in steam header pressure, FIC-110 would compensate, opening FCV-110A to increase steam flow back to the set point from primary controller TIC-115. If this was not a cascade loop and TIC-115 operated the steam valve directly, steam valve FCV-110A would likely go fully open and result in high temperature in E-100. In this example, the cascade controller provides improved temperature control without overshoot.

An important thing to note is that an increase in steam supply pressure would cause the valve to close more, whereas a decrease in steam supply pressure would cause the valve to open more. The flow loop will compensate for this problem. Level cascaded to flow

Figure 16.3 Heat exchanger (temperature/flow).
CREDIT: © Simulation Solutions.

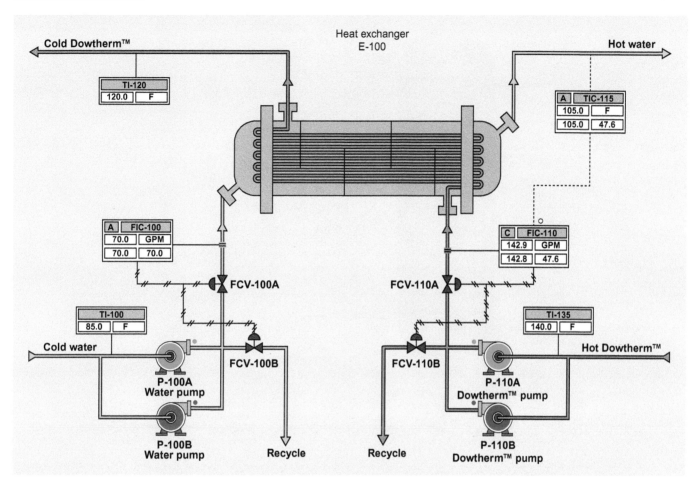

is better than plain level control. The cascaded control can control the vessel level more tightly, should the flow through the control valve change due to disturbances upstream of the control valve.

EXAMPLE 2: VESSEL "BOTTOMS" LEVEL CONTROL In Figure 16.4, the vessel level is controlled by resetting the flow rate of "bottoms" from the vessel. The level transmitter measures the vessel level and sends the measurement to the primary controller. The output of the primary controller is the set point of the secondary controller, which controls the flow rate from the vessel.

EXAMPLE 3: STIRRED TANK REACTOR In Figure 16.5, the temperature of the reactor is controlled by adjustments to the reactor jacket temperature. The temperature of the reactor contents can be controlled more tightly and quickly by allowing the secondary temperature controller to adjust for cooling water supply changes, rather than by waiting for disturbances to be propagated through the jacket and reactor contents.

The temperature transmitter measuring the reactor temperature provides the input to the primary controller. The output of the primary is the external set point to the secondary controller. In this example, the secondary loop is also a temperature loop. The reactor jacket is a contact cooling system that is built around the entire reactor wall, providing a large, consistent, and reliable direct contact cooling area. The reaction causes the reactor temperature to increase. The secondary temperature control loop responds by increasing the reactor cooling. If the reaction causes the reactor temperature to decrease, the reverse will happen.

Figure 16.4 Vessel (level/flow).
CREDIT: © Simulation Solutions.

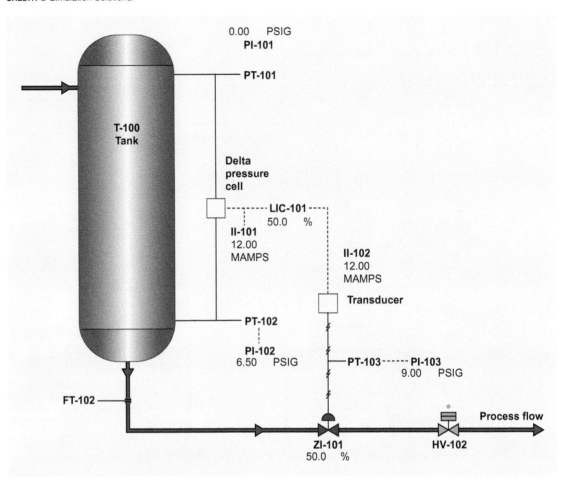

Figure 16.5 Stirred tank reactor (temperature/temperature).

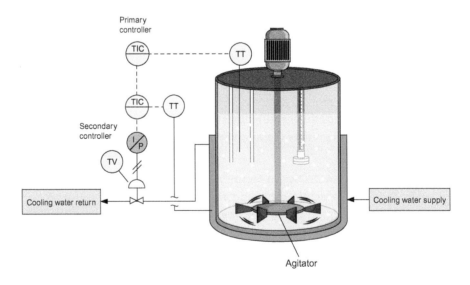

Ratio the proportion of two separate elements, such as the proportion of flow between two separate streams entering a mixing point.

Ratio control control used in blending raw materials to make a final product and in regulating the correct proportion of flow rates of reactants feeding into a reactor; also a term used to describe maintaining the proper fuel-air ratio feeding into a furnace.

Ratio control loop control loop designed to mix two or more flowing streams together while maintaining a quantitative ratio between them.

16.3 Ratio Control

The term **ratio** means proportion between the rates of flow of two separate flowing streams entering a mixing point. The diagram in Figure 16.6 shows an example of **ratio control**.

Ratio control loops are designed to maintain a fixed ratio of two or more flowing streams, regardless of flow changes in the uncontrolled stream. Ratio control is used when blending

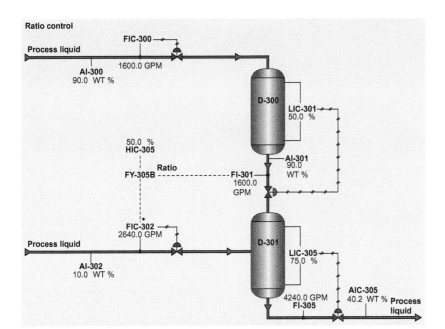

Figure 16.6 Ratio control.
CREDIT: © Simulation Solutions.

raw materials to make a final product and when proportioning the flow rates of reactants feeding into a reactor. It is also used to maintain the proper fuel-to-air ratio feeding into a furnace.

The hardware used to accomplish ratio control may look like a cascade loop between two flow loops. However, there are significant differences between the two. In a cascade control loop, the secondary loop operates in a dependent role to control the primary loop's process variable, whereas in a ratio control loop, the secondary control loop simply proportions its flow rate to the primary loop. In other words, the ratio loop does not receive feedback from the primary loop.

The ratio-producing component in a ratio loop may be a computing relay, such as a dividing relay, or it may be a specialized ratio controller. In either case, the ratio-producing component establishes the ratios of flow rates accordingly.

When working with ratio control schemes, there are a couple of points that should be considered. They are as follows:

- Let the flow ratio controller calculate the ratio PV. The process technician then adjusts the ratio set point within certain allowable limits.
- Ratio control is not recommended on streams whose compositions may vary. Ratio control on streams whose compositions may vary must use a proper analyzer and cascade the set point control to the flow ratio controller.

Examples of Ratio Control

The streams in ratio control can be controlled or uncontrolled. In order to better understand the difference between controlled and uncontrolled, it is important to look at several examples.

EXAMPLE 1: TWO CONTROLLED STREAMS In the diagram on Figure 16.7, two independently flowing streams (A and B) are mixed together in the proper proportions to produce a specific resulting mixture. Both A and B have controlled flow rates. Their flow rates are proportioned to one another by this control scheme and mixed in the mixing tee located downstream of both loops.

The flow transmitter on line A measures a flow and sends it to the two separate destinations, the flow controller and the ratio controller. The flow controller produces an output that works to keep the flow rate in line A at its predetermined set point. Concurrently, the ratio controller receives the same signal and produces an output that establishes a set point for the controller associated with the flow rate in line B. The control loop on line B controls the flow in line B at the predetermined proportion set on the ratio controller.

Figure 16.7 Two controlled streams.

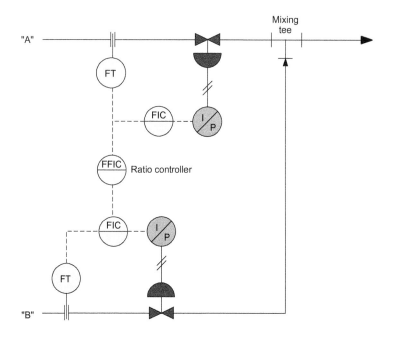

EXAMPLE 2: ONE CONTROLLED STREAM AND ONE UNCONTROLLED STREAM In the diagram on Figure 16.8, two independently flowing streams are mixed together in the proper proportions to produce a specific resulting mixture.

Figure 16.8 One uncontrolled stream and one ratio-controlled stream.

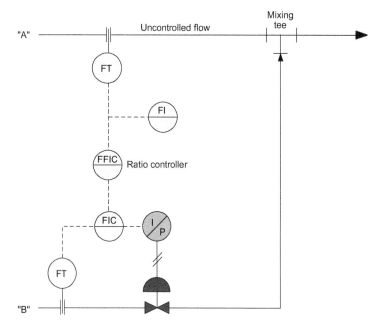

Stream A is considered to be an uncontrolled flow, while stream B is considered to be the controlled flow. Again, their flow rates are proportioned to one another by this control scheme and mixed in the mixing tee located downstream of both loops.

The flow transmitter on line A measures a flow and sends it to the ratio controller only. The ratio controller produces an output that provides a set point to the controller in the flow loop on line B. The control loop on line B then controls the flow in line B at the predetermined proportion established in the ratio controller.

EXAMPLE 3: FEEDFORWARD CONTROL Figure 16.9 shows a cascade control loop with the secondary controller adjusting steam flow. This setup is an improvement over a single-loop control system, but the problem remains that the system cannot correct for load changes until they cause a change in the product temperature from the heat exchanger.

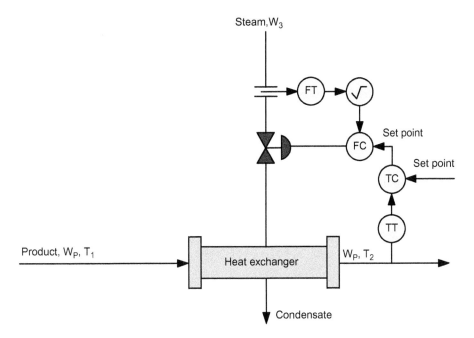

Figure 16.9 Cascade control.

Figure 16.10 shows an improved scheme with a secondary loop that adjusts for product load change. This loop responds to changes in load rate and temperature. The improved *bumpless* response is shown in Figure 16.11.

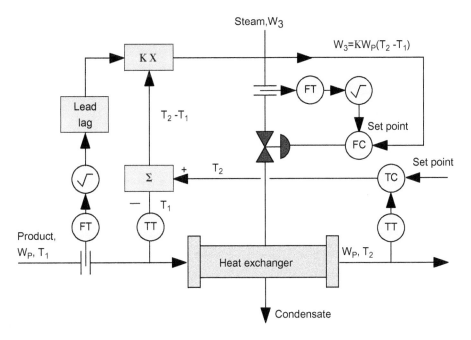

Figure 16.10 Feedforward control.

EXAMPLE 4: BOILER COMBUSTION CONTROL A common combustion control system employs a single variable, pressure, to control the two variables of fuel gas and air. In order to maintain steam header pressure under variable load conditions, fuel and air must be adjusted in parallel at a preset ratio. This does not necessarily provide the safest process operation. The control scheme shown in Figure 16.12 provides the following safety features:

- Air leads fuel on a load increase.
- Air lags fuel on a load decrease.
- If airflow falls, fuel flow is decreased.
- If fuel flow rises, airflow is increased.

Figure 16.11 Feedforward response.

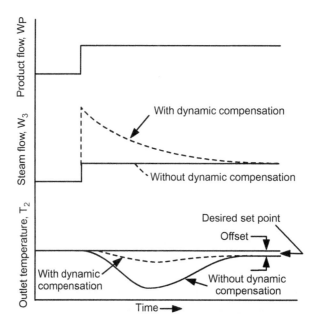

Figure 16.12 Boiler combustion control.

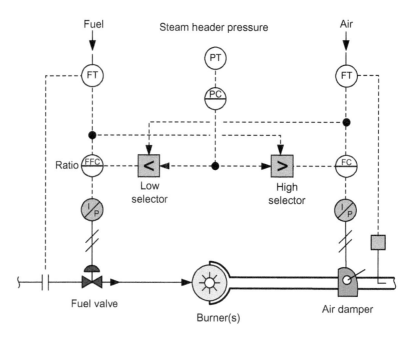

These requirements ensure that fuel gas flow does not exceed airflow output if the header pressure controller increases as steam load is increased. The load selector will not let the fuel flow set point increase until airflow increases. Air leads fuel on load increase, and air lags fuel on load decrease.

16.4 Split-Range Control

Split-range control is a control scheme in which the control signal is divided into several parts, each of which is associated with one of the manipulated variables. The final outcome of split-range control is a single process controlled through the coordination of several manipulated variables, all of which have an effect on the controlled output. The output of a split-range controller is divided between two final control elements. Examples of these control elements are shown in Figure 16.13.

Output of a split-range controller is divided between two final control elements. The most common scenario involves one final control element (valve A) responding to the lower

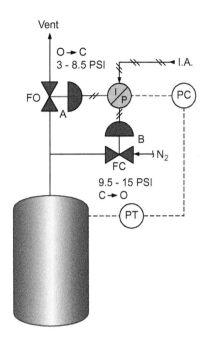

Figure 16.13 Split-range controller.

half of the output signal, while the other control element (valve B) responds to the upper half of the output signal. Consider this scenario:

A processing tank requires a constant vapor space pressure (space above the liquid) of 10 PSIG. During charging (tank filling) and heating cycles, the pressure in the vapor space increases, and a pressure release is required. After the heating cycle, the pressure drops so that increased pressure is required. To accommodate this, a controller with a split-range output applied to two separate control valves is required. One valve applies pressure to the tank, while another valve relieves pressure from the tank.

The most common way split-range valves would work together is to have one valve, valve B, respond by opening to an increasing signal from 9.5 PSI to 15 PSI and the other valve, valve A, respond by opening in an air to close arrangement from 8.5 PSI to 3 PSI. When the valves are receiving a signal of 8.5–9.5 PSI, they are both closed. As the pressure drops below 8.5 PSI in the tank, valve B opens, allowing pressure to enter the tank, while valve A remains closed. Conversely, when pressure rises above 9.5 PSI in the tank, valve A opens, allowing pressure to be relieved from the tank, while valve B remains closed.

16.5 Multivariable Control

Multivariable input refers to industrial processes that demonstrate an interaction between two or more process variables. When any one of these variables is changed, it affects the others in a swift and predictable manner. This is called multivariable control. In some highly interactive processes, such as fractionator control, a multivariable controller with multiple inputs and outputs can produce a level of control that is impossible with single-input or single-output controllers.

Multivariable control schemes are some of the most complicated and intricately designed control schemes in the processing industry. To understand and design a multivariable system fully requires extensive knowledge of process and control techniques.

Normally, single-input and single-output loops operate independently from other loops. When another variable in the same process changes, it will not affect other loops in a destabilizing manner.

In some cases, two or more loops may interact in a destabilizing manner in relationship to each other. If one manipulated variable changes, it will adversely affect one or more of the other variables. To illustrate this point, look at Figure 16.14, which contains two interacting loops.

Multivariable input industrial processes that involve an interaction between two or more process variables.

Figure 16.14 Multivariable control example.

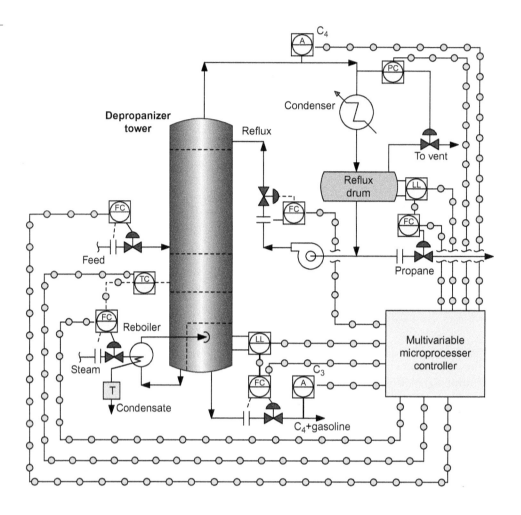

Multivariable controllers are most generally model-based controllers. In a multivariable controller, the inputs and outputs are custom "tailored" with a model (gain, time response, and delay) that is matched to the actual process being controlled. Being *model-based* versus standard PID–based is a distinguishing attribute of multivariable controllers.

Notice that a change in the output of loop 1 affects the measurement of loop 2, which in turn affects loop 1 again. This usually triggers the loops to start cycling with steadily increasing amplitude.

The need to implement a multivariable control system increases proportionally with the complexity of the interactions between instrument loops. The more complex the loop interaction, the more necessary a multivariable control system becomes. In industry today, multivariable control solutions have become more common due to the reliability and programmability of modern computerized control systems.

Some examples of places where multivariable control schemes can be used include the following:

- Chemical reactors
- Distillation and fractionation towers
- Heat exchangers

Figure 16.14 illustrates a depropanizer distillation system with all the variables that must be measured and controlled. These variables include the following:

- Feed rate
- Reflux rate
- Pressure

- Temperature
- Reboiler
- Overhead product (propane) rate, analysis
- Bottom product (C_4 + gasoline) rate, analysis

Conventional digital instruments are provided primarily for start-up, shutdown, and troubleshooting operations. The inputs and outputs to all instruments are connected to a microprocessor controller. Not only will a change in one input variable cause a change in several outputs, but changes in any output will also cause changes in an input variable. For example, a feed rate or composition change will trigger potential changes in reboil, reflux, and product rates. The microprocessor controller continuously updates its data and makes a series of adjustments, seeking the best attainable system performance. Continuous product analyses confirm the results of these adjustments. The computer controller adjusts all actuators in the same way as an analog controller.

The multivariable system allows continuous fine-tuning by the computer. In addition to maintaining all variables at conditions that provide desired results, the program can be modeled to perform helpful material balances and flag the approach to constraint conditions.

Override Controllers

An **override controller** is a device or program that remains inactive until a specific constraint (highest or lowest permitted extreme) on the measured variable is about to be reached. At that time, the override mechanism takes control of the manipulated variable (the final control element) and prevents the constrained variable from exceeding its limit. Only after the measured variable returns to a predetermined "safe" level does control of the manipulated variable returned to the normal controller.

In process refrigeration, override controllers are used to prevent an undesired condition. For example, in Figure 16.15, refrigerant liquid is admitted to a chiller so that a desired process outlet temperature can be attained. However, if the level in the chiller increases above a certain point, there is danger of liquid being carried over into the compressor, causing upset and potential damage to the compressor. Thus, the override controller limits the chiller level to a safe preset value (e.g., 80 percent) regardless of the action of the process temperature controller.

Override controller a device or program that remains inactive until a specific constraint (highest or lowest permitted extreme) on the measured variable is about to be reached.

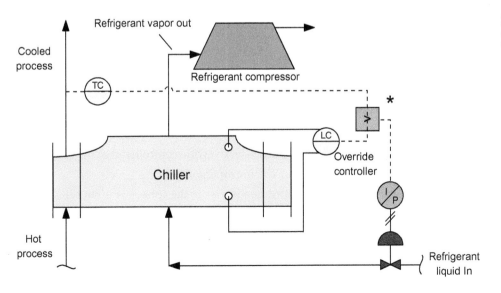

Figure 16.15 Override control scheme example.

★ High limiting selector. Output equals input or high level, whichever is lower.

Summary

Many applications require systems that are more complicated than single-loop, single-variable feedback controls. These types of systems are referred to as advanced control systems.

Advanced control refers to those systems that use the three basic control modes—proportional, integral, and derivative—to yield control benefits not possible with simple loops. Such benefits include reduced lag time, faster response to process changes, better control, disturbance minimization, high and/or low constant constraint controls, and ratio and mixing control.

Single-loop controllers operate on the equilibrium theory that a control device will be configured to manipulate a variable to bring the controlled variable to a predetermined set point and maintain that value.

Advanced control schemes go beyond the single-loop controller. There are many types of advanced control schemes, including cascade, split-range, and multivariable. Their configurations vary as widely as the processes they control. An override controller is a device or program that remains inactive until a specific constraint (highest or lowest permitted extreme) on the measured variable is about to be reached.

Checking Your Knowledge

1. Which of the following is a benefit of advanced control systems? (Select all that apply.)
   a. Reduced lag times
   b. Slower response to process changes
   c. Disturbance minimization
   d. Ratio and mixing control

2. Which controller always controls the final control element in a cascade control scheme?
   a. Primary controller
   b. Secondary controller
   c. Remote controller
   d. Local controller

3. The secondary controller should always set point track its process variable in _____ mode.
   a. slow
   b. process
   c. manual
   d. automatic

4. When is ratio control used? (Select all that apply.)
   a. When blending raw materials to make a final product
   b. When proportioning flow rates of reactants feeding into a reactor
   c. When maintaining the proper fuel-to-air-ratio feeding into a furnace
   d. When the flow from two separate streams needs to maintain a variable rate of flow

5. Who should calculate the ratio PV when working with ratio control schemes?
   a. The process technician
   b. The supervisor
   c. The flow rate controller
   d. The ratio PV does not need to be calculated in this scenario

6. A control system designed for a process in which the control signal is divided into several parts is called:
   a. Multivariable
   b. Split-range
   c. Override
   d. Cascade

7. Which of the following processes would be the most likely to require computer-assisted control?
   a. Multivariable
   b. Split-range
   c. Override
   d. Cascade

8. When is an override controller used?
   a. During emergencies
   b. When power to the controller is cut off
   c. When manually initiated by a process technician
   d. When a specific constraint on the measured variable is about to be reached

9. Identify the control scheme shown in the diagram to the right.
   a. Multivariable
   b. Split-range
   c. Override

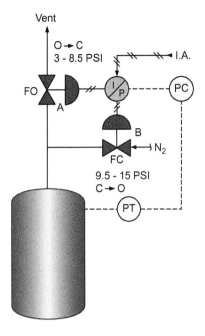

NOTE: Answers to Checking Your Knowledge questions are in the Appendix.

Student Activities

1. Write a paragraph to explain the purpose and function of cascaded control schemes.

2. Write a paragraph to explain the purpose and function of a split-range control scheme.

3. Write a paragraph to explain the purpose and function of a multivariable control scheme.

4. List the steps required to change instrument controllers without pumping the process.

5. Draw a cascade control system as a P&ID and then describe how the cascade loop operates.

6. Given a cascade control system with both an inner (secondary) and outer (primary) loop:
   A. Control the secondary (inner) loop at several different values in both manual and automatic control.
   B. Switch between local set point and remote set point on the secondary controller without significantly bumping the process.
   C. Control the process at several different set point values.

7. Given a ratio control setup configured according to the P&ID illustration used in the classroom, use a valve or rotameter with valve to adjust flow A and a ratio controller to adjust flow B.

Chapter 17
Introduction to Digital Control

Objectives

After completing this chapter, you will be able to:

17.1 Explain the development of digital control. (NAPTA Introduction to Instrumentation 1; Introduction to Control Loops 3*) p. 289

17.2 In general terms, describe the difference between analog and digital controllers. (NAPTA Introduction to Instrumentation 5; Introduction to Control Loops 3) p. 290

17.3 Define the following terms associated with process control schemes:
analog
discrete controllers
DCS
PLC
input
output
D to A
A to D
converter
multiplexer
demultiplexer (NAPTA Introduction to Instrumentation 5; Introduction to Control Loops 3) p. 291

---

*North American Process Technology Alliance (NAPTA) developed curriculum to ensure that Process Technology courses will produce knowledgeable graduates to become entry-level employees in process technology. Objectives from that curriculum are named here in abbreviated form. For example, "(NAPTA Introduction to Instrumentation 1; Control Loops 3)" means that this chapter's objective 1 relates to objective 1 of NAPTA's introductory course content on instrumentation and objective 3 of NAPTA's content on control loops.

17.4 In general terms, describe how a digital controller transmits an output signal to an analog final control element and how an analog controller transmits to a digital controller. (NAPTA Introduction to Instrumentation 5; Control Loops: Primary Sensors, Transmitters, and Transducers 3, 5) p. 292

Key Terms

Analog to digital (A/D)—the conversion of an analog signal to a digital signal so that it can be processed by a computer; also referred to as "A to D," **p. 293**.

Converter—a device that changes or converts a substance or a signal from one state to another (e.g., changing AC to DC), **p. 292**.

Data highway—a series of distributed computer processors, or nodes, that are all interconnected via a computer network, **p. 290**.

Demultiplexer (DEMUX)—a device that separates single stream, multiplexed signals back into their original multistream constituents, **p. 292**.

Digital to analog (D/A)—the conversion of a digital signal to an analog signal so that it can be interpreted and reported on an analog gauge or instrument (e.g., using an I/P converter to convert a digital signal to a pneumatic signal that triggers the opening or closing of a control valve); also referred to as "D to A," **p. 293**.

Discrete controllers—digital controllers that can accept and produce only digital input and output, **p. 292**.

Distributed control system (DCS)—a computer-based system consisting of a series of controllers used to monitor and control a process, **p. 290**.

Input—the process of entering data or information into a computer system or program; the data entered into a computer program or process, **p. 291**.

Interleave—to mix (two or more digital signals) by alternating between them, **p. 289**.

Multiplexer (MUX)—a device that merges or interleaves multiple signals into one stream or output signal in such a way that each individual signal can be recovered or separated out using a demultiplexer, **p. 292**.

Output—the number or value that comes out from a computer program or process, **p. 291**.

Programmable logic controller (PLC)—a computer-based controller that uses inputs to monitor processes and outputs to control processes, **p. 292**.

Remote processing unit (RPU)—devices that control remote nodes in a DCS, **p. 290**.

17.1 Introduction

In order to understand how information is transmitted and received, it is important to understand how digital control systems work and how they interrelate with analog components and converters.

Analog instruments, which were one of the earliest equipment types, use static gauges with an indicating pen that must move smoothly across all values. Digital instruments, however, provide a more accurate and detailed way of collecting, sending, and interpreting data.

In order to transfer signals between analog and digital equipment, converters must be used. There are several different types of converters, although D/A (digital to analog) and A/D (analog to digital) are the most common.

Once signals are converted and are ready to be transmitted, a multiplexer (MUX) is used to merge (**interleave**) the signals, so they can all be transmitted. Once received, a demultiplexer separates the signal back out into its original multistream constituents. Both multiplexers and demultiplexers are key components of modern distributed control systems (DCSs).

Interleave to mix (two or more digital signals) by alternating between them.

Brief History of Controllers

The concept of automated control is not new. As early as the mid-1920s, rudimentary forms of automatic control began appearing in industry, although a vast majority of processes were still controlled manually. In these manually controlled systems, process technicians were required to monitor gauges and make adjustments to variables in every tank throughout the workday. This was a very time-consuming and energy-consuming process.

By the 1950s and 1960s, analog electronics had begun to replace manually controlled systems. This significantly improved the process and reduced the amount of time and energy required for monitoring.

In 1971, the Intel Corporation introduced the model 4004 microprocessor, the world's first "processor on a chip." This technology allowed a new type of control system: *digital control*.

Digital control began as direct control. In direct control, a computer was used to process measurement data, calculate the required control output, and control final control elements (usually valves). However, using a single computer to control a process had one major weakness. If that governing computer failed, the whole unit was shut down. Improvements had to be made. As part of the improvement process, direct control systems evolved into supervisory control, and then further evolved into the **distributed control system (DCS)** we use today. The first-generation DCS was, basically, a series of microprocessor-based multiloop controllers built along a communications network called a **data highway**. Through this data highway and **remote processing units (RPUs)**, distributed controllers were able to manage multiple process loops simultaneously, as well as the digital signals associated with them.

Distributed control system (DCS) a computer-based system consisting of a series of controllers used to monitor and control a process.

Data highway a series of distributed computer processors, or nodes, that are all interconnected via a computer network.

Remote processing unit (RPU) devices that control remote nodes in a DCS.

17.2 Analog versus Digital Control

Digital signals are characterized by data that are represented with coded information in the form of binary numbers (a series of 0s and 1s). Measurements taken from a process are continuously variable quantities (analog) that must be converted into binary numbers and entered into a computer for processing (digital). The next sections discuss the differences and similarities of analog and digital equipment.

Analog

Analog describes something that is similar (analogous) to something else. In the case of a measured variable, that something may be a real-world indication of a variable such as pressure or temperature. In an analog system, pressure and temperature measurements change in a continuous manner or *wave* that mirrors the actual process variable. In an analog system, the readout shows that smooth change. Figure 17.1 shows an indicating pen on a pressure gauge that moves between all the values from one point to the next. Digital readout devices, discussed below, jump from one numerical value to the next.

Figure 17.1 Analog instrument.
CREDIT: © Emerson 2019.

Digital

Digital is a term applied to a device that uses binary numbers to represent continuous values or discrete (individual or distinct) states. Binary numbers are converted to decimal values before being displayed on a digital instrument like the ones shown in Figure 17.2. Unlike analog instruments, digital instruments jump from one value to the next, skipping values in between.

Figure 17.2 Digital instrument.
CREDIT: CI2004lhy/Shutterstock.

Computers and computer-based instruments are characterized by digital **input** and **output**. There are many advantages to using computer-based instruments, including programmability, self-diagnostics, and the ability to recalibrate from a remote location.

Industry uses both analog and digital systems because each has qualities desirable in specific situations. The true power of a digital system lies in the ability to manage far more complex information than an analog system can. Because of this, the use of digital instrumentation is on the rise, and analog systems are on the decline. As far as the near future is concerned, however, analog equipment will retain a significant role in the processing industry.

Digital instrumentation also has the capacity to manage devices from afar. The industrial Internet of Things (IoT) is one term used to describe the ability to monitor and manage equipment remotely. In the personal realm, IoT means being able to control lights, heat, microwaves, and so on while not at home. Applications in the industrial setting are developing rapidly.

Input the process of entering data or information into a computer system or program; the data entered into a computer program or process.

Output the number or value that comes out from a computer program or process.

17.3 Other Components of a Digital Control System

Remote Processing Units (RPUs)

RPUs are the heart of a distributed control system. An RPU is a microprocessor that is located close to the equipment that it monitors and controls. It is connected to workstations by the transmission subsystem or data highway. A large DCS can have hundreds of RPUs.

Originally, an RPU contained only 8 field terminations, or 8 *channels*. This limited the impact on process control in the event of a single RPU failure. If an RPU failed, only eight channels, out of potentially thousands, were affected. In addition, the abundance of channels made it easy and economical to provide redundancy or other backup strategies. Today, manufacturers offer modular I/O and expansion racks that can increase RPUs' I/O capability to hundreds of channels.

Programmable Logic Controllers (PLCs)

Programmable logic controller (PLC) a computer-based controller that uses inputs to monitor processes and outputs to control processes.

Programmable logic controllers (PLCs) are computer-based controllers that use inputs to monitor processes and outputs to control processes. They are used mostly in discrete manufacturing and were designed to replace electromechanical relay controls. (Relays, when used to control complex processes such as an assembly line, could fill rows of relay cabinets.) The use of PLCs made microprocessors and computers practical in industrial applications and enabled the development of the DCS.

The original PLCs were limited to on/off operations, but they did gain some analog capabilities over time. Present-day devices have complete digital and analog capabilities and are simply referred to as programmable controllers or PCs.

Discrete Controllers

Discrete controllers digital controllers that can accept and produce only digital input and output.

Discrete controllers, which tend to be smaller than their analog counterparts, are digital controllers that can only accept digital input and provide digital output. All computers and PLCs are considered discrete controllers.

Multiplexers/Demultiplexers (MUX/DEMUX)

The data highway in a plant DCS can accommodate multiple signals from field sources at one time. However, a controller must have a means to differentiate and process each individual signal.

Multiplexer (MUX) a device that merges or interleaves multiple signals into one stream or output signal in such a way that each individual signal can be recovered or separated out using a demultiplexer.

A **multiplexer (MUX)** is a device that merges, or interleaves, multiple signals into one stream or output signal in such a way that each individual signal can be recovered or separated out later. Multiplexing can be done with mechanical relays. However, digital methods are more common and are less likely to develop problems.

Demultiplexer (DEMUX) a device that separates single stream, multiplexed signals back into their original multistream constituents.

A **demultiplexer (DEMUX)** is a device that separates the single stream, multiplexed signals back into their original constituents. Demultiplexing is the opposite of multiplexing.

Multiplexers and demultiplexers are both key components of modern DCSs. See Figure 17.3.

Figure 17.3 Multiplexer/demultiplexer (MUX/DEMUX).

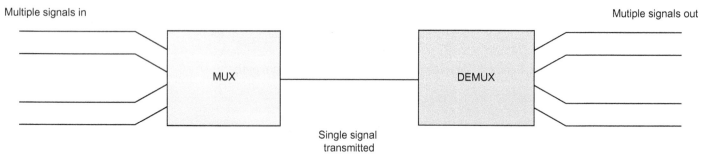

17.4 Signal Conversion and Transmission

As stated in the previous section, present-day process technology incorporates both analog and digital equipment. However, conversion is necessary for these two types of equipment to interact. In D/A conversion, digital signals are converted into analog signals. In A/D conversion, the reverse is true. Most devices perform these A/D and D/A conversions internally.

Converter a device that changes or converts a substance or a signal from one state to another (e.g., changing AC to DC).

While there are a number of different techniques used to convert a digital signal to its analog counterpart, the most common method is known as parallel-type conversion. These techniques require **converters**.

Digital to Analog (D/A)

There are several steps involved in a **digital to analog (D/A)** conversion. The following steps, which are diagrammed in Figure 17.4, are just one example:

1. The digital controller creates a digital (binary) signal. This signal is proportional to the difference between the set point and the actual value of the process variable.
2. The digital signal is then converted to an equivalent analog signal with the help of a D/A converter.
3. Next, the equivalent analog signal is converted into a pneumatic (air) signal in an I/P converter. (An I/P takes a 4–20 mA signal output from a transmitter, controller, DCS, or PLC, and creates a pneumatic signal to drive a device.)
4. The pneumatic signal is then sent to the control valve.

   NOTE: In some situations, there may be a positioner between the I/P and the control valve. The purpose of a positioner is to open the control valve very precisely. A positioner may be used when it is critical that the control valve opening be very precise or accurate.

In the example in Figure 17.4, the signal from the computer to the control valve would require two conversions in order to activate the control valve. Industrial control valves now have integrated communication to regulate their position per the requirements of the control strategy. This makes such conversions unnecessary.

Digital to analog (D/A) the conversion of a digital signal to an analog signal so that it can be interpreted and reported on an analog gauge or instrument (e.g., using an I/P converter to convert a digital signal to a pneumatic signal that triggers the opening or closing of a control valve); also referred to as "D to A."

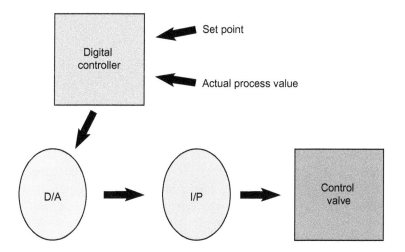

Figure 17.4 D/A conversion.

Analog to Digital (A/D)

Like D/A conversion, it is also important to convert signals in the other direction, from **analog to digital (A/D)**. Analog instruments such as flow transmitters, pressure transmitters, level transmitters, or thermocouples produce a 4–20 mA analog signal. This analog signal must be converted into its digital equivalent before it can be processed by a computer. There are numerous techniques for accomplishing A/D conversion. These include parallel, cascade, and indirect conversions. See Chapter 16, *Advanced Control Schemes*.

In the first diagram in Figure 17.5, an analog electrical transmitter is communicating with a discrete (digital) controller. An A/D converter is used to convert the signal.

In the second diagram in Figure 17.5, an analog pneumatic transmitter is communicating with a discrete controller. The following are the steps involved in this conversion:

1. The pneumatic signal is converted to an electrical signal through a P/I converter.
2. The converted electrical signal is sent through an A/D converter where it is changed from an analog signal to a digital signal.
3. The digital signal is sent to the digital controller.

Analog to digital (A/D) the conversion of an analog signal to a digital signal so that it can be processed by a computer; also referred to as "A to D."

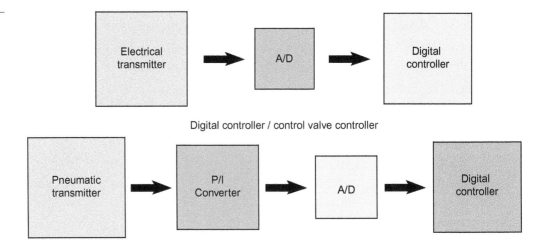

Figure 17.5 A/D conversion.

Stand-Alone Digital Controllers

Digital controllers are usually board mounted with 120-volt AC power. These units are able to take many different types of inputs and convert them internally to digital information. Some of the multiple inputs include resistance temperature detectors (RTDs), thermocouples, 4–20 mA or 1–5 volt, and of course digital. Some of the outputs can be sent as analog or digital signals.

Digital controllers usually have self-diagnostic and auto-tune capability. They can be programmed with limited logic and several different control schemes, such as cascade, ratio, feedforward, and feedback. The faceplate has a limited keypad for programming and shows set point (SP), process variable (PV), output, mode alarms, and tag number.

Smart Transmitters

In an attempt at streamlining the conversion process, transmitters called *smart transmitters* (microprocessor-based instruments) have been created and are currently being used in industry. Smart transmitters have an A/D converter built into them so that they can process information within the onboard microprocessor. It is important to note, however, that the output signal from a smart transmitter may be digital or may be both digital and analog. Some transmitters use high-frequency digital signals superimposed on the standard 4–20 mA output to communicate over the loop. The digital communication does not disturb the 4–20 mA signal since the net energy added to the loop is zero.

Summary

There are two main types of equipment signals, analog and digital. Analog, one of the earliest equipment types, uses static gauges with an indicating pen that must move smoothly across all values between point A and point B.

Digital, which was introduced in the 1970s, provides a more accurate and detailed way of collecting, sending, and interpreting data. Unlike analog equipment, digital instruments jump from one point to the next without passing all the values in between.

Analog and digital systems are both used in industry because each has unique qualities that are desirable in certain situations. While analog instrumentation has been the most prevalent in the processing industry, the use of digital is on the rise because of the amount of flexibility and accuracy it provides.

In order to transfer signals between analog and digital equipment, converters must be used. There are several different types of converters. However, A/D (analog to digital) and D/A (digital to analog) are the most common ones.

Once signals are converted and are ready to be transmitted, a multiplexer (MUX) is used to merge (interleave) the signals so that they can all be transmitted simultaneously. Once received, a demultiplexer (DMUX) separates the signal back out into its original multistream constituents. Both multiplexers and demultiplexers are key components of modern DCSs.

Checking Your Knowledge

1. Which type of instrument can jump from one value to the next without having to pass through all the values in between?
   a. Digital
   b. Analog
   c. Both digital and analog

2. If you wanted to convert a digital signal to analog, which type of conversion would you use?
   a. A/D
   b. I/P
   c. D/A

3. What is at the heart of a distributed control system?
   a. Converter
   b. PLC
   c. RPU
   d. CPU

4. Match the following terms with their definitions:

Term	Definition
I. Converter	a. A device that separates single stream, multiplexed signals back into their original multistream constituents.
II. Demultiplexer	b. Computers, designed to replace switches and push buttons of motor control centers.
III. Discrete controllers	c. Digital controllers that can accept and produce only digital input and output.
IV. Distributed control systems	d. The data or information entered into a computer system or program.
V. Input	e. A manufacturing system (usually digital) that consists of field instruments connected to multiplexer/demultiplexers and A/Ds (analog to digital) via wiring or buses and that terminate in a human-to-machine interface device or console.
VI. Multiplexer	
VII. Output	f. The number or value that comes out from a computer program or process.
VIII. Programmable logic controllers	g. A device that changes or converts signals from one state to another.
	h. A device that merges or interleaves multiple signals into one stream or output signal.

5. What is the most common method of converting a digital signal to its analog counterpart?
   a. Cascade
   b. Indirect
   c. Parallel
   d. Direct

6. Which type of instrument does the following graphic represent?
   a. Digital
   b. Analog
   c. Both digital and analog

CREDIT: © Emerson 2019.

7. You are a technician setting up a system. You have a pneumatic transmitter (A) and you need to send a signal to a digital controller (B). On the diagram below, identify and label all of the components (converters) that need to be added in order for the signal to be transmitted properly.

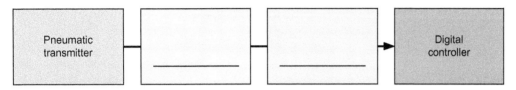

NOTE: Answers to Checking Your Knowledge questions are in the Appendix.

Student Activities

1. Draw a digital control loop and identify the parts. Discuss what each component is used for.

2. Given a digital control loop, identify the following components:
   a. Digital controller
   b. D/A
   c. Wiring
   d. I/P
   e. Positioner
   f. Control valve

3. Turn on power to the entire loop. Show the signal output from the controller and observe the response of the control valve. Record your observations.

4. Turn on power to the entire loop. Keeping the controller in manual, show the signal output from the controller and observe the response of the control valve. Record your observations.

5. Given a nonfunctioning D/A, determine what the problem is, and how it will affect the control loop and the control valve.

6. Given an I/P with a failed air supply, determine the mode of the failure and the impact the failure has on the control valve action.

Chapter 18
Programmable Logic Controls

 ## Objectives

After completing this chapter, you will be able to:

18.1 Explain the purpose of and terms associated with programmable logic control (PLC):
ladder logic diagram
sequential control
on/off
emergency shutdown (ESD) systems
integral to distributed control systems (DCS)
stand-alone capability (NAPTA Interlocks and Safety Features 1, 2*)
p. 298

18.2 Explain how ladder logic applies to programmable logic control. (NAPTA Interlocks and Safety Features 1, 2; Introduction to Instrumentation 1) p. 301

Key Terms

Boolean logic—in computer science, Boolean is a data type that has one of two possible values, intended to represent the two truth values of logic, **p. 301.**
And/or—a conditional logic statement used to control a process in which either one or both variables exist, **p. 301.**
If/then—a conditional logic statement used to control a process in which one condition leads to the next, **p. 301.**
Ladder diagrams—diagrams that guide electricians in the fabrication process, **p. 298.**

---

*North American Process Technology Alliance (NAPTA) developed curriculum to ensure that Process Technology courses will produce knowledgeable graduates to become entry-level employees in process technology. Objectives from that curriculum are named here in abbreviated form. For example, "(NAPTA Interlocks and Safety Features 1, 2)" means that this chapter's objective 1 relates to objectives 1 and 2 of NAPTA's course content on interlocks and safety features.

Motor control center (MCC)—hardwired relays, switches, or contacts, housed in large metal cabinets that control the starting, stopping, and sequencing of motors and other devices, **p. 298**.

On/off—a conditional logic statement used to control a process so that the process is either fully engaged or not engaged at all, **p. 298**.

Shelf position—the contact position of an electrical device when de-energized (e.g., a "normally open" switch is normally open when it is de-energized), **p. 300**.

18.1 Introduction

Control engineering has made many changes and advancements over the years. In the past, the main method for controlling a system was human intervention. Over time, however, fluid dynamics, electricity, and relays were employed to improve the process and reduce the dependence on manual intervention.

Relays were used to make simple logical control decisions and allow power to be switched on and off without a mechanical switch. However, once low-cost computers became readily available, a revolution occurred in industry that led to the computer-based programmable logic controller (PLC).

Modern electronic PLCs were first introduced during the 1970s. Today, PLCs represent the first and most widespread application of computers in process control, mainly because of the numerous advantages they offer over previous hardwired systems. Those advantages include the following:

- Cost effectiveness (a relatively inexpensive way to control complex systems)
- More sophisticated control (computational abilities and increased sophistication)
- Flexibility (modular construction that allows easy replacement and addition of units)
- Improved troubleshooting (programming that is easier and that reduces downtime)
- Reliability (components that tend to be more reliable and don't fail as frequently)

Historical Background of PLCs

In the precomputer PLC era, logic was accomplished by hardwired relays, switches, rotary contactors, or contacts housed in relay control cabinets (RCCs). These relay control cabinets were connected to electric motor control circuits contained in **motor control centers (MCCs)** or to electric solenoid valves for **on/off** control of control valves. MCCs, which are shown in Figure 18.1, controlled the starting, stopping, and sequencing of motors and other devices. Computer-based PLCs now perform these functions faster and with greater reliability than the older relay-based controls. MCCs, then and now, generally contain minimal numbers of control relays, switches, and so on. Their function is still the same as in the precomputer era, namely, to switch on/off higher powered equipment such as motors, fans, conveyers, and compressors.

Because logic had to be hardwired into the older relay-based controls, engineers were required to create diagrams to guide electricians in the construction process. These were called **ladder diagrams**. Figure 18.2 shows an example of a ladder diagram.

Having accurate ladder drawings was important because major modifications in control required time-consuming and expensive rewiring. Any time a change was made to the wiring or the control system, the ladder diagrams had to be updated for future reference and troubleshooting purposes. This, too, was an expensive and time-consuming process.

Motor control center (MCC) hardwired relays, switches, or contacts, housed in large metal cabinets that control the starting, stopping, and sequencing of motors and other devices.

On/off a conditional logic statement used to control a process so that the process is either fully engaged or not engaged at all.

Ladder diagrams diagrams that guide electricians in the fabrication process.

Programmable Logic Controls 299

Figure 18.1 Motor control centers (MCCs).

CREDIT: Lightpoet/Shutterstock.

Figure 18.2 Sample ladder diagram.

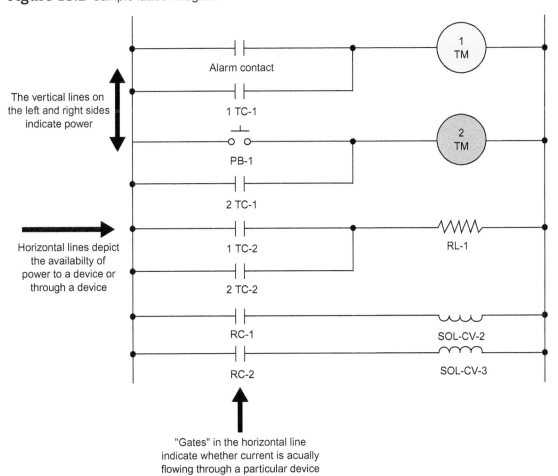

Reading Ladder Diagrams

Ladder logic diagrams derive their name from the ladder-like graphical symbolism used to depict PLC logic. The vertical lines, which look like upright ladder supports, indicate power to an electrical system. The horizontal lines, which look like the rungs of a ladder, depict the

Shelf position the contact position of an electrical device when de-energized (e.g., a "normally open" switch is normally open when it is de-energized).

availability of power to a device or through a device. "Gates" in the horizontal lines indicate whether current is actually flowing through a particular device.

Ladder logic diagrams are read from left to right and from top to bottom, just like a page of text would be read. Generally, devices are arranged in order of decreasing impact. So, devices that have a larger impact on the system are generally shown above devices with relatively minor impact. Contacts, relays, and other devices are shown in their **shelf position**. For instance, a normally open *(NO)* contact will be shown as open.

Consider switch A in Figure 18.3A. Switch A is considered normally open (NO) because there is no diagonal line intersecting it. When power is applied to this circuit relay, X remains de-energized because the circuit path is broken by switch A. However, if there had been a diagonal line, like in Figure 18.3B, the switch would be considered normally closed *(NC)*. When power is applied to this circuit relay, X is energized because the closed switch A enables current to flow in the circuit.

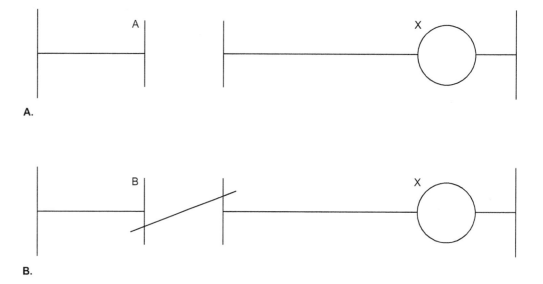

Figure 18.3 A. Ladder diagram (power on). **B.** Ladder diagram (power off).

Practical Tips

The following are practical tips:

- Instrument and electrical (I&E) technicians are typically quite good with PLCs and instrumentation. However, they may lack process knowledge. Process technicians provide a process liaison to work with the instrument technicians.
- Should it be necessary to bypass safety systems temporarily, management of change procedures must be strictly followed. A bypass, or *jumper* as it is called, should be documented and removed immediately after the completion of work.
- Before issuing a maintenance work permit to someone working on a PLC, plan the job thoroughly. It is not uncommon to see an entire plant shut down accidentally because a PLC is being repaired or modified. Process and maintenance technicians must work together when planning PLC modifications.

PLCs Today

Modern PLCs allow logic changes to be programmed easily using desktop personal computers. In addition, these PLCs allow changes to ladder diagrams to be recorded automatically during the programming process. This saves considerable amounts of time and money.

PLCs may be stand-alone units controlling only a single operation, or they may be integrated into a broader control scheme, such as modern distributed control systems (DCSs).

18.2 Logic in PLCs

The main programming method used for PLCs is ladder logic. Ladder logic was intentionally developed to mimic relay logic. By using a familiar logic structure, the amount of retraining needed for engineers and maintenance technicians was significantly reduced. While modern control systems do still use relays, they are seldom used for logic.

PLCs gained their name from the ability to use **Boolean logic** to program logic into these controls (e.g., programming **and/or**, **if/then**, or true/false functions like the one shown in Figure 18.4).

Boolean logic in computer science, Boolean is a data type that has one of two possible values, intended to represent the two truth values of logic.

And/or a conditional logic statement used to control a process in which either one or both variables exist.

If/then a conditional logic statement used to control a process in which one condition leads to the next.

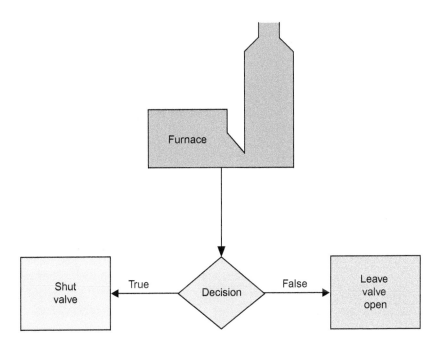

Figure 18.4 Controller logic.

In the following table, there are several decisions that can be made. These decisions, denoted here as "if" and "then" statements, could be as follows:

If . . . -	Then . . . -
The temperature in the furnace is <u>less than</u> 800 degrees Fahrenheit (427 degrees Celsius)	Leave the fuel flow valve open.
The temperature in the furnace is <u>above</u> 800 degrees Fahrenheit (427 degrees Celsius)	Shut the fuel flow valve off and shutdown the furnace.

Instructions such as these are expressed as binary code in the computer. Because of this functionality, PLCs are commonly used for safety-related applications. They make up the heart of emergency shutdown (ESD) systems.

PLC functionality can be expressed as a three-part process which does the following:

1. Scans inputs from switches or transmitters.
2. Executes logic or control within the PLC.
3. Sets outputs based upon results of logic.

Programming a PLC

There are a number of ways that a PLC can be programmed. These include function block diagram (FBD), ladder diagram (LD), structured text (ST; similar to the Pascal programming language), instruction list (IL; similar to assembly language), sequential function chart (SFC), or other computer languages. However, ladder diagrams tend to be the method most plant personnel are familiar with and are, therefore, the most commonly used method for interfacing with a PLC. An example of ladder logic can be seen in Figure 18.5.

In Figure 18.5, the vertical lines on the left and right are the power source "rails." In between the left rail and the right rail are horizontal lines, or "rungs." On these rungs are a combination of inputs and outputs. Inputs are entered into the system by sensors or switches. Outputs exit the system and are sent to devices such as lights and motors that are outside the PLC. If the inputs on the ladder diagram are opened or closed in the right combination, the power can flow from the left rail, through the inputs, to the power outputs, and finally to the right rail.

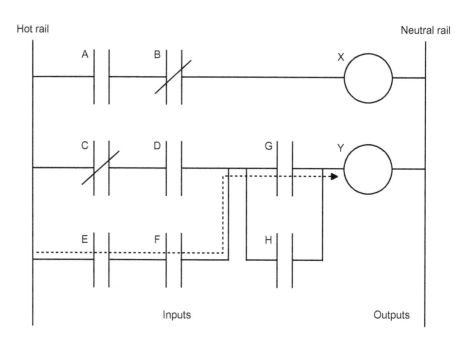

Figure 18.5 Ladder logic example.

Summary

Computer-based programmable logic controllers (PLCs), which made their introduction in the 1970s, are relatively small, stand-alone, easily programmed computers that can be used to control various processes.

Prior to the advent of PLCs, logic tasks were accomplished by hardwired relays, switches, or contacts housed in large metal cabinets called motor control centers (MCCs). MCCs controlled the starting, stopping, and sequencing of motors and other devices. However, PLCs now perform these functions faster and with greater reliability, more sophisticated control, more flexibility, less cost, and less downtime.

There are many ways to program PLCs. The most common method is through the use of ladder logic. Ladder logic, which was intentionally developed to mimic relay logic, is used because engineers and maintenance technicians were already familiar with it, and therefore required less retraining.

Ladder logic functions primarily as a series of and/or, if/then, or true/false statements and is represented through drawings called ladder diagrams. When reading ladder diagrams, it is important to note which devices are listed toward the top, since devices that have a larger impact on the system are generally shown above devices with relatively minor impact.

Checking Your Knowledge

1. Which of the following is an advantage of PLCs? (Select all that apply.)
   a. Cost effectiveness
   b. Less sophisticated control
   c. Modular construction aids in flexibility
   d. Increases downtime
   e. Components are more reliable

2. What type of diagrams did engineers need to create for older relay-based systems?
   a. Ladder diagrams
   b. Process diagrams
   c. Schematic diagrams
   d. Illustration diagrams

3. What symbol would be used on a ladder diagram to indicate a normally closed condition?
   a. Circle
   b. Straight line
   c. Diagonal line
   d. Diamond

4. What is the most common logic for programming PLCs?
   a. Structured text
   b. Instruction list
   c. Sequential function chart
   d. Ladder diagram

5. (*True or False*) In the following PLC diagram, switch A is considered "on."

6. (*True or False*) In ladder diagrams like the one below, devices at the bottom have more impact on the system than the ones at the top.

   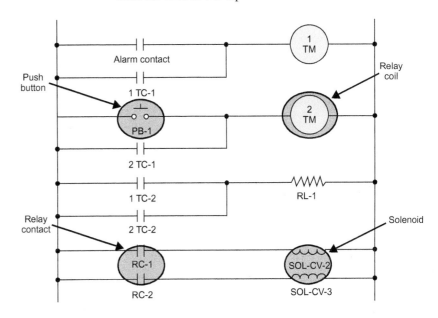

NOTE: Answers to Checking Your Knowledge questions are in the Appendix.

Student Activities

1. Describe what a PLC is.
2. Give an example of where you might use a PLC.
3. List the advantages of PLCs over relays.
4. Explain why you might use relays instead of a PLC.
5. Explain what a ladder logic diagram is, and how ladder logic applies to programmable logic control.
6. Explain the purpose of PLCs with regard to the following:
   a. Sequential control
   b. On/off
   c. Emergency shutdown systems (ESDs)
   d. Distributed control systems (DCSs)
   e. Stand-alone capability
7. Conduct a safety analysis of a simple process. Identify the type of ladder diagram that would be necessary to meet process safety requirements.
8. Consider effects of failure of a PLC on the plant safety. Emphasize the need for a separate power supply to such units.
9. Given a programmable logic controller (PLC):
   A. Load a simple circuit into a PLC with the use of a PC. You may consider programming a START–STOP button to start up and shut down a system.
   B. Change the input conditions and observe the response.

PART 4 DCS, Power Supply, Shutdowns, Malfunctions, and Troubleshooting

Chapter 19
Distributed Control Systems (DCSs)

Objectives

After completing this chapter, you will be able to:

19.1 Explain the purpose of a DCS. (NAPTA Introduction to Instrumentation 5*) p. 305

19.2 In general terms, describe the operation of a DCS. (NAPTA Introduction to Instrumentation 5) p. 306

19.3 Explain the function of a multiplexer/demultiplexer. (NAPTA Introduction to Instrumentation 5; Introduction to Control Loops 3) p. 307

19.4 Explain advantages of a DCS over an analog control system:
more precise
faster
more cost-effective
more trending operation
one location or space
bumpless process (NAPTA Control Loops: Controller 4, 5) p. 312

19.5 Explain how DCSs interact with analytics/trending, intrinsically safe display devices/tablets, and cyber security issues. p. 313

Key Terms

Data highway—a series of distributed computer processors or nodes that are all interconnected via a computer network, **p. 305**.

Drift—the change in the accuracy of an instrument signal or indication over time; it can be attributed to factors such as temperature and aging, **p. 312**.

*North American Process Technology Alliance (NAPTA) developed curriculum to ensure that Process Technology courses will produce knowledgeable graduates to become entry-level employees in process technology. Objectives from that curriculum are named here in abbreviated form. For example, "(NAPTA Introduction to Instrumentation 5)" means that this chapter's objective 1 relates to objective 5 of NAPTA's introductory course content on DCSs.

Node—a single computer terminal located on a computer network or data highway, **p. 305.**

Noise—disturbances that affect a signal and may distort the information carried by the signal, **p. 312.**

Trending—the analysis of a change in measured data over at least three data measurement intervals, **p. 313.**

19.1 Introduction

Distributed control systems (DCSs) are computer-based systems that are used to control processes (Figure 19.1). Because these systems are computer-based and interconnected, they are more fail-safe, reliable, and precise than previous systems. In addition, they can process and store data more quickly and efficiently, are more cost-effective, and require less control room space than their analog counterparts.

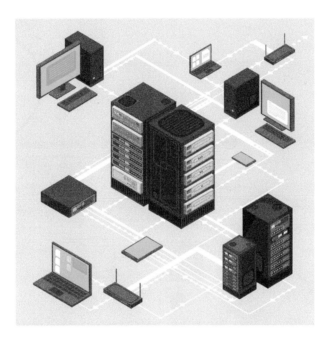

Figure 19.1 Distributed control system (DCS) with individual nodes.

CREDIT: MicroOne/Shutterstock.

Historical Background

In the 1970s, when use of computers became more prevalent in process control, single computers were used to process measurement data, calculate the required control output, and control final control elements. This was known as direct digital control. It was called *direct* because a single computer was directly connected to a particular part of the process. Although this type of control was effective, it had a major weakness. It lacked redundancy. If the computer failed, the entire unit was shut down. This weakness translated to increased safety risks and decreased productivity. Thus, direct control was forced to evolve into a more fail-safe system.

Over the years, digital control evolved from direct digital control, to supervisory control, and then to the distributed control systems (DCSs) we use today. The first-generation DCS was basically a series of microprocessor-based, multiloop controllers built along a communications network called a **data highway**.

A data highway holds a series of distributed computer processors, or **nodes**, that are all interconnected via a computer network. Because of this multinode architecture, DCSs are able to manage multiple process loops at one time. Furthermore, they are more fail-safe than traditional systems. If one node becomes inoperable or goes down, signals can be rerouted to one of the other nodes, and the process can continue. In risk management terminology, this is referred to as *distributed risk* since the risk of control system failure is no longer concentrated at

Data highway a series of distributed computer processors or nodes that are all interconnected via a computer network.

Node a single computer terminal located on a computer network or data highway.

one point. Because it is unlikely that all points will fail simultaneously, the overall reliability of the control system is increased. This translates to increases in both safety and productivity.

19.2 DCS Operation

A DCS divides production control in a plant into several subparts or nodes. The exact manner in which the DCS is divided depends on the degree of distribution desired and the cost and availability of the control systems.

Each node in a DCS is a local computer that is essentially independent of the other nodes (see Figure 19.2). A node may include a considerable number of control loops. The local computer for each node communicates with a central computer called the *host computer*. All nodes communicate with the host computer. Figure 19.3 shows an example of a typical DCS node.

Figure 19.2 DCS showing a node controlling part of a process.

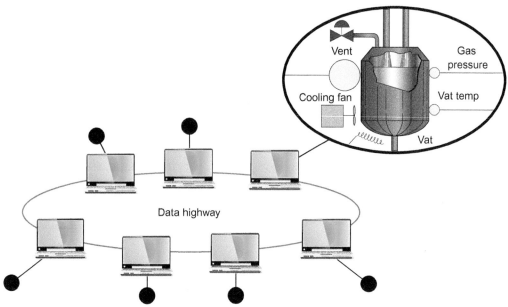

Figure 19.3 An individual node.

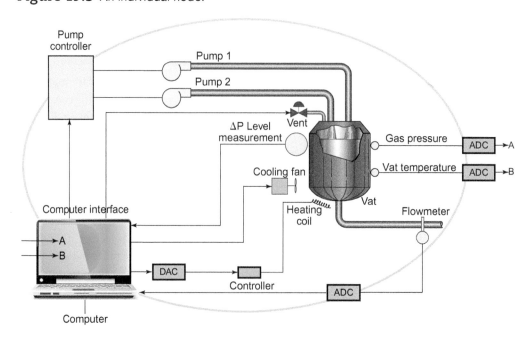

Function of a Typical DCS Node

The functionality of a typical DCS node can be explained as a series of steps as follows:

1. The DCS node processes signals received from the control loop, such as temperature, level, flow, pressure, and on/off digital signals.
2. The signals processed by the DCS are then converted from analog to digital through an analog-to-digital (A/D) converter.
3. Next, the digital signals produced by the A/D converter are forwarded to a local computer by a multiplexer (MUX), a device that sends many signals down a data highway that can later be recovered or separated out into the original constituents by a demultiplexer (DEMUX).
4. The local computer processes the incoming information and sends out control commands based on internally programmed logic and instructions.
5. The control commands generated by the local computer are forwarded by the demultiplexer to a digital-to-analog (D/A) converter.
6. The analog signals produced by the D/A converter are then forwarded to the appropriate valve or control element and the host computer.

19.3 DCS Architecture

Using the information above, let us take look at a typical distributed control system (DCS) installation. In a DCS, the amount of distribution of control and its location is determined by vendor selection and user requirements. Let us build a DCS starting with field instruments, programmable controllers, operating stations, and accessory devices.

Input/Output (I/O) Devices

Field instruments are usually connected to some type of input or output device (electronic card) that resides on an input/output (I/O) highway. This I/O highway may be located in the process unit or in a remote terminal building. The connections to the field devices are made by conventional electrical wiring or digital fiber optic technology (sometimes called bus technology). Each field instrument is wired to its own or representative electronic card. There are several different kinds of I/O cards. For example, analog input (AI), analog output (AO), digital input (DI), digital output (DO), frequency input (v), thermocouple input (T/C), and field bus cards. This interface can also be in the form of a computer (PLC or DCS) backplane circuit board with I/O cards inserted like a computer motherboard.

These cards contain programmable microprocessor chips and may perform many functions. For example, the AI card may perform A/D conversion and signal conditioning and may have current limiting in order to eliminate the use of fuses.

The power supply provides electrical power not only to the electronic cards but also to field devices such as transmitters and switches. The electrical power can be DC or AC depending on the type of card. Figure 19.4 gives an example of an I/O highway.

Usually each card has eight channels or eight input/output connections that are labeled 1 through 8. For example, the analog input card has eight input connections (+, −, and ground terminals), each wired to a field instrument that is given a unique tag number. For instance, in Figure 19.4, flow transmitter 1 (FT-1) is wired to an analog input (card 1, channel 2) and temperature transmitter 2 (TT-2) is wired to analog input (card 1, channel 3).

The field bus card is connected to one or two field terminal blocks that can contain as many as eight field instruments. These instruments are *smart* digital devices that contain programmable microprocessors such as advanced D/P transmitters, mass flowmeters, vortex meters, digital valves, and more. The field bus card can control two terminal blocks. The wiring is a single twisted pair from field instruments to the terminal block, and one pair of wires to the bus card, which is an obvious installation cost advantage.

Figure 19.4 I/O highway example (1).

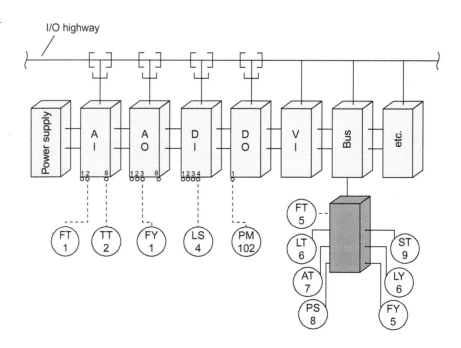

Network communication is performed digitally, with scan times of 100 milliseconds to 4 seconds, depending on the number and type of instruments. The communication contains control data, instrument status, diagnostic, and other information. A field bus transmitter or valve positioner can be programmed to perform PID control functions. This means PID control can occur autonomously entirely inside the field instruments without communication from the node or host computer.

It is important to note that the tag number is usually the same as the one shown on the unit P&ID. Each instrument tag number and where it is connected or located becomes part of a global database that is used by the DCS. This is configured (or programmed into the computer) when setting up the DCS architecture or when adding new controls.

Data Highway

The I/O highway with its devices is connected to the programmable controller that lies on an Ethernet or TCP/IP highway (in the example in Figure 19.5, highway 1 or node 1).

Figure 19.5 I/O highway example (2).

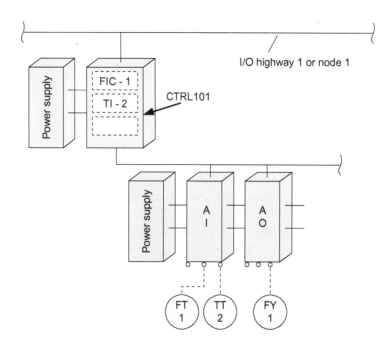

Ethernet and Transmission Control Protocol/Internet Protocol (TCP/IP) are networks set up for transmitting information. *Protocol*, in this context, is a set of rules that govern the transmission and makeup of binary data. Devices on the highway are assigned a unique node address (or tag number). They communicate by sending and receiving information through data packets. Examples of a packet might be start code, source address, control information, destination address, confirmation handshake, or stop code. Devices on the highway are polled or broadcast when they need to send or receive information.

The programmable controller is given a node address (or tag number) (e.g., CTRL-101, which translates to controller 01 on highway 1 or node one).

These programmable controllers can be configured to handle 500 to 1,000 loops. These loops can be a mixture of control loops, monitored inputs or outputs, startup or shutdown sequences, batch control, and logic functions. These controllers contain many microprocessors that can be configured to handle any type of control or unit functions. A loop could be input FT-1, output FY-1, and PID flow controller FIC-1 that is configured in CTRL-101. The system uses highway 1 or node 1/CTRL-101/FIC-1 for this control loop or a TCP/IP address. Most DCSs use tag number addresses to communicate with devices. Another loop shown might be column temperature indication TI-2 that is configured in CTRL-101 with its input TT-2: card 1, channel 8.

The programmable controllers have a built-in execution rate (usually 10 to 20 times a second or faster) to read all input signals, perform control functions, and send output signals. These controllers do not make control changes (e.g., auto to manual, set point, or output changes) unless directed by a supervisory computer or a process technician. In addition, each configured loop can be given its own scan time. For example, flow control loop FIC-1 could be configured with a scan time of 0.5 seconds, whereas column tray temperature indication TI-2 could be selected with a scan time of once every 10 seconds. Choosing various scan times, faster for fast loops or slower for slow loops, allows the programmable controller to run more efficiently and with more processor-free time. For example, an online analyzer AIC-123 sends a new signal only once per minute. Setting a scan time that is more frequent than the signal rate wastes processor operation. For example, if a 1-second scan time were configured, then for 59 scans, or seconds, the AIC-123 controller would calculate the same result, and the output may only change on the 60th scan. This scan time should be equal to when the analyzer has a new result (in this case, 1 minute).

Data highway 1 can have several more programmable controllers connected to it, each with its own I/O highway (e.g., CTRL-102 [highway 1, controller 02] in Figure 19.6). Thus, hundreds or thousands of loops can be connected to a single highway to control a large process unit.

Figure 19.6 I/O highway example (3).

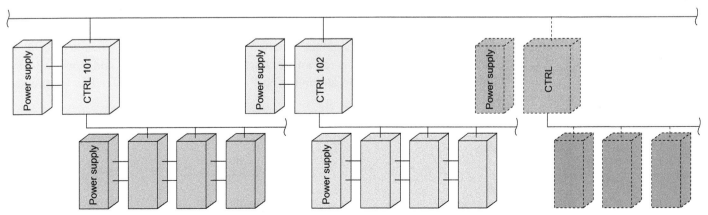

The programmable controllers and I/O may be located in several different places: in a control room, a remote terminal building, and/or in the field. The highways may be hundreds of feet long to miles long depending on the type and speed of the transmission system. The data highway speed ranges from coaxial cable (slowest), to twisted pair shielded wire, to fiber optic cable (fastest). Other transmission media may also be available, such as wireless communications for remote transmission and control (e.g., remote wireless communication to supervise or run an offshore oil platform from shore).

There can be more than one data highway throughout the plant; usually there are 8 to 12. These highways can communicate with one another through devices such as network traffic directors, bridges, gateways, and other types of communication hardware.

Operator Stations or Work Stations

So far, we have not discussed how to control the process. Process control is performed by a device called a *work station*, or *operator station* (Figure 19.7). A work station can be thought of as a window to the process. The main function of a work station is to allow the operator to control the process. The station not only collects process data but also monitors the DCS hardware. The data collected can be presented to the operator in the form of alarm lists, real-time or historical trends, unit graphics, material balance displays, interlock and help screens, and diagnostic information. Unit graphics have programmed control points that can be used to call up a controller faceplate. These stations can be connected to any data highway or transmission network.

Figure 19.7 Operator and work stations.

CREDIT: Alessia Pierdomenico/Shutterstock.

This type of work station used to be a proprietary device furnished by the DCS vendor. Now it is an off-the-shelf computer that must meet the DCS vendor's specifications. Hardware may consist of several hard drives for data collection, several monitors (with or without touch screens), keyboards, mice or trackballs, voice commands, and more. Software provided by the DCS vendor determines how the process data is displayed for the process technician.

Most process information can be configured as plant overview, unit graphics, faceplates, trends, alarm lists, interlock, and help displays. Special displays can also be configured. A digital picture of part of a process unit can be imported as a display with control points added. A mixture of these displays can be configured to represent the process unit to the process technician. The number of displays that can be configured may number in the thousands.

The stations can be selected to have a periodic update time that can be as fast as 1 second or as slow as several seconds. Any loop can be configured to report periodically or by function to the programmable controller and the operation station. A range of exception reporting values can be selected. For example, column temperature indication TI-2 can be selected to report if its PV changes from 1 percent of span. Alarms, however, are programmed to report when they occur and usually have a "first out" feature, which informs the user as to which alarm point was reached first.

Operator stations are usually located in main control rooms or satellite panels in a process area. There is considerable flexibility with regard to how these stations are configured (see Figure 19.7). For example, a particular operator ID can be configured to control system A but only monitor system B.

The main function of an operator station is to collect, monitor, and display process data; respond to alarm situations; record trends; and control the process. Our data highway now consists of an operator station and programmable controllers, each with its own I/O structure.

Developments in remote programming and control were first introduced in the private sector as the industrial Internet of Things (IoT). Applications for industry are clear but are balanced with concerns about the potential cybersecurity risks.

Redundancy

The redundant or backup feature of a DCS makes it extremely reliable. Any DCS device, such as I/O cards, programmable controllers, and highways, can be backed up with a secondary device (Figure 19.8). DCS redundancy is determined by the number of critical loops and the cost to install. Redundancy can be implemented like a "spare tire" setup where one device backs up four to eight like devices (AI cards) or a one-for-one backup, like a primary and secondary power supply.

Figure 19.8 Redundancy example.

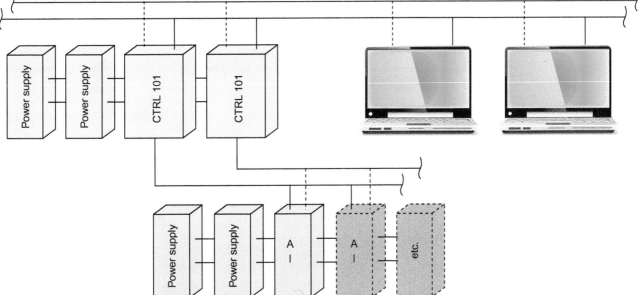

It should be pointed out that any device on the highway can be made redundant or backed up with a secondary device. If the primary device fails, the secondary takes over during the next scan and the process technician is alerted to the failure by an alarm. If more than one operator station is used, then all are online at the same time. If power were lost to the operator station screen, the programmable controllers would continue to run the unit until power was restored. Minimum redundancy should include primary and secondary highways, primary and secondary programmable controllers, primary and secondary power supplies, and the operator station.

Configuration Station

The configuration station stores the complete plant configuration on its hard drive. The primary function of a configuration station is to design the DCS with all its hardware devices and control schemes. A configuration station may be connected to any data highway and may even be located in a separate building. Changes can be made off-line and checked out. Following required coordination and approval (e.g., management of change), the changes can be downloaded to the unit control system. The changes can also be downloaded to a computer simulator. Simulators may be used to train process technicians in the new control schemes prior to implementation on the unit and to see how the changes worked.

Backup copies of the plant configuration should be stored in removable memory or other computers. It is a good idea to store a copy onsite and another copy offsite.

Accessory Devices

Some accessory devices connected to the data highway include printers, chart recorders, process information computers, data loggers, and advanced process control (APC) computers. Printers are used to print unit graphics and morning reports. Chart recorders are used to trend troublesome loops. A separate process information computer may be connected to the data highway to collect and store process data for historical trending. Data loggers are used to store alarms, sequential events, keystrokes, and more. The APC computer usually performs advanced control schemes and models predictive control. Some programmable controllers can perform the APC functions while the operator station contains the process information system and some of the data logger functions.

Plant Network

A local area network (LAN) is usually connected to the DCS through some type of gateway or security device that allows read-only data to be available to engineers, maintenance personnel, and operational personnel. The LAN may contain telephones, business computers, personal computers, and data servers. The operator should have access to a separate PC for items such as email, work orders, and internet. The operator should never have access to the internet from the work station.

19.4 DCS Advantages

Computer-based distributed control systems (DCSs) are preferred over analog-based systems for a number of reasons. These include: repeatability, accuracy, speed, cost-effectiveness, improved data management, space efficiency, and increased safety and reliability.

Noise disturbances that affect a signal and may distort the information carried by the signal.

Drift the change in the accuracy of an instrument signal or indication over time; it can be attributed to factors such as temperature and aging.

Repeatability and Accuracy

The overall repeatability and accuracy of a DCS is greater than that of the analog alternative. Digital systems can process more information involving signal checking and reduction of **noise**, and they can distinguish erroneous signals caused by **drift**.

Speed

Analog systems require a multitude of individual components to perform complex computations. Because of this, data processing speeds are slower than DCSs, which have the ability to process complex calculations quickly, thereby increasing the output speed. From a process technician's perspective, this also means the speed of response to an event is greatly improved because data is more readily accessible.

Cost-Effectiveness

The initial installation cost for a DCS may be less than or equal to that of an analog system. The cost of maintenance for a digital system, however, is considerably lower than for an analog system since digital components are significantly more reliable than their analog counterparts.

DCSs allow a more convenient means for configuring and changing the system to accommodate the needs of the given operation. Software programming changes are much more cost-effective and efficient than hardware changes. In addition, access to historical data, improved control schemes, graphics, complex calculations, predictive maintenance information, and other features are all inherent in the system and are significantly less expensive than hardware devices.

Improved Data Management

DCSs offer superior data storage, archiving, retrieval, and analysis capabilities. For example, if an alarm is indicated, a diagnostic message indicating which part of the system is not functioning properly may be displayed, and an alarm log created. These logs, along with the archiving capability and the ability for statistically analyzing, tracking, and **trending** data, are immensely valuable when it comes to troubleshooting process deviations and ensuring regulatory compliance.

Trending the analysis of a change in measured data over at least three data measurement intervals.

Space Efficiency

DCS displays require less physical space than their analog counterparts. This allows the size of the control room to be reduced while increasing operational productivity and troubleshooting capability. It also reduces maintenance costs.

Safety and Reliability

DCSs now incorporate functions that were traditionally handled by separate, hardwired PLCs for safety trips and interlocks. Because a DCS is computer-based, it allows for safer operation of a unit than analog counterparts. The redundancies inherent in this type of system ensure that signals can be rerouted and operations can continue should a specific local computer (node) fail, thereby increasing reliability and mean time between failures and reducing the potential for unsafe conditions.

Bumpless Transfer

In DCSs, configuration choices (e.g., set point or PV tracking) eliminate bumpless transfer issues. *Bumpless transfer* is the act of changing the controller from manual to automatic (or vice versa) without a significant change in controller output.

19.5 Processes, Equipment, and Safety for DCSs

Analytics/Trending

Operations technicians have relied on real-time and historical data trend charts for years. However, those data were only observed by board operators for troubleshooting, and any communication of that information to other divisions was slow and often in paper format.

What is changing in Industrial 4.0 is that all analytics can now be immediately available to all departments in any location.

This new access to trending of the process parameters in a plant allows very precise quality control and increased throughput of the operation. Process technicians need to understand that every bit of operational data is critical and needs to be accurate. Many decisions, from feedstock supply to material/energy balance and product delivery, will rely on this "big data" to evaluate the health and performance of the plant. Data-based decisions will allow increased operational efficiency by improving planning and scheduling and by minimizing process errors. The ability to track assets and allow quick location of all personnel, equipment, and instrumentation will help plants to manage preventive maintenance. It will help eliminate costly unplanned maintenance, accidents, and lost production time.

Intrinsically Safe Display Devices/Tablets

We have all experienced the Internet of Things (IoT) in our lives, from the ability to adjust a home thermostats from a cell phone to a refrigerator that can order your groceries. These same control systems are now utilized in the industrial Internet of Things (IoT) through Human Machine Interface (HMI), handheld devices such as computer tablets and cellular smart phones.

These units need to be intrinsically safe (a design technique applied to electrical equipment and wiring for hazardous locations) so that we can input and acquire information safely from anywhere in the plant. Plant operators with these input tools can effectively perform simple or complex tasks anywhere. They can also access data and applications including work permits, communicate over voice and video, and participate in just-in-time training from any location. These interfaces will improve asset management; they will also increase productivity, worker safety, troubleshooting accuracy, and remote operations.

Cybersecurity

We have all heard of or experienced data breaches and cyberattacks in networks, which occur partly because systems need to be very "open" so users can access information and perform transactions. As evolution continues in technologies of the new Industrial 4.0 evolution and connectivity of IoT communications that use wireless and fiber optic pathways, the security of operations will become critical.

Sound security policies and procedures will be needed to lock down the industrial Operational Technologies (OT) and tunnel the information to the Informational technologies (IT) business operations in a very secure environment. This is accomplished initially with traditional firewalls and password policies but is further enhanced by data diodes. Data diodes enforce an actual physical "air-gap" separation between the OT and IT networks, providing an absolute secure environment. Perimeter fences, secured access gates, and employee security entrance badges have been a common means of safeguarding a physical plant location. Similarly, the security of the data networks and associated control capabilities will be the responsibility of all personnel, and the operator will be a key person in enforcing this security.

Summary

Distributed control systems (DCSs) are computer-based systems that are used to control processes. These computer-based systems first entered into process technology in the 1970s. Prior to the 1970s, processes were controlled by analog controllers, which were effective but did not contain the same level of functionality as their digital counterparts.

A DCS consists of a series of components that are interconnected via a computer network or data highway. DCSs are more fail-safe than analog systems because of the redundancy built into the systems and the reliability of digital components (e.g., if a node in a DCS becomes inoperable or goes down, data signals are automatically rerouted to a different node and the process continues). This translates to increases in both safety and productivity.

DCSs are also more user-friendly and more accurate. They provide more flexibility in choosing control station locations. Furthermore, they can process and store data more quickly and efficiently, are more cost-effective, and require less control room space than their analog counterparts.

As the advanced technologies of Industry 4.0 and the communication processes of the IoT networks are implemented into process operations, technicians will need to continually advance their knowledge. This information will be collected and analyzed utilizing HMIs allowing real-time data to be accessed to improve asset management; they will also increase productivity, worker safety, troubleshooting accuracy, and remote operations Additionally, the security of the data networks and associated control capabilities will be the responsibility of all personnel, and the operator will be a key person in enforcing this security.

Checking Your Knowledge

1. A single computer failure using a redundant DCS would likely cause:
   a. Environmental violation.
   b. Catastrophic fires.
   c. No operational interruption.
   d. Unit shutdown.

2. A node in a DCS consists of:
   a. A single control loop.
   b. A host computer and several local computers.
   c. A local computer and several control loops.
   d. A host computer and several control loops.

3. How many terminal blocks can a bus card control?
   a. 2
   b. 4
   c. 8
   d. 10

4. Which of the following provides the highest data transmission capacity?
   a. Coaxial cable
   b. Twisted pair shielded wire
   c. Fiber optic cable

5. (*True or False*) A DCS reduces the noise and drift that can make an analog system less accurate.

6. The reduced space requirement for DCS controls may result in: (Select all that apply.)
   a. Greater productivity in operation.
   b. All controls now being located in the field.
   c. Ease in troubleshooting.
   d. Greater flexibility in control station locations.

NOTE: Answers to Checking Your Knowledge questions are in the Appendix.

Student Activities

1. Given a diagram of a DCS, identify and label the components and describe the function of each.
2. Draw a schematic diagram of a typical DCS and identify its parts.
3. Given a series of "what-if" analysis of failures:
   a. Explain which sections will be affected.
   b. Identify the steps that will ensure that the affected parts will shut down safely.
   c. Discuss the effect of noise (crosstalk) or loose connections on the safety and performance of a DCS.
4. Given a DCS:
   a. Identify the components and describe the function of each.
   b. Trace the system from the sensor to computer (node), from the computer to the host, and then back to the final control element.
   c. Deliberately cause failure of a thermocouple and observe the results.

Chapter 20
Instrumentation Power Supply

 ## Objectives

After completing this chapter, you will be able to:

20.1 Provide an overview of plant power supply systems. (Introduction to Instrumentation 5*) p. 318

20.2 Identify the purpose and components of uninterruptible power supply (UPS) systems. (Switches, Relays, Alarms 5) p. 321

Key Terms

Alternating current (AC)—an electrical current that reverses direction periodically, 60 cycles per second in North and South America and 50 cycles per second in other parts of the world, **p. 318**.

Automatic transfer switch (ATS)—an electromechanical device that switches an electric load between two separate power sources (primary and secondary) depending on power availability from these sources; it is typically configured to be in the normal position when primary power is available, **p. 322**.

Battery—an electrical device consisting of two or more cells that convert chemical energy into electrical energy and provides a steady state of DC voltage, **p. 321**.

Breaker—a switchlike device in electrical panel boxes, used to keep electrical current from exceeding the recommended load, **p. 318**.

Charger—a device used to restore batteries to a proper electrical charge, **p. 322**.

Direct current (DC)—electrical current that flows in a single direction through a conductor, **p. 322**.

Generator—a device that converts mechanical energy into electrical energy, **p. 322**.

Inverter—a device used to change direct current (DC) into alternating current (AC), **p. 322**.

---

*North American Process Technology Alliance (NAPTA) developed curriculum to ensure that Process Technology courses will produce knowledgeable graduates to become entry-level employees in process technology. Objectives from that curriculum are named here in abbreviated form. For example, "(Introduction to Instrumentation 5)" means that this chapter's objective 1 relates to objective 5 of NAPTA's curriculum on instrumentation.

Uninterruptible power supply (UPS)—a backup power unit, usually consisting of large batteries, rectifier, inverter, battery charger, and transfer switch that provides continuous auxiliary power when the normal power supply is interrupted, **p. 319.**

UPS alarm—an audible or a visual signal used to draw attention to problem situations, **p. 322.**

20.1 Introduction

Electrical power is a very important part of industrial processes. It is used to power instrumentation, provide lighting, drive motors, power computer equipment, and much more. However, in order for us to use electrical power, we must harness it, convert it, and then route it so that it is usable. Harnessing, converting, and routing can be accomplished using devices such as automatic transfer switches, batteries and battery chargers, inverters, generators, and uninterruptible power supplies (UPSs).

Plant Power Supply Overview

Most high-voltage power lines that you see stretching across the landscape carry 138 kilovolts (kV) of **alternating current (AC)**. Alternating current is used in industrial applications because it can be stepped up or stepped down to meet different voltage requirements with the use of transformers (Figure 20.1).

Alternating current (AC) an electrical current that reverses direction periodically, 60 cycles per second in North and South America and 50 cycles per second in other parts of the world.

Figure 20.1 High-voltage power lines.

CREDIT: Zhao jiankang/Shutterstock.

When power reaches a plant substation, transformers step the voltage down from 138 kV to 13.8 kV before it enters the substation. Inside the substation, power is routed to various units within a plant. Figure 20.2 shows an example of power routing.

Once power reaches the unit, it is then routed to the unit's main **breaker** and disconnect switches located in an electrical equipment room, like the one shown in Figure 20.3. Generally speaking, the power from substation breakers (13.8 kV) is sent either to 13.8 kV switchgear or to step-down transformers that in turn feed the 480 volt MCCs. Then power is distributed to 480 volt three-phase plant equipment or to other step-down transformers that supply 240/120 VAC single-phase power.

Breaker a switchlike device in electrical panel boxes, used to keep electrical current from exceeding the recommended load

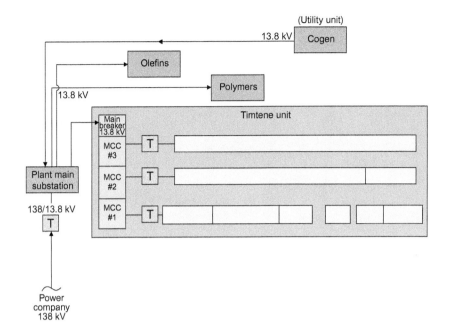

Figure 20.2 Power plant scheme example.

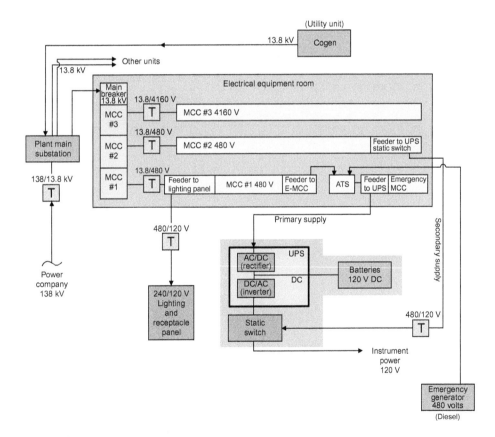

Figure 20.3 Power routing example.

Motor control centers (MCCs), which are shown in Figure 20.4, are rows of steel cubicles that contain electrical switching equipment. This equipment supplies the power needed by the many motors and devices that are used in the unit, as well as the power needed to run the emergency motor control centers (E-MCCs) and the **uninterruptible power supply (UPS)**.

Uninterruptible power supply (UPS)
a backup power unit, usually consisting of large batteries, rectifier, inverter, battery charger, and transfer switch that provides continuous auxiliary power when the normal power supply is interrupted.

Figure 20.4 Motor control center (MCC).

CREDIT: ETAJOE/Shutterstock.

Figure 20.5 shows an example of power flow from the main 138 kV power line, through the breakers and MCCs to the UPS.

Figure 20.5 Sample power flow.

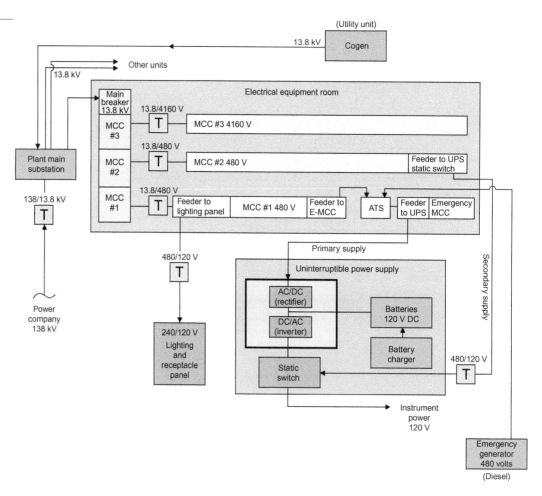

When studying electricity, it is important to know that every piece of electrical equipment in the unit is supplied by a cubicle in one of the MCCs. Each cubicle has its own breakers. Breakers are the primary point at which to connect or disconnect power from a piece of equipment (Figure 20.6). The following are the three states of a breaker and their descriptions.

Breaker Handle Position	Description
ON	The handle is in the "up" position; power to the equipment is on and conditions are normal.
TRIPPED	The handle is in the "middle" position; the breaker has been tripped and power to the equipment is off; the breaker will have to be reset in order to restore power to the equipment.*
OFF	The handle is in the "down" position; power to the equipment is off and conditions are normal.

*NOTE: Some breaker switches may be oriented horizontally. To reset a tripped breaker, pull the handle to the off position and then return it to the on position. Different plants will have different regulations on the procedures for resetting tripped breakers. For safety reasons, never stand directly in front of a breaker when resetting it.

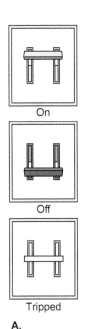
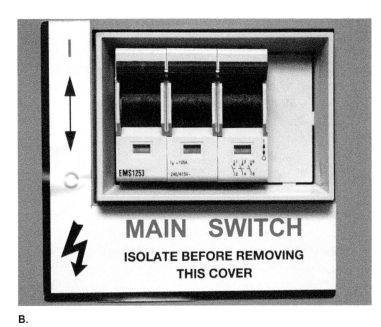

Figure 20.6 A. Breaker switch positions (ON, OFF, and TRIPPED). **B.** One example of an industrial breaker.

CREDIT: R.Moore/Shutterstock.

A.   B.

Most operations personnel do not normally throw breakers. However, process technicians may be expected to throw breakers that energize or de-energize individual pieces of equipment. Use of arc flashing protection is crucial whenever working with breakers. Training and proper PPE are required by law (OSHA) to operate certain electrical breakers, MCC cubicles, disconnects, and so on. In some plants, operations personnel do manipulate circuit breakers for lockout/tagout purposes. Be sure to follow your facility's policies and training protocols.

20.2 Uninterruptible Power Supply (UPS) Systems

Under normal operating conditions, equipment receives power from a traditional power source. In scenarios where there is no uninterruptible power supply (UPS), a power source failure would result in a loss of instrumentation and an abrupt system shutdown. With a UPS, however, an auxiliary power source is available to maintain the functionality of critical systems. Thus, the purpose of the UPS is to ensure an uninterrupted power flow to unit instrumentation.

A UPS uses **battery** power to replace a normal power source should the normal power source fail or become unavailable. UPS systems may range from very small (enough to power one computer) to very large (enough to power instrumentation for an entire process unit). It is important to note, however, that power supplied by a UPS is generally short-lived (30 minutes to 3 hours) since batteries can hold only a finite amount of energy. Thus, the intent of a UPS is to provide temporary power until an alternate power supply can be implemented (either house power is restored, or an emergency generator is employed) or a safe shutdown condition can be achieved.

Battery an electrical device consisting of two or more cells that convert chemical energy into electrical energy and provides a steady state of DC voltage.

Automatic transfer switch (ATS) an electromechanical device that switches an electric load between two separate power sources (primary and secondary) depending on power availability from these sources; it is typically configured to be in the normal position when primary power is available.

Direct current (DC) electrical current that flows in a single direction through a conductor.

Inverter a device used to change direct current (DC) into alternating current (AC).

Charger a device used to restore batteries to a proper electrical charge.

Generator a device that converts mechanical energy into electrical energy.

UPS alarm an audible or a visual signal used to draw attention to problem situations.

UPS Components

BATTERIES In a UPS system, batteries provide backup power to instrument systems. If normal power is lost, an **automatic transfer switch (ATS)** detects the outage and automatically switches the power source from normal to battery backup. (See Figure 20.5)

Because most computer systems and instruments require AC power, the power coming from the battery system must be converted from **direct current (DC)** to AC by an **inverter** located within a UPS. A battery **charger** begins charging when power is restored and is an integral part of the battery backup system. Once normal power is restored, the ATS switches the power source from battery backup to normal power.

AUTOMATIC TRANSFER SWITCHES During normal operation of the unit electrical system, the ATS plays no special role. Uninterrupted and unaltered electricity flows from the motor control center (MCC), through the ATS, to the emergency MCC (E-MCC).

The ATS monitors the presence of power from an MCC feeder. If that power fails, the ATS will start an emergency generator. The purpose of the ATS is to always supply 480 volt power to the E-MCC. However, the ATS cannot maintain absolutely continuous power to the E-MCC. There will be a temporary loss of power from the moment that the ATS senses a loss of feed and provides a start signal to the **generator**, until the generator gets up to speed and the ATS switches to it.

EMERGENCY GENERATOR The primary purpose of emergency generators is to provide electrical power at the desired voltage, phase, and frequency. The emergency generator system consists of a driver (diesel engine or gas turbine), AC generator, and necessary switches and instruments. Generation capacity design is based on the power demand of the electrical load. Many of these systems are designed to come on automatically with a loss of normal power. To ensure operability of these systems when needed, they should be operated and tested at prescribed intervals.

UPS ALARMS AND INDICATORS Due to the critical role played by the UPS, process technicians need to be familiar with **UPS alarms** and indicators, especially those used to indicate abnormal conditions. The following is a list of typical status indicators that might be found on a small computer UPS system:

Indicator	Description
Online	A light-emitting diode (LED) that illuminates when the unit is running nominally online power. **NOTE:** This indicator will be off if the "On Battery" LED is on.
On battery	An LED that illuminates when the UPS is running on battery power. **NOTE:** This indicator will be off if the "Online" LED is on.
Overload	A light that illuminates if someone tries to power up more equipment than the UPS system can accommodate. **NOTE:** If this light appears, the amount of equipment attached to the unit needs to be reduced, or the size of the UPS needs to be increased.
Site wiring fault	An LED, often located on the back of the unit, which illuminates if there is a problem in the circuit feeding the UPS (not the UPS itself). **NOTE:** A qualified electrician should examine the circuit if this LED illuminates.
Replace battery	An LED that illuminates if the battery is not charging properly or staying within normal operating parameters. **NOTE:** Battery charging and operating parameters are determined automatically by the UPS through periodic internal tests.
Low battery	An LED that illuminates when the UPS detects the battery is almost exhausted and is close to a shutdown. **NOTE:** UPS systems will normally shut down before the battery is completely drained as a fail-safe measure, since a complete battery discharge could damage the unit.
Battery status	A series of LEDs, often in a vertical "bar graph" configuration, that indicates the amount of battery power still remaining. **NOTE:** These types of indicators may not be found on lower-end (i.e., less expensive) UPS systems.
Load status	A series of LEDs, similar to battery status LEDs, that indicate how much of the unit's capacity is currently being drawn by the equipment attached to it. **NOTE:** These LEDs are good indicators as to whether or not the UPS can support additional equipment.

Be aware that, in addition to these visual indicators, some UPS systems also have audible indicators or alarms that are used to draw attention to problem situations. To learn more about indicators, especially those specific to your UPS system, refer to the user manual that came with the equipment.

Troubleshooting Tips and Warnings

Few, if any, process technicians will be responsible for servicing or maintaining a UPS system. However, process technicians may be asked to assist an I/E technician with preventative maintenance activities, so it is important to be aware of some of the features and hazards associated with UPS systems:

- **Risk of Electrical Shock**—UPS systems produce live AC current at the output even if the unit is disconnected from the primary power source. Therefore, personnel should always be aware of the presence of AC and the potential for electrical shock (see Figure 20.6).
- **Battery Maintenance**—The status of the battery charger and the batteries must be checked on a routine basis. It is important to maintain the proper operating conditions specified by the manufacturer.

Summary

Electrical power is used in industrial process for instrumentation, lighting, motors, computers, and equipment. A/C power is generally delivered via high-voltage wires and is converted from 138 kV to 13.8 kV to enter a substation. In the substation, it is generally converted to 480 volts to power MCCs. It can also be converted to 240 kV as needed.

Harnessing, converting, and routing are accomplished with such devices as automatic transfer switches, batteries and battery chargers, inverters, and generators. The goal is for plants to maintain an uninterruptible power supply (UPS) to allow continuous operation.

In some plants, operations personnel do manipulate circuit breakers for lockout/tagout purposes. Be sure to follow your facility's policies and training protocols. Process technicians must be aware of risks involved in working with electrical power. They need to be familiar with alarms and indicators that signal a problem with electrical supply.

Checking Your Knowledge

1. Most high-voltage power lines that you see stretching across the landscape carry _____ kilovolts (kV) of alternating current (AC).
   a. 13.8
   b. 40
   c. 128
   d. 138

2. To reset a tripped vertical breaker after investigating the issue that caused the breaker to trip:
   a. pull the breaker handle down.
   b. pull the breaker handle up.
   c. pull the breaker handle down and then up.
   d. pull the breaker handle up and then down.

3. What does a UPS use to replace normal power if the normal power source should fail?
   a. A set of batteries
   b. A generator
   c. An automatic transfer switch
   d. A secondary UPS

4. The _____ will signal the backup power source to turn on in the event of power failure from the primary power supply.
   a. main breaker
   b. transfer switch
   c. automatic transfer switch
   d. backup relay switch

5. The _____ ensures clean and continuous power to instrumentation.
   a. motor control center
   b. uninterruptible power supply
   c. transfer safety switch
   d. emergency generator

6. When does a site wiring fault indicator for a UPS light display?
   a. When the UPS is running on battery power
   b. When someone tries to power up more equipment than the UPS can handle
   c. When the battery is almost exhausted and close to shutdown
   d. When there is a problem in the circuit feeding the UPS

NOTE: Answers to Checking Your Knowledge questions are in the Appendix.

Student Activities

1. Given a drawing of a UPS, identify the location of each of the following indicators, explain when each one would be illuminated, and explain what each one indicates.
   a. Battery status
   b. Load status
   c. Low battery
   d. Replace battery
   e. Online
   f. On battery
   g. Overload
   h. Site wiring fault

2. In the failure scenario diagrammed below, power to motor control center (MCC) #1 has failed.
   a. Describe the problem(s) this scenario will cause.
   b. Explain why this could mean major trouble for the entire unit.
   c. Explain how this failure might impact the UPS.
   d. Explain why you should check to see that generator has started and that the UPS is not running off batteries.

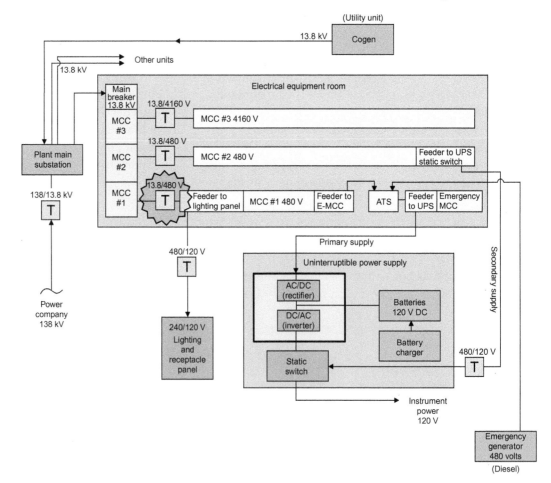

3. In the failure scenario diagrammed below, a breaker trips in the MCC #1 feeder to the ATS.
   a. Describe the problem(s) this scenario will cause.
   b. Identify what corrective action you should take if the breaker is tripped.
   c. Explain why you should check to see that generator has started and that the UPS is not running off batteries.

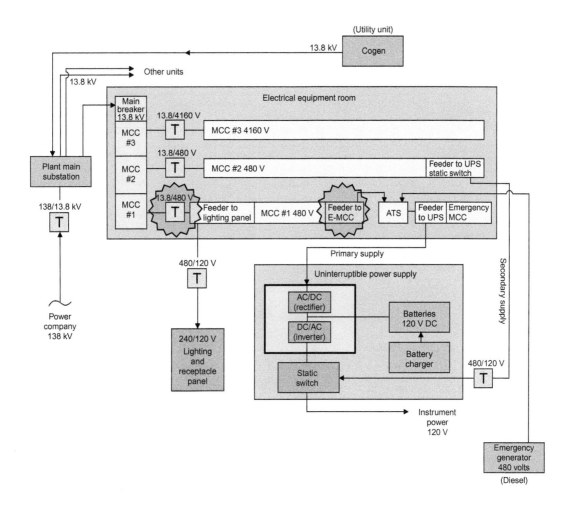

4. In the failure scenario diagrammed below, either power to motor control center (MCC) #1 has failed or a breaker has tripped in the MCC #1 feeder to the ATS and the generator has not started.
   a. Describe the problem(s) this scenario will cause.
   b. Identify what corrective or investigative action(s) you should take with regard to UPS battery voltage levels, the emergency generator, and all UPS alarms.
   c. Explain why you should check to see that generator has started and that the UPS is not running off batteries.

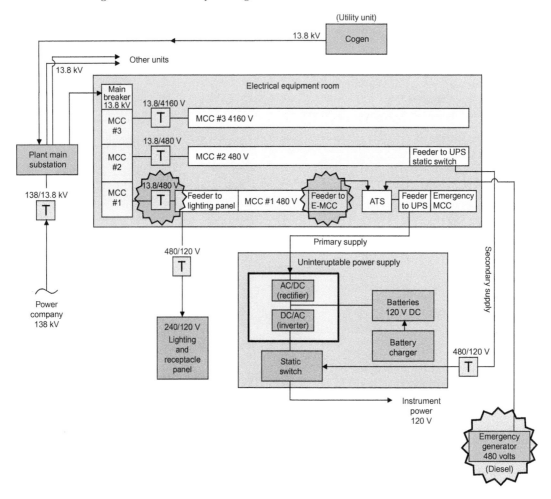

Chapter 21
Emergency Shutdown (ESD), Interlocks, and Protective Devices

 Objectives

After completing this chapter, you will be able to:

21.1 Identify terms associated with emergency shutdown systems:
startup permissives
interlocks
emergency shutdown. (ESD). (NAPTA Interlocks and Safety Features 1, 2, 3*) p. 329

21.2 Describe common types of alarms:
high alarm
high high alarm shutdown
low alarm
low low alarm shutdown
deviation
vibration
gas detection. (NAPTA Switches, Relays, and Alarms 6) p. 333

21.3 Identify ways of avoiding shutdowns:
redundancy
voting logic. (NAPTA Interlocks and Safety Features 2) p. 336

---

*North American Process Technology Alliance (NAPTA) developed curriculum to ensure that Process Technology courses will produce knowledgeable graduates to become entry-level employees in process technology. Objectives from that curriculum are named here in abbreviated form. For example, "(NAPTA Interlocks and Safety Features 1, 2, 3) means that this chapter's objective 1 relates to objectives 1, 2, and 3 of NAPTA's course content on interlocks and safety features.

21.4 Explain methods for testing and resetting ESDs and issues that are associated with bypassing an ESD logic. (NAPTA Interlocks and Safety Features 3) p. 338

Key Terms

Bypass—a short patch cable or wire used to complete an electrical circuit, usually temporarily, for testing or diagnostic purposes (also called a *defeat* or *jumper*), **p. 340.**

Defeat—*See* bypass, **p. 340.**

Deluge system—an emergency shutdown (ESD) system component that, when a leak is detected, sprays the area of leakage with large quantities of water droplets to dilute the concentration of vapors so they are below the toxic limit for personnel or the lower explosive limit (LEL), thereby reducing danger to personnel and the potential for fire, **p. 339.**

Deviation alarm—an alarm that is activated when the difference between two variables exceeds a set limit (e.g., in a particular catalyst bed, the difference or deviation between the outlet temperature and the inlet temperature should not exceed 50 degrees Fahrenheit); in this type of alarm, the actual temperature value is not important, what is important is the difference between the two values), **p. 335.**

Emergency shutdown (ESD) system—a system consisting of interlocks, breakers, sensors, and other equipment responsible for automatically shutting down equipment in an extremely abnormal situation in order to protect people and equipment and reduce the risk of catastrophe, **p. 330.**

High (H) alarm—the first alarms triggered when a process variable (e.g., fluids in a tank) rises above a predetermined high level; the purpose of this alarm is simply to notify an operator that the level is abnormal, **p. 333.**

High high (HH) alarm—alarm triggered when a process variable (e.g., fluids in a tank) continues to rise above a predetermined maximum level after a high (H) alarm has already been triggered. The purpose of this alarm is to notify a process technician that the alarm condition of a process variable is becoming increasingly abnormal and to initiate a shutdown or other corrective action regardless of any action by the process technician, **p. 333.**

Interlock—a system for connecting mutually independent equipment that can stop upstream equipment when downstream equipment is shut down, **p. 330.**

Jumper—*See* bypass, **p. 340.**

Low (L) alarm—the first alarm triggered when a process variable (e.g., fluids in a tank) drops below a predetermined low level; the purpose of this alarm is to notify a process technician that the level is abnormal, **p. 334.**

Low low (LL) alarm—alarm triggered when a process variable (e.g., fluids in a tank) continues to drop below a predetermined minimum level after a low (L) alarm has already been triggered. The purpose of this alarm is to notify a process technician that the alarm condition of a process variable is becoming increasingly abnormal and to initiate a shutdown or some other corrective action (e.g., shutting off power to a pump to prevent cavitations, which could result in pump damage), **p. 335.**

Permissives—a set of conditions designed to ensure safe operations that must be met before a piece of equipment can be turned on, **p. 329.**

Redundancy—a design feature that provides more than one operation for accomplishing a given task, so the failure of one operation does not impair the system's ability to operate, **p. 337.**

Vibration alarm—an alarm designed to be triggered if the level of vibration increases above an acceptable level in order to protect vibrating equipment, **p. 335.**

Voting logic—computer logic that analyzes the signals from several devices, all of which are monitoring the same condition; it will initiate shutdown if a majority or a single high priority variable of the monitoring devices signals a dangerous condition, **p. 337.**

21.1 Introduction

The process industry uses equipment and processes that can be extremely hazardous. In order to ensure a safe work environment and reduce the likelihood of equipment damage or catastrophe, procedural and equipment safeguards must be implemented. These safeguards include safe startup procedures, equipment designed to detect and warn of dangerous conditions, automatic devices without operator input, and equipment and procedures necessary for shutdown in the event dangerous conditions occur.

Some of the main devices responsible for ensuring system safety include **permissives**, interlocks, sensors, and alarms. In addition to these devices, specific operating procedures, such as those used when installing a bypass or jumper, are also employed.

Permissives a set of conditions designed to ensure safe operations that must be met before a piece of equipment can be turned on.

Permissives and Interlocks

When operating equipment, it is imperative that safeguards be implemented prior to startup and during operation. Two of the most common types of safeguards include permissives and interlocks.

Startup Permissives

Before a piece of equipment can be turned on, a set of conditions must be met to ensure safe operation. This set of conditions is called the startup permissives for that piece of equipment.

For example, startup permissives for a furnace may include verification of fuel pressure, LEL check (a check for the presence of hydrocarbons), and verification of an adequate steam flow purge. Permissives may be checked manually in a step-by-step manner or by programmed computers. The primary purpose of startup permissives is to ensure that it is safe to put a particular piece of equipment into operation (Figure 21.1).

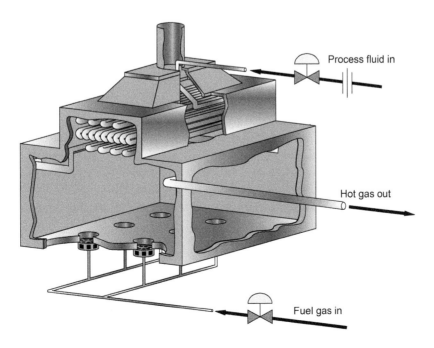

Figure 21.1 Furnace permissives.

Permissives should be clearly documented in plain, easy-to-understand language. They should never be tampered with or bypassed. In addition, their duration should never be shortened without written management approval because doing so could jeopardize safety.

Interlocks

Although process emergencies are rare, the potential for danger to lives and equipment is high. Therefore, a means for detecting dangerous conditions and for safe automatic shutdown of a process or a piece of equipment is imperative. All critical equipment is designed with a shutdown circuit for accomplishing this task. This shutdown circuit is called an **interlock**. Figure 21.2 shows an example of an interlock.

Interlock a system for connecting mutually independent equipment that can stop upstream equipment when downstream equipment is shut down.

Figure 21.2 Distillation and reboiler.

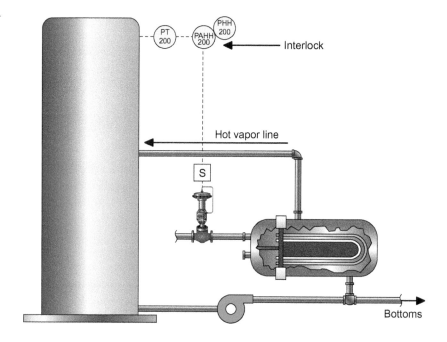

In this example, an interlock prevents the pressure within a distillation column from exceeding a safe limit by executing the steps below:

1. A monitoring device signals that a dangerous pressure has been reached.
2. The interlock action closes the steam valve going into the reboiler.
3. The product pressure from the reboiler into the distillation column begins to drop.
4. Once pressure has fallen to a safe level, the steam valve may be manually reset.

A basic interlock consists of a process signal activating a circuit (including relays or contacts) that causes a solenoid to shut off or completely open a control valve. Interlocks may be hardwired, not PC controlled, or programmed into a programmable logic controller (PLC) like the one shown in Figure 21.3.

Because interlocks play such a critical role in process safety, it is imperative that process technicians have access to the most up-to-date information on the specific interlocks used in their plant. Process technicians need to also be aware that bypassing interlocks typically requires management approval.

Emergency shutdown (ESD) system a system consisting of interlocks, breakers, sensors, and other equipment responsible for automatically shutting down equipment in an extremely abnormal situation in order to protect people and equipment and reduce the risk of catastrophe.

Emergency Shutdown (ESD)

Emergency shutdown (ESD) systems play a critical role in protecting lives and equipment. Some of the most common items in an ESD system include interlocks, breakers, and sensors.

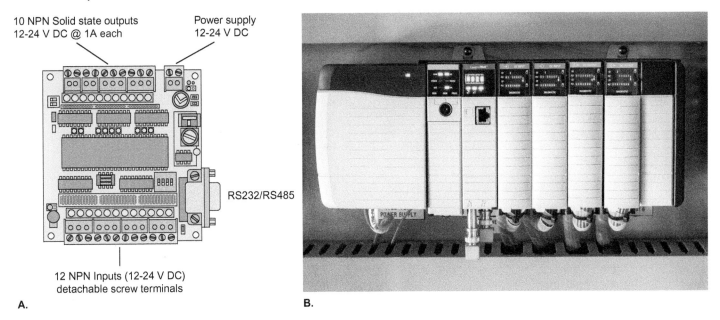

Figure 21.3 Programmable logic controller (PLC). **A.** Illustration. **B.** Photo of PLC.
CREDIT: B Xmentoys/Shutterstock.

Interlocks in Emergency Shutdown

An emergency shutdown (ESD) for a piece of equipment or process may involve a number of related interlocks.

For example, a furnace, like the one shown in Figure 21.4, could be equipped with interlocks designed to shutdown fuel flow in response to any one of several conditions such as:

- Excessively high stack temperature
- Low or no flow of the process fluid
- Low fuel gas inlet pressure or flameout
- Loss of draft.

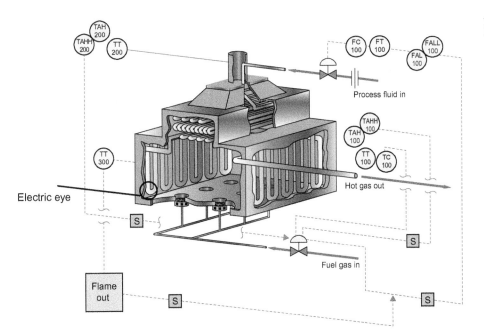

Figure 21.4 Furnace ESD.

In another example, a large compressor, like the one shown in Figure 21.5, may include interlocks to shut down the compressor in the event of:

- High suction drum level
- Excessively low suction pressure
- High discharge temperature
- Excessive vibration
- Lube or seal oil failure.

Figure 21.5 Compressor ESD.

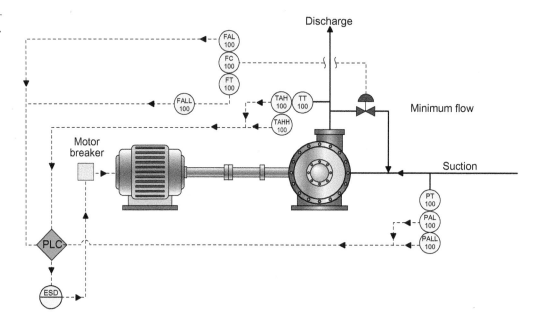

Finally, a third example of emergency interlocks is a lube oil/seal system like the one shown in Figure 21.6. Lubrication is critical for rotating or reciprocating equipment. Lube oil system permissives or interlocks are designed to ensure that the equipment will not operate unless the lube oil system is functioning properly.

Figure 21.6 Lube system interlocks.

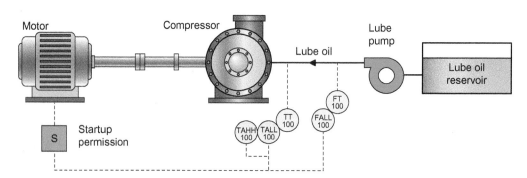

In many types of equipment, lube oil must achieve a certain minimum temperature to ensure the viscosity range appropriate for lubrication. Conversely, if the lube oil temperature exceeds a certain value, the oil will disintegrate and "carbon up." Therefore, lube oil systems contain both high- and low-temperature alarms.

Breaker Switches in Emergency Shutdown

A breaker switch is a safety device used to provide protection against overcurrent. In the event of high temperature, a bimetallic strip expands inside the breaker, causing the breaker to trip. Once a breaker is tripped, it must be manually reset (**NOTE:** Many large motors have restrictions regarding how many times they can be started).

Electric Eye Sensors in Emergency Shutdown

Some boilers and furnaces are equipped with sensors called *electric eyes* (also referred to as *red eyes* or *flame scanners*) that monitor the flame burning in the combustion zone. Absence of flame means that the fuel is pouring out of the burner and is not being burned completely. If all burners, or at least a large number of them, have ceased burning, the condition is called *flameout*. Because of the dangers associated with gas that has not been properly combusted, flameout immediately activates the shutdown of fuel gas to equipment.

21.2 Alarms

ESD monitoring equipment is designed to (1) automatically shut down equipment if a dangerous condition occurs, and (2) warn operations personnel in such a circumstance. These warnings can be either auditory, visual, or both and may vary by severity and cause. Four common types of alarms include high (H), high high (HH), low (L), and low low (LL). See Figure 21.7.

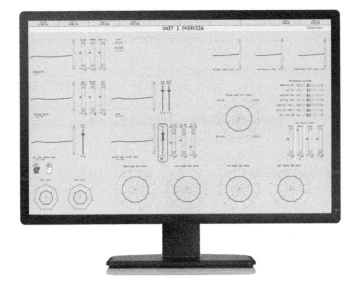

Figure 21.7 Alarms example on DCS screen.
CREDIT: © Emerson 2019.

High Alarm (H)

(**NOTE:** In the following example, the process variable used for illustration purposes is level. However, any process variable can have high, high high, low, or low low interlocks and shutdowns.)

High (H) alarms are the first alarms triggered when a process variable (e.g., fluid in a tank) rises above a predetermined high level. The purpose of this alarm is to notify a process technician that the level is not normal (high).

Consider the example in Figure 21.8. In this example, a tank is equipped with a high level (H) alarm that will come on if the fluid level in the tank reaches 70 percent of tank capacity. If this alarm is sounded, the process technician must acknowledge the alarm to silence it, and then look into the cause of the alarm (e.g., the control valve needs to be closed more). As long as the level stays above the preset value (70 percent for this example), the alarm light stays lit. As the level goes below 70 percent, the alarm light turns off.

High High Alarm Shutdown (HHSD)

High high (HH) alarms are triggered when a process variable (e.g., fluid in a tank) continues to rise and reaches a predetermined maximum level after a high (H) alarm has already been triggered. The purpose of this alarm is to notify a process technician that a shutdown has been initiated and to automatically take corrective action to variable level from rising.

High (H) alarm the first alarms triggered when a process variable (e.g., fluids in a tank) rises above a predetermined high level; the purpose of this alarm is simply to notify an operator that the level is abnormal.

High high (HH) alarm alarm triggered when a process variable (e.g., fluids in a tank) continues to rise above a predetermined maximum level after a high (H) alarm has already been triggered. The purpose of this alarm is to notify a process technician that the alarm condition of a process variable is becoming increasingly abnormal and to initiate a shutdown or other corrective action regardless of any action by the process technician.

Figure 21.8 Tank alarms (H and HH).

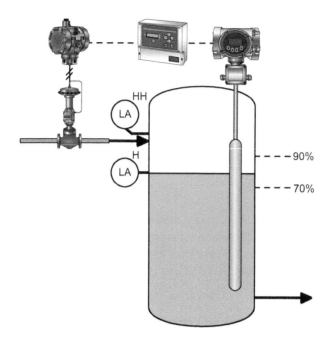

Again, consider the example in Figure 21.8. In this example, a high (H) alarm would have been triggered when the fluid level reached 70 percent. If the tank level continued to rise above 90 percent, the controller would provide another audiovisual alarm (an HH alarm) and would initiate a safe shutdown of the system (typically the actuation of a solenoid that shuts off the inlet valve). Once shutdown is initiated, safety mechanisms prevent the system from being restarted until the level has fallen below the high high (HH) preset values determined by safety considerations and process needs.

Low Alarm (L)

Low (L) alarm the first alarm triggered when a process variable (e.g., fluids in a tank) drops below a predetermined low level; the purpose of this alarm is to notify a process technician that the level is abnormal.

Low (L) alarms are the first alarms triggered when a process variable (e.g., fluid in a tank) drops below a predetermined low level. The purpose of this alarm is to notify a process technician that the level is not normal (low). Whether a given process needs a low (L) alarm depends on the process requirements. A tank, like the example shown in Figure 21.9, may

Figure 21.9 Tank alarms (L and LL).

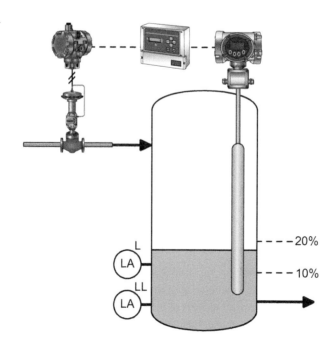

include a low level (L) alarm because a very low fluid level may cause cavitation problems in the outlet pump. Similarly, certain heat exchangers may also need a low temperature (L) alarm if there is a risk that low temperatures may lead to solidification or freezing of certain chemicals.

Low Low Alarm Shutdown (LLSD)

Low low (LL) alarms are triggered when a process variable (e.g., fluid in a tank) continues to drop and reaches a predetermined minimum level after a low (L) alarm has already been triggered. The purpose of this alarm is to notify a process technician that the level is becoming increasingly abnormal (low) and initiate a shutdown or some other corrective action (e.g., shutting off power to a pump to prevent cavitation, which could result in pump damage).

Deviation Alarm

In certain processes, like the one shown in Figure 21.10, the absolute value of a variable may not be important, but a certain differential is important. For instance, in a particular catalyst bed, the difference (deviation) between the outlet temperature and the inlet temperature should not exceed 50 degrees Fahrenheit (27 degrees Celsius). In this case, signals from two thermocouples are sent to a different unit or a computer. If the difference is greater than 50 degrees Fahrenheit (10 degrees Celsius), an audiovisual **deviation alarm** is activated. This type of alarm can also be used with other variables, such as the pressure found in a distillation column.

Low low (LL) alarm alarm triggered when a process variable (e.g., fluids in a tank) continues to drop below a predetermined minimum level after a low (L) alarm has already been triggered. The purpose of this alarm is to notify a process technician that the alarm condition of a process variable is becoming increasingly abnormal and to initiate a shutdown or some other corrective action (e.g., shutting off power to a pump to prevent cavitations, which could result in pump damage).

Deviation alarm an alarm that is activated when the difference between two variables exceeds a set limit (e.g., in a particular catalyst bed, the difference or deviation between the outlet temperature and the inlet temperature should not exceed 50 degrees Fahrenheit); in this type of alarm, the actual temperature value is not important, what is important is the difference between the two values).

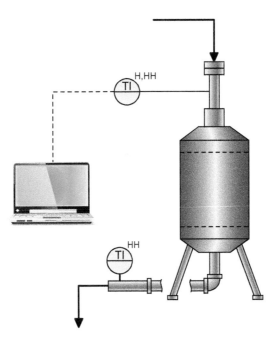

Figure 21.10 Reactor with deviation alarm.

Vibration Alarm and Shutdown

Large rotating machinery is often equipped with vibration monitoring since running such machinery with high vibrations could damage it and/or cause catastrophic failure resulting in danger to nearby personnel.

Typically, radial and axial vibration monitoring is performed by instruments known as proximity probes, located at several strategic points (e.g., radial and thrust bearings and gear units). If vibrations exceed a certain value, an audiovisual alarm is activated. If the level of vibration increases once the alarm has been triggered, a second alarm will initiate shutdown to protect personnel and the equipment. Figure 21.11 shows an example of a compressor with a **vibration alarm**.

Vibration alarm an alarm designed to be triggered if the level of vibration increases above an acceptable level in order to protect vibrating equipment.

Figure 21.11 Compressor with vibration alarm.

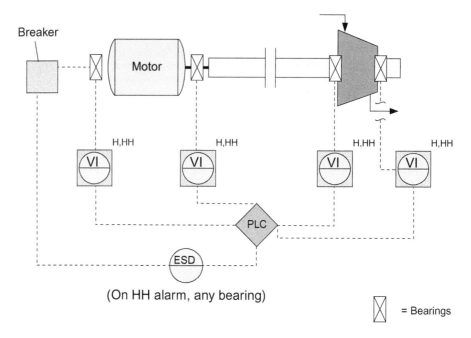

Gas Detection Alarm and Suppression Systems

Although modern pumps and sealing systems are designed to minimize leaks, the possibility of leaks still exists. A large spontaneous leak of hydrocarbons, or even a small leak over an extended period of time, can lead to fire or explosion and threaten the loss of lives and equipment. Therefore, facilities that handle flammable materials almost always employ several gas detection alarms located strategically throughout the plant. Depending on the type of flammable vapors involved, the monitors may employ several different gas detection techniques including infrared (IR), ultraviolet (UV), or cloud chamber detection.

Typically, large units have several gas detection monitors that provide an audiovisual alarm to the control room. The alarm from each of these monitors identifies the area or location of the leak.

Many gas detection alarms are also tied to deluge systems, like the one shown in Figure 21.12. Deluge systems spray the area with large quantities of water to help fight the fire, and to keep metal pipes, metal tanks, and metal structural members cool so that they retain their strength. This reduces the potential for explosion and/or fire and also protects lives and equipment.

To ensure continual reliability and avoid nuisance tripping of deluge systems, monitor systems should be checked periodically.

Figure 21.12 Compressor with deluge system triggered by a gas detection alarm.

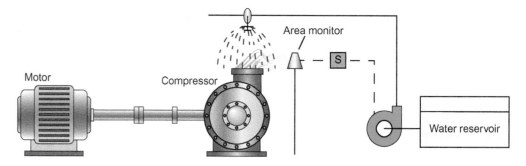

21.3 Avoiding Unnecessary Shutdowns

Unnecessary shutdowns are time consuming and expensive. In order to reduce the likelihood of a shutdown, a variety of hardware, software, and techniques are employed. Two of the most common techniques used are the incorporation of redundant hardware and computer logic called voting logic.

Redundancy

Previous chapters have used the term **redundancy** to describe functional duplication of plant components to ensure uninterrupted operations. Two examples of redundancy are found in distributed control systems (DCSs) and uninterruptible power supplies (UPSs).

In a DCS, more than one computer is capable of handling any particular operation. Therefore, if one computer fails, the data can be rerouted and another computer in the DCS can assume the task. This linking is routinely called a network.

An operating unit may have its data backed up routinely by an adjoining unit so that data will be available in case of an emergency. Some plants also back up their entire network once a day, for example at midnight, on either the main frame of the plant or at an offsite location like a corporate headquarters.

In a plant UPS system, batteries are used as a backup power source in the event of a power failure or power decrease. This ensures that power flow to critical instrumentation will continue, thereby reducing the likelihood of safety concerns for plant personnel and unnecessary shutdown.

Redundancy a design feature that provides more than one operation for accomplishing a given task, so the failure of one operation does not impair the system's ability to operate.

Voting Logic

Redundancy can be designed into an alarm system to minimize nuisance or spurious shutdowns. A common method of redundancy involves a computer function called **voting logic**, which is shown in Figure 21.13.

Voting logic computer logic that analyzes the signals from several devices, all of which are monitoring the same condition; it will initiate shutdown if a majority or a single high priority variable of the monitoring devices signals a dangerous condition.

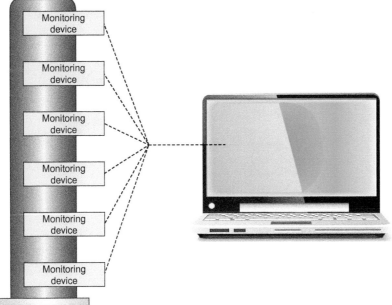

Figure 21.13 Voting logic.

With voting logic, a computer program analyzes the signals from several devices, all of which are monitoring the same condition, and will initiate shutdown only if a predetermined number of the monitoring devices signal a dangerous condition.

Consider temperature control in a reactor. In this scenario, thermocouples are used to monitor reactor temperature. If the temperature exceeds a certain value, shutdown is activated. However, if a faulty thermocouple is installed, the result might be a costly and unnecessary shutdown. Therefore, a voting logic approach is often the best solution.

In a voting logic approach, several thermocouples are installed, and each sends readings to a microprocessor. The microprocessor then compares the incoming readings and initiates a shutdown only if a predetermined number of the thermocouples are registering unacceptable values.

21.4 Emergency Shutdown (ESD) Resetting and Testing

The purpose of any emergency shutdown (ESD) system is to shut down a system or piece of equipment in the event of an unsafe condition. Because it is a safety shutdown, an automatic reset would not be feasible. Thus, ESD systems must be reset manually. Resets can be either simple (single push-button type) or multistage. They also may include a manual relink (reconnection) of the field device.

Simple Reset

Relatively simple ESD systems, such as those involving only one or two related loops, can be reset simply by pushing a button (e.g., some lube or burner management systems are designed with a single push-button configuration). Figure 21.14 shows an example of a simple reset.

Figure 21.14 Simple reset.

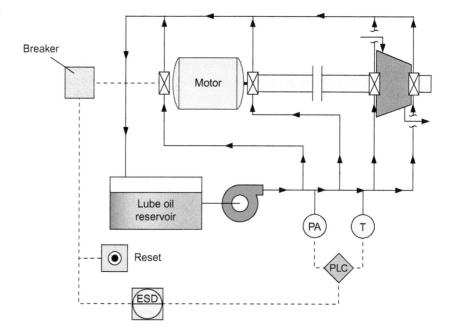

Multistage Reset

Typically, ESD systems that involve many loops utilize multistage resets like the one shown in Figure 21.15. In this example, an ESD system on a distillation column has shutdown product feed and steam to the reboiler. In order to get the system up and working again, each component must be reset individually and in a particular order as well.

Reset of a multiple-stage system should take place only after each component of the system is carefully checked and the cause of the shutdown has been determined. Knowing the cause of the shutdown may be useful in avoiding a recurrence.

While ESDs require manual reset, some alarms can be reset automatically. This is allowed because efficiency concerns outweigh the minimal risk involved in the auto reset of an alarm.

Testing an ESD

ESD testing is necessary to ensure the system will function properly when needed. A typical method of checking involves the testing of a measurement instrument in order to activate the ESD system and alarms. Once an instrument has been tested, one of two results may occur: either (1) the system will perform an actual shutdown of a valve or a final control element with all systems activated (full test); or (2) the system will perform a shutdown with some systems modified or disengaged (dry testing).

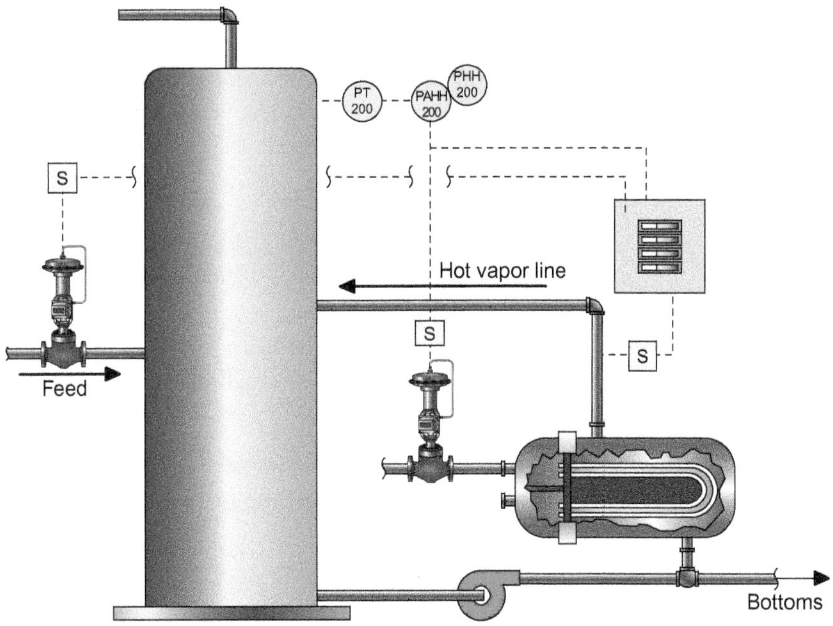

Figure 21.15 Multiple reset.

Full Testing

Full testing, which is almost always done when a plant or process is being commissioned for the first time or following a turnaround, involves the application of a specific electrical charge to an ESD instrument in order to induce the shutdown of a valve or final control element.

While full testing provides a high degree of confidence in the system, it does have a downside, and that is the potential for equipment damage. For this reason, many individuals opt to perform a dry test, which is less conclusive but is also less likely to damage critical and expensive equipment.

Dry Testing

Dry testing is similar to full testing in that the alarm and shutdown devices are activated to induce a shutdown. However, in dry testing, the actual shutdown system is disengaged, so the test cannot be fully executed (Figure 21.16).

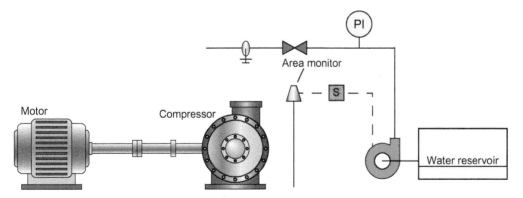

Figure 21.16 Dry test.

Consider the full test mentioned earlier. In a full test, the **deluge system** is activated, and water is dispensed just as if the actual problem were occurring. If this same system were to be tested using a dry test method, the signal would still be sent to the deluge system, but the water would not be dispensed.

A dry test can allow personnel to observe whether or not a deluge system mechanism is activated. However, since the system is unable to execute completely, dry testing cannot give any assurance that the system will function as specified in the design if it is needed.

Deluge system an emergency shutdown (ESD) system component that, when a leak is detected, sprays the area of leakage with large quantities of water droplets to dilute the concentration of vapors so they are below the toxic limit for personnel or the lower explosive limit (LEL), thereby reducing danger to personnel and the potential for fire.

Practical Tips

ESD system testing and calibration should be done when the plant, or section of the plant, is down. Close coordination between process technicians and instrumentation personnel is essential. All test findings should be carefully documented for regulatory, legal, and technical reasons.

ESD Bypassing

Bypass a short patch cable or wire used to complete an electrical circuit, usually temporarily, for testing or diagnostic purposes (also called a *defeat* or *jumper*).

Jumper See bypass.

Defeat See bypass.

There may be times when it is necessary to override an interlock. In order to perform this override, a piece of electrical wire, called a **bypass** (also referred to as a **jumper** or **defeat**), can be installed.

NOTE: While bypasses can be a convenient and necessary expedient, misuse may result in damage to expensive equipment. Therefore, top management approval is mandatory before bypasses can be installed.

Bypass systems can be designed to permit testing of alarms and trips during operation. Close coordination between process technicians and instrument personnel is essential in order for these tests to be successful. Upon completion of the work, the bypass system must be deactivated and documented for regulatory, legal, and technical reasons.

Consider the example shown in Figure 21.17. In this example, it has become necessary to perform repairs on a compressor while the compressor is running. However, the compressor is equipped with an interrupt designed to initiate a shutdown if a certain temperature is exceeded. It is considered likely that this limit will be exceeded during the repair. In this situation, a bypass may be installed to prevent a high-temperature shutdown.

Figure 21.17 Compressor with interlock defeats.

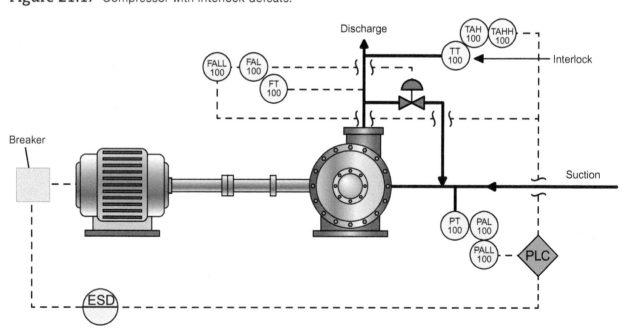

ESD Bypass Management

ESD systems are designed to protect lives and equipment. In the event that a bypass is necessary, companies must have established procedures that are followed carefully in order to minimize risk.

Summary

The process industries use equipment and processes that can be extremely hazardous. In order to ensure a safer work environment and reduce the likelihood of equipment damage or catastrophe, procedural and equipment safeguards must be implemented and maintained. These safeguards include the use of safe equipment startup procedures, called permissives, and emergency sensing and shutdown equipment, called interlocks. Other safety devices, such as breakers, sensors, and alarms, may also be incorporated.

Within many systems, there are several types of alarms that may be sounded if conditions become off normal. These include high (H), high high (HH), low (L), and low low (LL) alarms. High alarms indicate when process variables rise above predetermined levels. Low alarms indicate when process variables drop below predetermined levels. Both of these alarm types may trigger a shutdown or other corrective action if the levels go too far beyond the threshold. Low low and high high alarms can be incorporated into systems to respond automatically when levels are more extreme in order to protect plant personnel, equipment, and the process.

In addition, many systems also contain deviation, vibration, and gas detection alarms. Deviation alarms notify the process technician when the difference between two variables exceeds acceptable levels. Vibration alarms notify the process technician if the amount of vibration generated by rotating equipment exceeds acceptable limits (excessive vibration can damage equipment). Gas detection alarms notify process technicians of gas leaks and may trigger various suppression systems, such as a deluge system which douses the area with water in order to dilute the concentration of vapors, so they are below the lower explosive limit (LEL).

In addition to alarm and suppression systems, many systems also include components that help protect the equipment and the process by avoiding unnecessary shutdowns. Two of these components include system redundancy (e.g., a UPS system supplying backup power to critical systems during a power failure) and computer logic (which measures inputs from multiple sensors and only initiates a shutdown if a majority of the sensors are registering unacceptable values).

If shutdown is initiated, many systems require a manual reset in order to make them operational again. Resets can either be simple (e.g., a single push button) or multistage (in which each component must be individually checked and reset before the system will be operational).

In order to ensure an alarm or suppression system will function properly during abnormal conditions, performance testing is needed. Testing can either be full testing or dry testing. Full testing is almost always done when a plant or process is being commissioned for the first time or following a major turnaround. During a full test, the emergency shutdown system (ESD) is triggered and suppression system (e.g., a deluge system) is activated. While this is more accurate than a dry test, there are limitations, namely the likelihood of equipment damage. Dry testing is similar to full testing in that the ESD is triggered and a signal is sent to the suppression systems. However, the suppression system in this case is disengaged, so it does not dispense.

Finally, when working with systems, there may be times when it is necessary to override an interlock using a bypass, or jumper. If a bypass is to be used, management must first approve the use of the bypass and document their approval on a bypass form. Once the work is complete, bypasses must be deactivated and the removal documented for regulatory, legal, and technical reasons.

Checking Your Knowledge

1. Which of the following is a set of conditions required for equipment startup?
   a. Interlocks
   b. Proximity probe
   c. Bypass
   d. Permissives

2. _____ is (are) critical for rotating or reciprocating equipment.
   a. Interlocks
   b. Lubrication
   c. Permissives
   d. Safety

3. What is the first alarm triggered when a process variable rises above a predetermined high level?
   a. High
   b. Low
   c. High high
   d. Low low

4. A _____ is an instrument used to detect vibration.
   a. proximity probe
   b. vibratect
   c. vibro-indicator
   d. contact probe

5. What redundancy method involves the use of a computer?
   a. UPS
   b. DCS
   c. Voting logic
   d. Proximity probe

6. In order to restart equipment after an ESD, _____ of each piece of equipment will be necessary.
   a. interlock
   b. bypass
   c. reset
   d. defeat

7. (*True or False*) Multiple reset means that a single button reset of each piece of equipment will be necessary.

8. A _____ test allows some systems to be disengaged. This type of testing is less disruptive to operations, but does not completely verify that the full ESD is operational.
   a. dry
   b. wet
   c. full
   d. partial

NOTE: Answers to Checking Your Knowledge questions are in the Appendix.

Student Activities

1. Given a diagram of a relatively simple loop, identify the instrumentation and the alarm and shutdown systems.

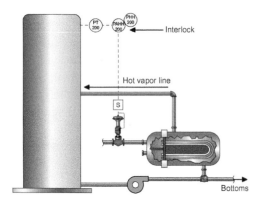

2. Take a single interlock system and identify components that can fail and can cause spurious shutdown.

3. Provide students with a P&ID illustrating several pieces of common plant equipment. Have students identify each interlock system along with its components. Have students discuss what each ESD is designed to protect and to protect against. For example:
   - Level applications: Can the tank overflow? What are the safety and environmental consequences of such an overflow? In many applications, you may provide an H level alarm and a HH level shutdown.
   - Pressure applications: When would you use a high-pressure alarm, followed by an HH pressure shutdown? Although relief valves (RVs) provide a protection against overpressure, an HH pressure shutdown gives an added assurance of protection.
   - Temperature applications: If a potential exists for a runaway reaction (where the temperature goes out of control), an HH temperature shutdown is worthy of consideration.
   - Flow applications: Some situations may require you to shut down the process. A severe flood in a distillation column would necessitate that the feed be shutdown.

Chapter 22
Instrumentation Malfunctions

 Objectives

Upon completion of this chapter, you will be able to:

22.1 Recall the methods used for determining if a sensing or measuring device is malfunctioning. (NAPTA Instrumentation Troubleshooting 1, 3*) p. 344

22.2 Describe the failure modes of the following:
temperature elements
thermocouples
RTDs
level floats
flow elements
pressure elements
analytical elements. (NAPTA Instrumentation Troubleshooting 7) p. 349

22.3 Explain how a control loop will respond to typical malfunctions in the following:
primary sensing elements
transmitters
controllers
final control elements. (NAPTA Instrumentation Troubleshooting 9) p. 352

---

*North American Process Technology Alliance (NAPTA) developed curriculum to ensure that Process Technology courses will produce knowledgeable graduates to become entry-level employees in process technology. Objectives from that curriculum are named here in abbreviated form. For example, "(NAPTA Instrumentation Troubleshooting 1, 3)" means that this chapter's objective 1 relates to objectives 1 and 3 of NAPTA's course content about instrumentation malfunctions and troubleshooting.

Key Terms

Noisy signal—a signal that fluctuates dramatically and that is most likely the result of a loose sensor connection, **p. 348.**

22.1 Introduction

Process technicians rely upon instrumentation to monitor all process variables and ensure safety. It is important that all instrumentation be accurate and functioning properly since malfunctioning instruments may fail to indicate an actual problem or may indicate process problems where they do not exist.

In the event of an instrument malfunction, process technicians need to know how to identify the malfunction, isolate it, and determine the possible cause. In order to do this, technicians must be familiar with various types of instrumentation (e.g., pressure, temperature, level, flow, and analytical elements) and the symptoms associated with their failure.

Identifying Instrument Malfunction

Process technicians work closely with instrumentation technicians during instrumentation troubleshooting. A basic understanding of typical instrumentation problems and troubleshooting methods is essential. Technicians need to know how to determine the source of the problem. This chapter will identify the types of failures and their causes.

Isolating the Problem Source

When troubleshooting, technicians should always use the process of elimination to systematically identify and isolate faulty components (Figure 22.1). Take, for example, a control loop.

A control loop is an integrated system consisting of a sensor, a controller, a transmitter, and a final control element. A malfunction in any of these components affects the loop performance and stability. Therefore, it may be necessary to use the process of elimination to verify signals into and out of each instrument in turn in order to isolate the faulty component.

A number of problems could cause the control loop to be unstable. Valve failure could be caused by such factors as worn elastomers and seat seals; debris in the pipeline; operating in excessively high-temperature situations; and improper valve installation, maintenance, and assembly.

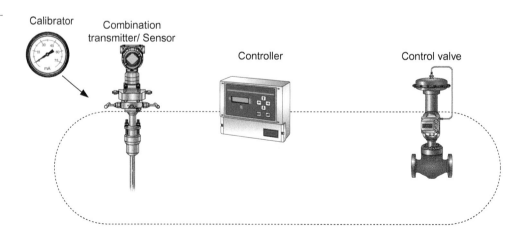

Figure 22.1 Process of elimination.

Air leaks on the pneumatic valve operators or positioners could cause the valve to be erratic. The instrument air regulator is a key to determining if there is an instrument air header pressure drop that is causing an inaccuracy in the control valve. These regulators are usually in the 0 to 120 PSIG range. They have an input pressure air indicator and an output instrument air indicator (nominally 0–100 or 120 and 0–30). See Figure 22.2.

Figure 22.2 A simple air regulator.

CREDIT: Yuthtana artkla/Shutterstock.

If the control valve is set at 50 percent for operations, the output signal then should be 50 percent of 30 PSIG, or 15 PSIG. If it is not within a small variance of that reading (i.e., 1 or 2 PSIG), the problem may exist with the air regulator.

These regulators are equipped with a drain valve. The operator can isolate the regulator to avoid header pressure and open the drain to see if water and/or debris falls out of the drain. The operator can then restore header pressure to this regulator and verify whether the control valve has a better response.

If the control valve is not performing close to its specifications, the operator should inspect for bends, kinks, leaks, or disturbed fittings in the pneumatic tubing that connects the control valve to the regulator. This check is especially needed if other plant personnel or outside contractors have been working around the control valve site. A simple use of a liquid at the connection points can help the technician spot leaks to and from the regulator or control valve.

Calibration of Control Valves

This is a fairly easy procedure to verify.

Step 1:

The outside operator watches the control valve output signal from the PI, as the inside (control room) operator "strokes" the control valve (i.e., moves it from the current reading to another reading that is at least 20 percent different).

For example, inside operators may move the control valve from 50 percent open *up* another 20 percent. The outside operator sees that the control valve readout in the control room would be 70 percent. Thus, the reading on the air regulator PI would go from 15 PSIG up 20 percent, or 6 additional PSIG to 21 PSIG.

The inside operator then returns the control valve back to 50 percent, allowing the outside reading to reach steady state.

The inside operator may then stroke the control valve 20 percent *down* to verify that this moves the control valve the same amount in a decreasing mode. The air regulator should go from 15 PSIG down 20 percent to 9 PSIG.

This simple test would verify that all is okay with the regulator and the control valve.

If this does present the expected results *and* if the process can withstand a momentary/brief disturbance, proceed to Step 2.

Step 2:

The outside operator *first* opens the bypass valve system and then closes the inlet and outlet valves to the control valve.

The inside operator then strokes the valve slowly *up* from its present reading to 90 percent (27 PSIG), then slowly *down* to 10 percent (3 PSIG), and then returns it to the proper set point of 50 percent (15 PSIG). If there is a small blockage in the line, this step may cause the blockage to pass, thus restoring the system to acceptable control.

Calibration of Sensors

One of the first things that should be investigated when troubleshooting instrumentation is calibration (Figure 22.3). Calibration is a common method used to determine whether a sensor is working. Take, for example, a sensor in a control loop.

In a control loop, a sensor may be removed or isolated from the process and subjected to rigorous testing. For example, a thermocouple can be placed in boiling water and the output observed with the use of a handheld calibrator to see if the output temperature is within the manufacturer's published accuracy for that thermocouple at the boiling point of water. If the output is inaccurate, one can assume the sensor (thermocouple) is faulty. This method can be applied to both local and remote indications. Although it may be practical to calibrate a temperature-measuring instrument this way, it may be difficult to apply this method to other types of sensors (e.g., those that measure flow, level, and pressure).

Once it has been determined that an instrument is faulty, the next step is to determine the cause of the failure. Once the faulty component is identified, it must be repaired or replaced.

Failure Types and Possible Causes

Instrument failures can be of many types and causes. The change in output, which could be attributed to signal noise, may be abrupt, gradual, or a fixed deviation.

Figure 22.3 Calibration. **A.** Illustration. **B.** Photo of calibration device.

CREDIT: B. Xmentoys/Shutterstock.

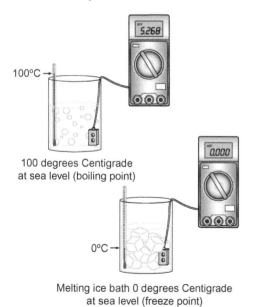

ABRUPT CHANGE Abrupt changes in the indicated value of a process variable with no corresponding change in the process are reliable indications that the sensor or transmitter is faulty (Figure 22.4). For example, if a level indicator registers an abrupt change in a tank level, but the sight glass does not show a corresponding change, the problem might be a transmitter wire with a short in it. Simple observation may identify the most likely source of the problem in both remote and local indications.

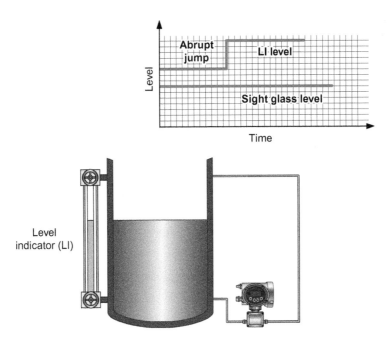

Figure 22.4 Abrupt indication change.

GRADUAL CHANGE Gradual change in instrument readings is somewhat harder to diagnose (Figure 22.5). However, if there is gradual change in the indication, a freshly calibrated instrument should be installed and the values should be checked again. If the replacement shows a different value from the original instrument, this confirms that a gradual change in the original instrument's indication has occurred. This is true for both remote and local indications.

Gradual change problems that were common with early analog instruments have been largely eliminated with the introduction of digital instrumentation. However, most thermocouples do show a gradual change with age. Plugged lines to a D/P cell on a flowmeter can show gradually increasing or gradually decreasing signals, depending on which leg is plugging faster (the responses are discussed in more detail later in this chapter). Gradual changes in flow and level indications may be caused by leaks in the impulse lines to the transmitter. If the impulse lines are fluid filled and the fluid is not replaced or is replaced with a different fluid, such as process fluid, the transmitter output may show a gradual change. (The time scale, illustrated in Figure 22.5, could be minutes, hours, days, or weeks.) The slower the drift, the more difficult the problem is to detect and analyze.

FIXED DEVIATION A fixed deviation may be more difficult to recognize and troubleshoot than abrupt or gradual changes. For example, the improper installation of an orifice plate, like the one installed backward in Figure 22.6, will give flow indications that are in error by a fixed margin. Troubleshooting this type of problem requires the use of engineering calculations. This technique can be applied to both remote and local indications.

NOISY SIGNALS Most process measurements have a small amount of inherent noise; noise itself does not signify a problem. Excessive noise on signals that are normally "clean,"

Figure 22.5 Gradual deterioration and drifting measurements.

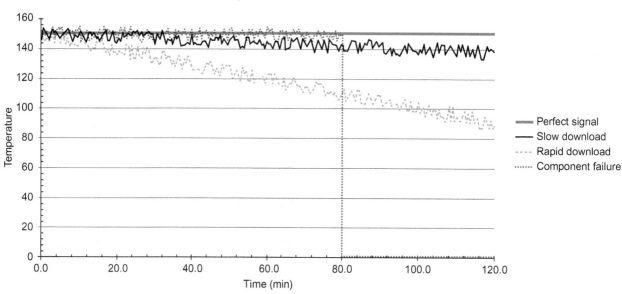

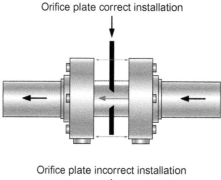

Figure 22.6 Fixed deviation can be caused by an improperly installed orifice plate.

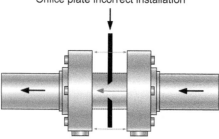

however, does indicate a problem that must be fixed. Excessive noise (signals that fluctuate dramatically), as shown in Figure 22.7, can be generated when signal lines are not properly shielded or are placed too close to electrically noisy equipment, such as large motors or high-voltage electrical lines. For this reason, it is wise to avoid running temporary power and extension cords in the vicinity of instrument signal wiring. A **noisy signal** may also indicate instrumentation failure. For example, the source of this noise may be loose sensor connections. If that is the case, tightening the connections should eliminate the problem. This solution applies to both remote and local indications.

Noisy signal a signal that fluctuates dramatically and that is most likely the result of a loose sensor connection.

NO INDICATED CHANGE The fact that a process variable does not change at all can also be an indication of a failure. Rarely do process variables remain completely unchanged for very long. This is particularly true for faster processes such as flow. This type of failure

Figure 22.7 Noisy signal measurements.

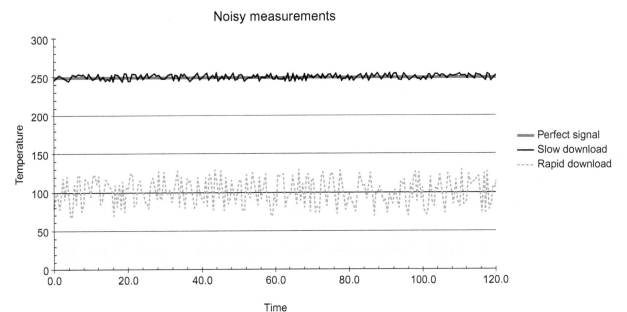

is more apparent with digital readouts, which often are accurate up to one or two decimal points. The process technician should, for example, be suspicious of an absolutely steady flow rate reading of 52.46 GPM. This could well be a false reading caused by a transmitter failure or a software calculation that has locked up and stopped functioning.

22.2 Instrument Failure Modes

Different types of sensor or transmitter elements have different failure modes. This means that each type of equipment will exhibit certain behaviors and characteristics when it begins to fail. The following paragraphs describe some of the different failure modes for various types of instrumentation.

Temperature Elements

A resistance temperature device (RTD), also called a resistance temperature detector, is composed of both a resistance source and a voltage source. The current allowed to flow through the circuit is inversely proportional to the temperature being measured. An open RTD would have a high resistance, which would provide a false indication of a high temperature. A shorted RTD would be a low resistance, which would provide a false indication of a low temperature.

Environmental conditions may also cause the gradual deterioration of an RTD and, consequently, a gradual decline in the RTD output signal (Figure 22.8). In addition, RTD circuits can show fluctuating output if the wiring connections are loose or if there is cross-talk along the signal path. Although twisted pair wires, shields, and grounding are effective means for preventing interference from cross-talk, defective workmanship in any of these items can result in a fluctuating signal.

Thermocouple failure modes are comparable to those of an RTD, at least from an electrical point of view. Although a thermocouple may be protected by a thermowell, an improper environment could cause instrument failure. RTDs and thermocouples are the most common types of temperature sensors, but other types are sometimes used. These include thermal bulb, bimetallic, and infrared.

Figure 22.8 RTD failure.

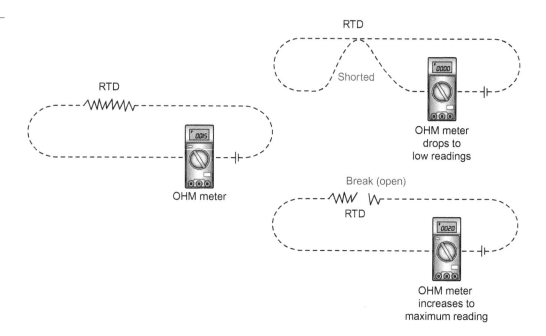

Examples of thermocouple materials are: Type T (copper constantan), Type J (iron constantan), Type E (chromel constantan), and Type K (chromel alumel). These thermocouple types are arranged here in order of increasing temperature applications (i.e., Type T being the lowest, and Type K being the highest).

Flow Elements

Flow may be measured by a variety of devices and methods including: orifice, Venturi, pitot, Doppler, transit time, vortex, magnetic meter, Coriolis, turbine meter, elbow meter, and thermal. Use of an orifice meter is the most common of these methods (Figure 22.9). An orifice meter typically includes an orifice plate connected to a differential pressure (D/P) cell. An orifice meter may malfunction as a result of incorrect installation, orifice plate corrosion, plugged pressure lines, incorrect valve position, or component failure.

Figure 22.9 Orifice meter. **A.** Illustration. **B.** Photo of an orifice meter.
CREDIT: B. © Emerson 2019.

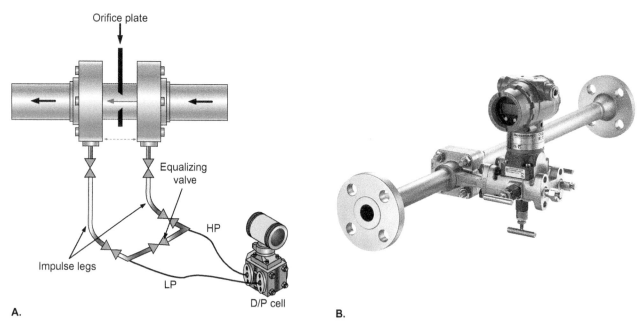

Incorrect installation of the orifice plate can cause problems. If an orifice plate is installed backward, the flow indication will be inaccurate by a fixed, or constant, margin. Similarly, if an orifice plate is installed without a sufficient length of straight pipe upstream and downstream from the plate, a fixed erroneous error signal will result. These errors can be diagnosed rather quickly simply by checking the field installation of the orifice plate.

Corrosion of the orifice plate can result in a gradually declining flow indication. This type of failure is rare and should not occur if the plate is constructed from the proper material.

The impulse lines connecting the flow pipe with the D/P cell may be the source of instrument failure. Impulse lines are short lengths of metal tubing. One is called the high-pressure (HP) line and the other is called the low-pressure (LP) line (low pressure leg). If either of the HP or LP taps become plugged, the result will be an erroneous flow indication. Failure to winterize (freeze preparation) the impulse lines, when necessary, can also result in an erroneous flow indication. If the equalizing valve is inadvertently left open, the flow indication will continue to stay at zero. If the equalizing valve is opened during normal operation, the freeze protection fluid will be flushed from the HP line to the LP line and could cause an imbalance and an erroneous output. Failure of any component of the D/P cell will also result in erroneous or zero-flow indications.

Pressure Elements

If a D/P cell is used as a pressure sensor or transmitter, error can occur. Some pressure transmitters contain diaphragms to isolate the process fluid from the internal transmitter components. If the pressure transmitter is used in a viscous service or slurry service, the diaphragm may become coated with thick deposits. This will dampen the response of the pressure sensor. In extreme cases, the sensor will register a zero reading. Rapidly pulsating pressures can destroy the pressure sensor. This can often be corrected with the use of a restrictive device (pulsation dampener).

Analytical Elements

It may be necessary to test the composition of the product at a given stage in the process. To accomplish this, a variety of analysis techniques may be used including: flame ionization detection (FID), thermal conductivity (TC), photo ionization detection (PID), infrared (IR) detection, and ultraviolet (UV) detection. Although analyzers associated with each technique have their own specific problems, all types share common problems related to sample conditioning. See Figure 22.10.

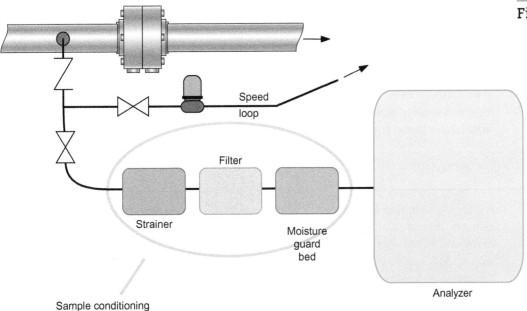

Figure 22.10 Analyzer.

Analyzers typically have a system for sample conditioning. Sample conditioning may utilize strainers, filters, and guard beds (adsorbents) to remove moisture. Samples containing impurities in the form of particulates are passed through a fine filter. A heavy load of trash may plug the filter and result in zero output from the analyzer. For this reason, filters must be changed at regular intervals. When samples are captured for analysis, note should be made of the analyzer field reading so that it can be cross-checked with sample lab results.

Excessive moisture can quickly saturate a guard bed and then seep into the analyzer. Most moisture traps have a color indication. When a color change occurs, it is necessary to change out the guard bed.

Another source of error may be improper use of reference solutions. Some chemicals, such as phosgene and benzene, have stringent regulations for how they must be tested. If the reference solution is old or chemically incorrect, the resulting analytical reading will be "off" as well.

Increasingly, in some industries, flare stack readings from any gaseous or "hot" stream are required to measure nitrous oxides and the destruction capability of the particular chemical being burned.

22.3 Loop Response to Instrumentation Failure

So far in this chapter, failures in different components of a control loop have been discussed in isolation. However, these components interact, making it difficult to determine which component is responsible for a particular failure. Furthermore, more than one component may contribute to the problem. For example, the unstable behavior of a control loop may be the result of an improperly tuned controller, a noisy signal from the transmitter, a hesitation in valve response (e.g., the packing may be too tight), or a combination of any or all of these causes. See Figure 22.1.

Primary Sensing Elements

Primary sensing elements may affect a loop in different ways. A fixed signal deviation or a gradual signal deviation may result in a corresponding deviation in the process variable. For example, an incorrectly installed orifice plate may result in a consistently low flow indication. This will not make a loop unstable. If, however, the orifice plate shows fluctuating flow due to corrosion, the controller response will be unstable, and the valve will constantly open and close. Similarly, a poor ground on a thermocouple may produce a fluctuating temperature signal, which will make a temperature loop unstable.

Transmitters

Transmitters convey information from a sensor to a controller. A faulty transmitter may be responsible for a signal that is consistently in error. In addition, a fluctuating transmitter signal due to loose wires or interference may cause a control loop to be unstable.

Controllers

The most common problem associated with controllers stems from improper tuning. A sluggish controller (i.e., proportional band that is too high) will fail to respond to process changes, possibly leading to unsafe conditions. For example, if a sluggish control loop fails to respond to changes in pressure and the relief valve fails, a vessel can rupture. If a control loop is too sensitive (i.e., the proportional band is too small), small variations in the process variable may result in instability that is known as *cycling*. Cycling can produce wild swings, which can reach dangerous levels.

Figures 22.11 and 22.12 show stable and unstable responses of a controller to a sizeable step change in the temperature it is trying to control. Both figures show responses of a controller using proportional, integral, and derivative (PID) modes (see Chapter 14, *Controllers*). When controllers are set for "tight" control, they are tuned to respond rapidly to small deviations from their set point. If deviations become large for unexpected reasons, instability can occur in an otherwise stable control loop. Figures 22.11 and 22.12 illustrate the difference between stable control, in which the controller is set to a reasonable response, and unstable control, in which the control is too "tight" and responds rapidly to small deviations from its set point. If deviations become large for unexpected reasons, instability can occur in an otherwise stable control loop. Figure 22.11 shows the stable response of a correctly set controller, and Figure 22.12 shows the unstable response of a too tightly set controller.

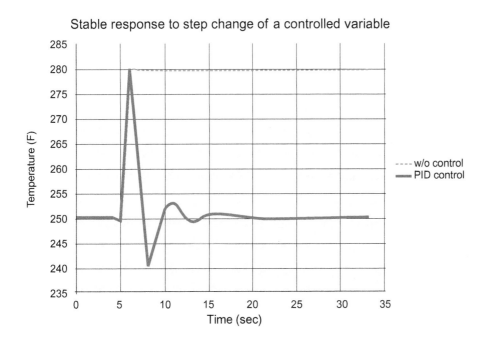

Figure 22.11 Stable response to step change of a controlled variable.

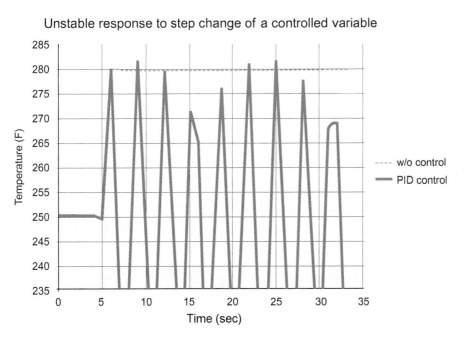

Figure 22.12 Unstable response to step change of a controlled variable.

Control Valve

A control valve may be the source of loop instability. Highly viscous liquid service or slurry service may cause plugging problems. This may force the control valve to open more than would be required under normal conditions. Severe plugging may create the illusion that a control valve is too small. This may not affect the stability of the control loop. However, in extreme cases, a control valve may not be able to respond adequately to the needs of the control loop. In such cases, the loop will be off set point all the time and the controller will be unable to control the process. In addition, if plugging or overtightened packing causes the valve stem to stick, the valve will constantly be sticking and freeing itself. Air leaks on the pneumatic valve operators or positioners often cause the valve to be erratic. Any of these problems could cause the control loop to be unstable.

A useful rule of thumb says that control valves should normally operate between 15 percent and 85 percent open. If a valve is constantly at 90 percent open, it is plugged, or it is too small for the service. At the other extreme, if a valve is constantly at 10 percent open, it is too large for the service, or the trim has worn severely. In both cases, the valve flow characteristics should be changed. This may require a complete, new valve or possibly a simpler (and cheaper) change of the internal trim package.

Summary

Process technicians rely upon instrumentation to monitor and control all process variables in order to maintain desired targets and ensure safety. Improperly functioning instrumentation may fail to indicate a problem when one actually exists or may send up false indications when no problems exist.

In the event of an instrument malfunction, process technicians need to know how to identify the malfunction, isolate it using a process of elimination, and determine the possible cause.

Failures can be of many types and causes. The changes indicated during a failure may be abrupt, gradual, fixed deviation, or highly fluctuating (as is the case with noisy signals).

Different types of sensor or transmitter elements (those that measure pressure, temperature, level, flow, and analytical data) have different failure modes, each with its own unique characteristics. It is important for technicians to be familiar with the types of process deviations being encountered and how these deviations might relate to various components such as transmitters, impulse lines, controllers, and final control elements.

Checking Your Knowledge

1. Troubleshooting often utilizes a process of _____.
   a. supplication
   b. elaboration
   c. elimination
   d. alternation

2. _____ is a common method used to determine whether a sensor is working.
   a. Calibration
   b. Elaboration
   c. Investigation
   d. Adjudication

3. Gradual change is less likely to occur with _____.
   a. thermocouples
   b. digital instruments
   c. pneumatic instruments
   d. analog instruments

4. A _____ is a common temperature-measuring device.
   a. resistance temperature device
   b. temperature transmitter
   c. heat indication device
   d. temperature variance recorder

5. The backward installation of an orifice plate will result in:
   a. A gradual deviation in the meter output signal
   b. An erratic output signal
   c. No change in the meter output signal
   d. A fixed error in the meter output signal

6. If the diaphragm in a pressure transmitter becomes severely coated, a sensor may register:
   a. A zero reading
   b. A high reading
   c. An erratic output signal
   d. A gradually increasing out signal

7. Controller failure most often stems from:
   a. improper tuning
   b. improper atmosphere
   c. electrical spikes
   d. a plugged HP leg

8. (*True or False*) A sticking control valve can cause instability in a loop.

NOTE: Answers to Checking Your Knowledge questions are in the Appendix.

Student Activities

1. Look around your home or workplace and think of all the different instrument and control malfunctions you have experienced in either of these places (e.g., heating or air conditioning system failures, clothes dryer failures, and automobile system failures such as the exhaust gas sensor or oxygen sensor). Record these failures on a piece of paper and provide a description of each.

2. Utilizing a flow control loop, artificially cause the orifice plate to show erroneous flow. With the controller in AUTO, observe the valve response.

3. Discuss what could go wrong with a displacer.

4. Loosen the wiring connections to and from a transmitter and observe the loop response.

Chapter 23
Instrumentation Troubleshooting

 Objectives

After completing this chapter, you will be able to:

23.1 Explain the extent of an operator's role when troubleshooting problems with process instruments and the importance of process knowledge in troubleshooting. (NAPTA Instrumentation Troubleshooting 1, 4*) p. 357

23.2 Identify typical malfunctions found in primary sensing elements and transmitters. (NAPTA Instrumentation Troubleshooting 2, 3, 5) p. 358

23.3 Explain the methods used for determining if a sensing or measuring device is malfunctioning. (NAPTA Instrumentation Troubleshooting 5, 7) p. 358

23.4 Explain the importance of communication between the board technician and the outside process technician when troubleshooting a control loop problem. (NAPTA Instrumentation Troubleshooting 7, 8, 9) p. 362

23.5 Explain the proper use of tools and equipment related to process troubleshooting. (NAPTA Instrumentation Troubleshooting 5, 8) p. 362

23.6 Discuss safety and environmental issues related to troubleshooting process instruments. (NAPTA Instrumentation Troubleshooting 6) p. 364

Key Terms

Calibration—the act of applying known input spans to an instrument and adjusting the device so that it provides an indication or output corresponding to the known values, **p. 363**.

North American Process Technology Alliance (NAPTA) developed curriculum to ensure that Process Technology courses will produce knowledgeable graduates to become entry-level employees in process technology. Objectives from that curriculum are named here in abbreviated form. For example, "(NAPTA Instrumentation Troubleshooting 1, 4)" means that this chapter's objective 1 relates to objectives 1 and 4 of NAPTA's course content on troubleshooting.

Over range—a condition that exists when the value of a measured variable exceeds the upper limit of an instrument, **p. 358.**

Smart instruments—instruments that have one or more microprocessors or "smart chips" included in their electronic circuitry so that they may be programmed and have diagnostic capability, **p. 364.**

Troubleshooting—the systematic search for the source of a problem so that it can be solved, **p. 358.**

23.1 Introduction

Due to their extensive knowledge regarding the equipment on their unit and their familiarity with its operation and characteristics, process technicians are relied upon to recognize abnormal operating situations. They should be aware of general equipment condition and be prepared to address problems, such as the occasional failure of control instrumentation as well as process problems those failures can cause (Figure 23.1).

Generally, process technicians are expected to report all maintenance problems, and in some cases, take the action required to repair the problems. To do either requires a certain level of knowledge associated with normal plant operations.

Figure 23.1 Process technician checking unit instrumentation.

CREDIT: Oil and Gas Photographer/Shutterstock.

A technician must first know what is right in order to recognize what is wrong. If a technician's only responsibility is to report instrumentation problems in a general sense, then this technician must be able to accurately describe the problem from a symptomatic standpoint. However, if a technician is expected to go one step further and identify the malfunctioning component of the instrumentation loop, then additional training on instrumentation system troubleshooting must be made available.

The level of troubleshooting involvement expected of the process technician varies from one company to another and may even vary between plants in the same company and from one unit to another in the same plant. With so much variation in the industry, the process technician's role as an instrument troubleshooter is less defined than other broadly accepted tasks. Therefore, in this chapter, troubleshooting is discussed in general terms so that it can be applied to any processing plant.

The importance of understanding the troubleshooting process cannot be overstated. The ability to identify and correct a problem in its early stages, before it causes a major unit upset, can prevent many hours of unit instability and lost production. A process technician who possesses good troubleshooting skills will recognize a problem as it occurs, or soon after, depending upon the effect it has on process operations. Some problems are obvious and easily diagnosed and corrected, while others can only be identified when some other part of the process or another controlled variable becomes abnormal. The more a process technician knows about the process and the equipment associated with it, the better equipped that technician will be when troubleshooting a malfunction.

23.2 Typical Malfunctions

Equipment failures are most likely to occur soon after a new piece of equipment is installed or at the end of the equipment's expected life span. Failures influenced by environmental factors such as exceeding temperature limits, external corrosion, or humidity can happen at any time. Problems such as a gradual accumulation of solids in the impulse tubing connecting the pressure transmitter to the process can be avoided through a regular preventive maintenance program.

Preventive maintenance is specifically intended to avoid instrumentation failures by addressing some aspect of maintenance that prevents malfunctions. This is important because scheduled maintenance reduces unscheduled downtime and improves profitability and operating efficiency.

The following are some common factors, including those attributable to environmental conditions, which can cause instrument malfunctions:

- Excessive temperatures
- Corrosion
- Sudden change in temperature
- **Over range** (exceeding the maximum value of a device or system)
- Inclement weather
- High humidity
- High vibration
- Mechanical damage

Over range a condition that exists when the value of a measured variable exceeds the upper limit of an instrument.

Table 23.1 provides a list of various instrument components and the malfunctions normally associated with each.

23.3 Troubleshooting Methods

Troubleshooting is a skill based on sound reasoning and technical knowledge. A number of different troubleshooting methodologies are used today. While each has its own unique strengths, they all share some similarities in their approach to problem solving. The following troubleshooting steps provide a general framework of thought and action that a technician can use to find and repair an instrument problem (Figure 23.2).

Troubleshooting the systematic search for the source of a problem so that it can be solved.

1. Gather information and data—Look at historical trends and associated operational variables.
2. Interpret and analyze the data—Use the results of your analysis to decide if a problem exists and if it is an instrument problem or an operational problem.
3. Evaluate and infer—If you believe the problem is caused by an instrument malfunction, consider which specific piece of equipment could be causing the problem (sensor, transmitter, controller, etc.).
4. Repair—If the problem is one that a process technician can repair, fix the problem. If not, make the proper notifications to obtain assistance from an instrumentation technician.
5. Reevaluate—Once repairs have been completed, observe the operation of the equipment to ensure that the problem has been corrected.

6. Document—Document the repair or change for historical purposes and regulatory reasons.

Table 23.1 identifies some of the most common instrumentation malfunctions.

Table 23.1 Instrument Components and Common Malfunctions

Sensor-Related Instrument	Common Malfunctions
Impulse tubing	• Plugged or partially plugged tubing • Leakage • Improperly sized tubing • Excessive vibration • Heat tracing too hot • Heat tracing too cold • Mechanical damage
Thermocouples	• Burnout • Corrosion • Bad connection • Broken wire • Thermal aging
Signal wire	• Bad connection • Broken wire • Excessive loop resistance • Shielding problems
Transmitters	• Low voltage from power supply (or pressure decrease in air supply with pneumatic instruments) • Loss of power • Environmental problems (e.g., corrosion or excessive humidity) • Calibration shifts
Controller	• Plugging • Moisture in the air supply • Power loss (in electronic controllers) • Moisture in the device itself • Air supply system failure • Change in transmitter span, necessitating retuning
Final control element	• Process plugging • Corrosion or erosion in the valve • Packing leaks • Sticking valve • Mechanical failure of tubing or valve components • Moisture or contaminants in the air line • Moisture or contaminants in the device itself • Air or electrical power failure • I/P transducer or positioner out of calibration

CREDIT: Yellow Cat/Shutterstock.
CREDIT: AlexLMX/Shutterstock.
CREDIT: Nordroden/Shutterstock.
CREDIT: © Emerson 2019.
CREDIT: © Emerson 2019.

Figure 23.2 **A.** Technician interprets and analyzes the data. **B.** Technician repairing the problem.
CREDIT: **A.** Gorodenkoff/Shutterstock. **B.** Reggie Lavoie/Shutterstock.

A.

B.

Dividing the Loop in Half

One of the best ways to identify a malfunctioning instrument in a loop is to divide the loop into halves until the problem has been isolated to one component. Typically, only one device fails at any given time, although, there are times when multiple problems occur simultaneously. Since most loops have a sensor, a transmitter, a controller, and a final control element, the logical place to divide the control loop is in the middle, at the controller. See Figure 23.3.

Place the controller in manual and observe the response. Based on your observations, you should be able to determine that the problem is in one of these locations:

1. In the controller itself
2. In the instrument(s) before the controller
3. In the instrument(s) beyond the controller

Figure 23.3 Typical pressure control loop. **A.** Illustration of pressure control loop. **B.** Photo of pressure control loop.
CREDIT: **B.** Frantic00/Shutterstock.

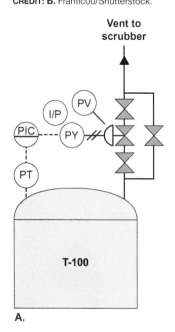
A.

B.

Checking the Controller

In an analog loop, the controller (Figure 23.4) is a stand-alone device that may exhibit any number of problems normally associated with hardware. If the controller is a part of a DCS, however, the problem may be associated with the analog or digital interface (PLC) components rather than the hardware itself.

Software, unlike hardware, has a failure curve that is rather steep initially and then diminishes over the remainder of its useful life. Therefore, outside of a problematic programming change, the problem is not likely to be in the process computer. If it is, other instrumentation loops will probably be affected as well.

When troubleshooting a controller:

1. Make sure that the PV indication on the controller corresponds to the true output from the transmitter by having an instrument tech simulate signals on the input to the controller or from the transmitter.
2. Place the controller in manual and see if the process stabilizes.
3. With the controller still in manual, and the final control element bypassed, determine if the controller moves the final control element properly (e.g., 0 percent, 50 percent, and 100 percent).
4. Confirm that the final control element movement is smooth.
5. Ensure that the loop is properly tuned (have tuning parameters checked).

Figure 23.4 Controllers. **A.** Field controller. **B.** Integrated controller.
CREDIT: **A** and **B**. © Emerson 2019.

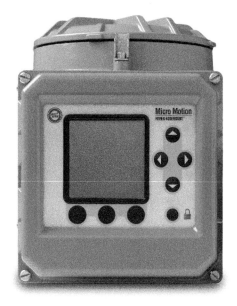

A.

B.

Process to Controller (Before the Controller)

If a determination is made that an instrument or sensor on the upstream side of the controller, located between the process and the controller, is the problem, checking the output of the sensor or transmitter is the next step.

A transmitter is an input/output device, so if the input is known then the expected output should be known as well. If both the input and output are different from what is expected, a technician would need to do the following:

1. Check the transmitter zero and span.
2. Check process connection or piping.

3. Disconnect and check the sensor and calibrate or replace as needed.
4. Reconnect the transmitter to the controller and check for proper reading.

Controller to Process (Beyond the Controller)

If a determination is made that an instrument located on the downstream side of the controller, between the controller and the process, is the problem, then checking the final control element, usually a control valve, is the next step. The most common instruments on the output side of the controller are the I/P converter (transducer) and the control valve or final control element. The I/P converter and control valve can be checked by having a field technician look for problems with the valve as the control room technician moves the valve open and closed. The outside technician should look for smooth stem motion and a complete stroke as the valve is driven open and closed.

23.4 Communication during Troubleshooting

During the troubleshooting process, good communication between the control room technician and the field technician is essential. These two technicians must work as a team, usually communicating over two-way radios, as one technician instructs the other technician to either initiate a change or watch for a response. Each technician has a different perspective of the process, which only he or she can see. The control room technician can see the signal coming into the control room from the sensor or transmitter as well as the outgoing corrective action signal being sent to the final control element. The outside technician can read a local indication of the process variable, and in many cases, can see the true process variable to verify the reading. The outside technician can also view the actual response of the final control element to the correction sent by the controller, which can be compared to the signal being sent by the controller. In the case of local instruments, the control room technician may only have an indication of the process variable, with the outside technician having the responsibility for observation and control of the variable.

23.5 Troubleshooting Equipment

At many sites, the most important piece of troubleshooting equipment that process technicians possess is their brain. In many cases, using their knowledge of and experience with an operating unit allows process technicians to quickly identify and correct process problems. At sites where process technicians are expected to carry out advanced instrumentation troubleshooting and repair, they will commonly use the following tools during a typical troubleshooting procedure (Figure 23.5):

- Slotted and Phillips head screw drivers
- Adjustable wrench
- Multimeter to test the electrical side of the loop
- Pressure and/or temperature testing gauges

Failure to use the proper tools can make solving problems more difficult or create new problems. For example, channel-locks are a convenient tool; however, they are usually not the proper choice for the job since channel-locks tend to round off the edges of hex nuts or bolt heads making them difficult to retighten. Furthermore, using the wrong sized screwdriver is also a problem. If a screwdriver is too large or too small, it may strip or damage the slot or mating surface and compromise the mechanical integrity of the screw.

Figure 23.5 Basic troubleshooting tools.

CREDIT: (top row, left to right) MichaelJayBerlin/Shutterstock; Jose Gil/Shutterstock; Lotus_studio/Shutterstock; DnDavis/Shutterstock. (bottom row, left to right) Yuri Shebalius/Shutterstock; Dmitry Kalinovsky/Shutterstock.

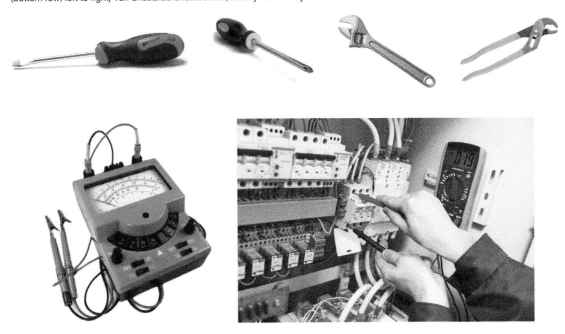

Calibration and Troubleshooting

Over time, even under normal conditions, instrument readings will drift from their original calibrated spans. Because of this, instrumentation must be checked and adjusted periodically against a standard. This action is called **calibration** (Figure 23.6), the act of applying known input values to an instrument and adjusting the device so that it provides an indication or output corresponding to the known values.

While the proper use of calibration equipment requires a substantial amount of training, using a meter to check a signal value is a less complex task, which process technicians may receive training to do. The technician needs to be familiar with the testing equipment and the signal type used in the testing procedure (e.g., a 4–20 mA transmitter may have a millivolt test signal). The best way to determine which type of signal to expect is to read the instrument's instruction manual.

If instruments are calibrated according to a regular preventive maintenance schedule, the overall number of malfunctions and instrument failures will decrease, and the need for troubleshooting will decline as well.

Calibration the act of applying known input spans to an instrument and adjusting the device so that it provides an indication or output corresponding to the known values.

Figure 23.6 Instrument calibration.

CREDIT: Xmentoys/Shutterstock.

Troubleshooting "Smart" Instruments

Smart instruments instruments that have one or more microprocessors or "smart chips" included in their electronic circuitry so that they may be programmed and have diagnostic capability.

Smart instruments are instruments that have one or more microprocessors (or smart chips) included in their electronic circuitry. This means that the device is programmable and has diagnostic capability. In smart instruments, the microprocessor can compute and store analog and digital signals as well as additional pieces of digital data pertaining to the status of the instrument. Some of these devices include digital control valves and transmitters, along with motor control circuitry.

Smart calibrators are used to calibrate and check out the status of smart instruments. An instrument technician can call up this data from anywhere in the loop at the (junction box, at the device, or from any set of loop screw terminals). An added feature of these instruments is that all the data can be sent to a DCS operator station. This allows the process or instrument technician to analyze the data when the loop is in the diagnostic mode. (Note: Loop control functions have priority over diagnostics.)

Some of this digital data includes manufacturing specifications, materials of construction, sensor calibration graphs, and online status. For example, a smart flow transmitter can report a plugged impulse line, reverse flow, an empty pipe, or a bad sensor, as well as many other conditions. A smart temperature transmitter can detect the difference between a bad sensor, open loop circuitry, RTD drift, and calibration error.

While smart field devices have made instrument troubleshooting easier, it is still important for process and instrument technicians to be able to analyze this data correctly.

23.6 Troubleshooting Safety and Environmental Issues

Troubleshooting usually involves the manipulation of one or more instruments. If a transmitter or control valve is to be taken out of service, then safety and environmental precautions must be addressed. Prior to opening any process line, proper personal protective equipment (PPE) must be donned and proper energy isolation and safety procedures must be followed. Once work has been completed, all lines must be secured and returned to their normal safe operating condition.

Summary

Process technicians are the first line of defense against unit-related problems, especially the occasional failure of control instrumentation. Depending on the particular facility requirements, process technicians may be expected to repair instrumentation problems or simply report them so that maintenance can perform the repairs. In either event, technicians must understand normal operation in order to recognize abnormal operation of the instruments in their area of responsibility.

Awareness of the factors that increase malfunction and failure rates helps to minimize or prevent the occurrence of those problems. Issues attributable to environmental conditions such as excessive temperatures, corrosive atmospheres, or excessive humidity may happen spontaneously, while other problems, such as the accumulation of solids in pressure transmitter tubing, may happen over time and may not be as noticeable in the short term. To prevent excessive wear and tear on process instrumentation, it is important to conduct daily inspections and perform routine preventative maintenance.

Troubleshooting is a skill based on sound reasoning and technical knowledge. There are many different approaches that can be used to troubleshoot a process, but they all share the common goal of identifying and correcting a problem in order to return the process to normal operating conditions.

When troubleshooting control loops, the controller is a logical place to start the process, because it is the main component in the control loop and is the component in the middle of the loop. After ruling out the controller as the source of the problem, the loop can be divided in half (upstream of the controller and downstream of the controller) and the various instruments checked until the specific problem is identified. This process usually requires good communication between control room and field personnel and coordination of various activities.

Various diagnostic equipment and tools may be used during the troubleshooting process. These include screwdrivers, multimeters, and pressure and temperature-testing gauges. It is important to use the proper tools for the job since instruments are sensitive pieces of equipment and easier to damage than many other types of process equipment.

When troubleshooting instrumentation associated with hazardous process streams, additional safety and environmental procedures must be observed. These procedures may require the wearing of additional PPE, the use of special tools, or special material handling techniques.

When instruments are installed, they are calibrated to operate within certain ranges and at specific set points. Over time, these instruments may lose their calibration due to vibration or other processing or environmental conditions. Therefore, each instrument should be tested and calibrated periodically to ensure that the reading indicated by the instrument matches a standard known value.

Checking Your Knowledge

1. (*True or False*) Process technicians have a clear set of duties in the area of troubleshooting instrumentation that are universal from plant to plant and company to company.

2. When are equipment failures most likely to occur? (Select all that apply.)
   a. Soon after installation
   b. Immediately after the initial testing of the equipment
   c. In the middle of the equipment's life
   d. At the end of the equipment's expected life span

3. Which of the following is a common malfunction for thermocouples?
   a. Burnout
   b. Shielding problems
   c. Low voltage from power supply
   d. Loss of power

4. Which is usually the first step in the troubleshooting process for an instrumentation loop?
   a. Evaluate and infer
   b. Interpret and analyze
   c. Repair
   d. Gather information and data

5. During troubleshooting, radio communication usually takes place between the field technician and the _____.
   a. control room technician
   b. unit engineer
   c. I&E technician
   d. maintenance supervisor

6. Which of the following tools may be used when troubleshooting instrumentation? (Select all that apply.)
   a. Channel locks
   b. Multimeter
   c. Slotted screwdriver
   d. Temperature-testing gauges
   e. Sledgehammer

7. (*True or False*) Technicians must wear proper PPE (hard hat, safety glasses, gloves, safety boots) when troubleshooting all types of instrumentation.

8. When calibrating an instrument, the technician applies known _____ values to the instrument in order to adjust indication correspondingly.
   a. temperature
   b. pressure
   c. variable
   d. level
   e. input

9. Match the instruments below with the malfunctions normally associated with each.

Instrument	Common Malfunctions
 CREDIT: Nordroden/Shutterstock. I. Signal wire	a. • Burnout • Corrosion • Bad connection • Broken wire • Thermal aging
 CREDIT: Yellow Cat/Shutterstock. II. Impulse tubing	b. • Bad connection • Broken wire • Excessive loop resistance • Shielding problems
 III. Transmitters	c. • Plugging • Moisture in the air • Power (in electronic controllers) • Moisture in the device itself • Air supply system failure • Change in transmitter span, necessitating retuning

Instrument	Common Malfunctions
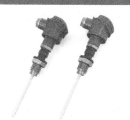 CREDIT: AlexLMX/Shutterstock. IV. Thermocouples	d. • Low voltage from power supply (or pressure decrease in air supply with pneumatic instruments) • Loss of power • Environmental problems (e.g., corrosion or excessive humidity) • Calibration shifts
 CREDIT: © Emerson 2019. V. Final control element	e. • Plugged or partially plugged tubing • Leakage • Improperly sized tubing • Excessive vibration • Heat tracing too hot • Heat tracing too cold • Mechanical damage
 CREDIT: © Emerson 2019. VI. Controller	f. • Process plugging • Corrosion or erosion in the valve • Packing leaks • Sticking valve • Mechanical failure of tubing or valve components • Moisture or contaminants in the air line • Moisture or contaminants in the device itself • Air or electrical power failure • I/P transducer or positioner out of calibration

NOTE: Answers to Checking Your Knowledge questions are in the Appendix.

Student Activities

1. In your library or online, research various types of troubleshooting methodology. Apply the techniques to instrumentation as used in this chapter and discuss how they would or would not be applicable.

2. Given a control loop malfunction scenario, correctly apply a deductive troubleshooting methodology. Successfully deduce the source(s) of the malfunction and explain what actions are necessary to return the variable to within acceptable limits.

3. Role-play an inside technician and an outside technician communicating during the troubleshooting of a control valve action.

4. Given a simulation trainer that utilizes a flow and level loop, and instructor-induced malfunctions, determine whether the problem is an instrumentation problem or process-related problem.

5. Perform minor instrumentation calibrations.

Appendix

Answers to Checking Your Knowledge Questions

Chapter 1	Answer	LO #
1.	b, d, a, c	1.1
2.	I. Flow—b II. Level—c III. Differential pressure—e IV. Temperature—d V. Pressure—a	1.3
3.	b	1.2
4.	c	1.3
5.	c	1.4
6.	a	1.5
7.	a	1.5

Chapter 2		
1.	c	2.1
2.	c	2.1
3.	c	2.2
4.	I. Manometer—d II. Capacitance transducer—b III. D/P cell—c IV. Strain gauge—a	2.3
5.	d	2.2
6.	d	2.2
7.	a	2.4

Chapter 3		
1.	b	3.1
2.	a	3.1
3.	a, d	3.2
4.	c	3.2
5.	d	3.3
6.	c	3.4
7.	b	3.4

Chapter 4		
1.	d	4.1
2.	I. Ullage—c II. Innage—b III. Indirect level measurement—a IV. Direct level measurement—e V. Interface level—d	4.1
3.	b	4.2
4.	a	4.2
5.	b	4.2
6.	d	4.3
7.	a	4.3

Chapter 5		
1.	a, b, c, d	5.1
2.	b	5.1
3.	d	5.1
4.	c	5.2
5.	I. Orifice plate—c II. Flow nozzle—d III. Venturi tube—e IV. Pitot tube—b V. Rotameters—a	5.2
6.	d	5.2
7.	I. Mass flow meter—c II. Magmeter—b III. Turbine meter—a IV. D/P transmitter—d	5.2
8.	c	5.2
9.	b	5.2
10.	c	5.3
11.	I. Orifice plate—d II. Flow nozzle—b III. Venturi tube—a IV. Pitot tube—e V. Multiport pitot tube (Annubar®)—c	5.2
12.	I. Electromagnetic flow meter (magmeter)—c II. D/P transmitter—e III. Turbine meter—d IV. Rotameter—b V. Mass flow meter—a	5.2

Chapter 6		
1.	I. pH—c II. ORP—b III. Conductivity—e IV. Chromatography—d V. Combustion—a	6.1
2.	b	6.2
3.	c	6.2
4.	b	6.2
5.	c	6.2
6.	a	6.2
7.	c	6.3
8.	c	6.3
9.	d	6.3

368 Appendix

Chapter 7	Answer	LO #
1.	b	7.1
2.	d	7.1
3.	b	7.1
4.	b	7.2
5.	a	7.2
6.	c	7.2
7.	a	7.3
8.	c	7.4
9.	b	7.4

Chapter 8		
1.	a	8.1
2.	c	8.1
3.	d	8.2
4.	a	8.2
5.	c	8.4
6.	d	8.4

Chapter 9		
1.	a	9.1
2.	b	9.2
3.	a, b, d	9.2
4.	a	9.3
5.	c	9.3
6.	I. P/I—c II. I/P—a III. E/I—b IV. I/E—d	9.4
7.	b	9.5
8.	37.4%	9.6
9.	d	9.7
10.	a. 0–100 b. 0 c. 20–50 (or 30) d. 100	9.6

Chapter 10		
1.	d	10.1
2.	a, b, c, d	10.1
3.	True	10.2
4.	Light switch—OPEN Toilet flush—OPEN Water faucet—OPEN Automobile cruise control—CLOSED	10.2
5.	a	10.3
6.	I. b II. d III. c IV. a	10.4
7.	d	10.4
8.	b, c	10.4
9.	a. Open loop b. Closed loop	10.2

10.

INSTRUMENT CONTROL LOOP				10.3
Component	Element Type	Process Variable Being Controlled	Component Function	
FE	Orifice plate	Flow	Sensing	
FT	Transmitter	Flow	Transmitting	
FC	Controller	Flow	Comparing	
FY	Transducer	Flow	Converting	
FCV	Control/Valve	Flow	Manipulating	

11.

INSTRUMENT CONTROL LOOP				10.3
Component	Element Type	Process Variable Being Controlled	Component Function	
LE	Differential pressure	Level	Measuring	
LT	Transmitter	Level	Transmitting	
LC	Controller	Level	Comparing	
LY	Transducer	Level	Converting	
LCV	Control/Valve	Level	Manipulating	

12.

INSTRUMENT CONTROL LOOP				10.3
Component	Element Type	Process Variable Being Controlled	Component Function	
TE	Sensor	Temperature	Measuring	
TT	Transmitter	Temperature	Transmitting	
TIC	Indicator controller	Temperature	Comparing	
TY	Transducer	Temperature	Converting	
TCV	Control/Valve	Temperature	Manipulating	

13.

INSTRUMENT CONTROL LOOP				10.3
Component	Element Type	Process Variable Being Controlled	Component Function	
PT	Transmitter	Pressure	Transmitting	
PIC	Indicator controller	Pressure	Comparing	
PY	Transducer	Pressure	Converting	
PCV	Control/Valve	Pressure	Manipulating	

Chapter 11		
1.	c	11.1
2.	b	11.2
3.	b, d, e	11.2
4.	b	11.3
5.	c	11.3
6.	175	11.3
7.	c	11.4
8.	b	11.4
9.	575°F 8 mA 150 PSI	11.4
10.	d	11.5
11.	b	11.6

Appendix

Chapter 12	Answer	LO #
1.	I. Controller responds to the rate at which the variable changes.—e II. Controller responds to an increasing input with a decreasing output.—b III. Controller responds to reset the measurement back to the set point.—d IV. Controller responds to an increasing input with an increasing output.—a V. Controller takes the full range of input to drive the output.—c	12.1
2.	d	12.1
3.	b	12.1
4.	a	12.2
5.	I. Two of the same type of process variables must be controlled proportionally.—e II. One controller is needed to manipulate two final control elements.—c III. The controller and the final control element are remotely located and their actions are isolated and unrelated to other processes.—a IV. The output of one controller serves as the set point for another controller.—d V. The controller is removed from the process area.—b	12.4
6.	d	12.5
7.	a	12.5
8.	a, b	12.5
9.	I. Split-range—c II. Ratio—e III. Remote—b IV. Local—a V. Cascade—d	12.4

Chapter 13		
1.	a	13.1
2.	b	13.2
3.	a	13.2
4.	d	13.3
5.	a	13.3
6.	a	13.3
7.	a	13.3
8.	c	13.4
9.	c	13.5
10.	b	13.5
11.	a, b, d	13.5
12.	b	13.6
13.	I. Body—H II. Actuator—L III. Diaphragm—B IV. Seat—I V. I/P transducer—D VI. Stem—E VII. Bonnet—G VIII. Plug or disc—J IX. Handwheel—A X. Spring—C XI. Valve positioner—K XII. Packing and packing box—F	13.1
14.	I. Globe body valve—a II. Butterfly valve—c III. Three-way valve—b	13.5

Chapter 14		
1.	I. Dead time—d II. Load change—c III. Over range—a IV. Process equilibrium—e V. Spanning—b	14.1
2.	b	14.1
3.	b	14.1
4.	I. Proportional gain—c II. Windup—g III. Offset—e IV. Tuning—f V. Trend—d VI. Hunting—a VII. Steady state—b	14.1, 14.2, 14.3
5.	c	14.3
6.	b, d	14.3

Chapter 15		
1.	c	15.1
2.	a	15.1
3.	c	15.1
4.	a	15.1
5.	a, d	15.2
6.	a	15.2

Chapter 16		
1.	a, c, d	16.1
2.	b	16.2
3.	c	16.2
4.	a, b, c, d	16.3
5.	c	16.3
6.	b	16.4
7.	a	16.5
8.	d	16.5
9.	b	16.4

Chapter 17		
1.	a	17.1, 17.2
2.	c	17.1
3.	c	17.3
4.	I. Converter—g II. Demultiplexer—a III. Discrete controllers—c IV. Distributed control systems—b V. Input—d VI. Multiplexer—h VII. Output—f VIII. Programmable logic controllers—e	17.2, 17.3, 17.4
5.	c	17.4
6.	b	17.2
7.	P/I converter A/D	17.4

Chapter 18	Answer	LO #
1.	a, c, e	18.1
2.	a	18.1
3.	c	18.1
4.	d	18.2
5.	True	18.1
6.	False	18.2

Chapter 19		
1.	c	19.1
2.	c	19.2
3.	a	19.3
4.	c	19.3
5.	True	19.4
6.	a, c	19.4

Chapter 20		
1.	d	20.1
2.	c	20.1
3.	a	20.2
4.	c	20.2
5.	b	20.2
6.	d	20.2

Chapter 21		
1.	d	21.1
2.	b	21.1
3.	a	21.2
4.	a	21.2
5.	c	21.3
6.	c	21.4
7.	False	21.4
8.	a	21.4

Chapter 22		
1.	c	22.1
2.	a	22.1
3.	b	22.1
4.	a	22.2
5.	d	22.2
6.	a	22.2
7.	a	22.3
8.	True	22.3

Chapter 23		
1.	False	23.1
2.	a, d	23.2
3.	a	23.2
4.	d	23.3
5.	a	23.4
6.	b, c, d	23.5
7.	True	23.6
8.	e	23.7
9.	I. Signal wire—b II. Impulse tubing—e III. Transmitters—d IV. Thermocouples—a V. Final control element—f VI. Controller—c	23.2

Glossary

Accelerometer a vibration measuring device (e.g., piezoelectric type).

Acceptable limits the operating range within which a piece of rotating equipment can operate without causing excessive wear to the bearings or other types of catastrophic failure.

Accuracy how close a measurement corresponds to its true value.

Actuator an electrical, pneumatic, or hydraulic device that manipulates a control device, usually a valve.

Alarm switch a switch used to notify an operator when a process variable enters an abnormal range (e.g., high or low) by triggering an alarm (in the form of a light, a horn, or both).

Alarm system a series of alarms associated with one process unit or equipment system.

Algorithm preset mathematical function calculated in a controller that can be mechanical, analog, or digital. The three most common output functions deal with proportional (P), integral (I), and derivative (D) tuning.

Alternating current (AC) an electrical current that reverses direction periodically, 60 cycles per second in North and South America and 50 cycles per second in other parts of the world.

Analog a continuous signal that changes to mirror the actual process variable.

Analog to digital (A/D) the conversion of an analog signal to a digital signal so that it can be processed by a computer; also referred to as "A to D."

Analog transmission a signal transmission in which process variable measurements change in a continuous manner mirroring the actual process variable.

Analytics use of a logical technique to perform an analysis; in instrumentation, a measurement of a physical or chemical property.

Analyzer measurements qualitative and quantitative.

And/or a conditional logic statement used to control a process in which either one or both variables exist.

Atmosphere(s) the pressure at any point in the atmosphere due solely to the weight of the atmospheric gases above the point concerned; 14.7 PSIA, which equals one atmosphere, is the pressure at sea level.

Atmospheric pressure the pressure at the surface of the earth (14.7 PSIA at sea level).

Automatic to manual and manual to automatic switching the shift of controller action from automatic to manual or vice versa by adjusting the set point of the controller to the actual controlled point and then switching the mode; or switching from automatic to manual by simply repositioning the mode selector.

Automatic transfer switch (ATS) an electromechanical device that switches an electric load between two separate power sources (primary and secondary) depending on power availability from these sources; it is typically configured to be in the normal position when primary power is available.

Automatic/manual switch a switch that allows a process technician to select either automatic or manual control from the front of a controller.

Back-pressure regulator a device used to regulate and/or control the pressure of a process fluid upstream of the device location.

Balloon the basic instrumentation symbol.

Bar measurement of pressure equal to 0.987 atmospheres.

Basic control functions sensing, measuring, comparing, calculating, correcting, and manipulating.

Basic equipment symbol common equipment such as pumps, towers, furnaces are basic pieces of equipment for most processing facilities and have commonly recognizable, or basic, equipment symbols.

Battery an electrical device consisting of two or more cells that convert chemical energy into electrical energy and provides a steady state of DC voltage.

371

Bimetallic strip two dissimilar strips of metal bonded together that expand and contract at different rates when exposed to temperature change, causing a rotating effect; used as the primary element in a temperature gauge or bimetallic thermometer.

Block flow diagram (BFD) a simple illustration that shows a general overview of a process, indicating its parts and their interrelationships.

Body the housing component of a valve.

Bonnet a bell-shaped dome mounted on the body of a valve where the valve disc is held when the valve is opened.

Boolean logic in computer science, Boolean is a data type that has one of two possible values, intended to represent the two truth values of logic.

Breaker a switchlike device in electrical panel boxes, used to keep electrical current from exceeding the recommended load.

British thermal unit (BTU) the amount of heat energy required to raise the temperature of 1 pound of water 1 degree Fahrenheit.

Bubbler a level measurement system used in open or vented containers that measures head pressure and that then converts the measurement to a level reading; this allows the measurement of head pressure in a liquid without the pressure sensor coming in contact with the process fluid.

Bump a process upset occurring when a controller is switched from automatic to manual mode.

Bumpless transfer the act of changing the controller from manual to automatic (or vice versa) without a significant change in controller output.

Bypass a short patch cable or wire used to complete an electrical circuit, usually temporarily, for testing or diagnostic purposes (also called a *defeat* or *jumper*).

Bypass switch a switch used to override the normal operation of a system or device for equipment startup operations or for routine check of alarms.

Calibration the act of applying known input spans to an instrument and adjusting the device so that it provides an indication or output corresponding to the known values.

Calorie the amount of heat energy required to raise the temperature of 1 gram of water 1 degree Celsius (4.1868 joules).

Capacitance transducer an electrical device that contains a measurement diaphragm and capacitor plates; it converts an applied pressure to an electronic signal measurement that is proportional to the process being measured.

Cascade control loop control loop characterized by the output of one controller becoming the set point of another.

Cascade control employing two controllers so one process variable is controlled by controlling another; the output of one controller is the remote set point of another.

Celsius/centigrade scale of measurement to determine temperature; freezing point of water is 0 degrees Celsius (C) and boiling point of water is at 100 degrees C.

Charger a device used to restore batteries to a proper electrical charge.

Chart recorders devices that physically or electronically record data that process and instrument technicians can use to troubleshoot process trends and instrumentation problems.

Chromatogram a graphic record of the separated components.

Chromatography a process where molecular components of a liquid or gas are separated and identified by means of a tube, called a column.

Closed control loop when a control loop has feedback (e.g., controller in automatic mode).

Color/optical analyzers photometers that operate in the visible light spectrum (400–800 mμ); two types include the visual color analyzer and the photometer, or spectral color analyzer.

Combustion the rapid consumption of fuel resulting in its conversion to heat, light, and gases.

Comparator a component of a controller that compares the measurement to a predetermined set point.

Comparing, calculating, and correcting element the control loop component that receives the appropriate signal from the transmitter and compares the signal to a desired value (set point); if there is a difference, then the output of the comparison causes a calculation to be performed to cause a corrective response by the controller output signal to the final control element.

Conduction the transfer of heat through matter via vibrational motion; exchange media must be in direct contact.

Conductivity analyzer a device that measures all ions in an aqueous solution.

Conductivity meter a device that measures the conductivity of a process sample by comparing it to a known value or standard cell.

Conductivity a measurement of the ability of a material to conduct an electrical current; the measure of the ability of a liquid (or any solution) to conduct electricity.

Continuous level measurement monitoring of all level points in the tank from the 0 percent level (bottom of the measuring device) to 100 percent full; used in processes with control loop instrumentation to provide ongoing measurement.

Control loop   the series of instruments that work together to monitor and control a single process variable.

Control mode   the control action or the control algorithm response such as PID or a programmed function.

Control valve   the final component of a control loop, operated either remotely or by hand, that receives its signal from the controller and manipulates the process variable in response to a system disturbance or change.

Controlled variable   a process variable that is sensed to initiate the control signal.

Controller   an instrument that receives a signal from the transmitter, compares it to a set point, and produces an output to a final control element.

Controlling   keeping a variable at a specific quantity.

Convection   the transfer of heat through the circulation or movement of a liquid or a gas.

Converter   a device that changes or converts a substance or a signal from one state to another (e.g., changing AC to DC).

Converting and transmitting element   the control loop component that converts the sensed process variable and transmits the measured signal.

Converting device   a device that receives information in one form of an instrument signal and changes it into another form of an instrument signal.

Cycling   the moving or shifting of a process variable above and/or below the set point.

Data highway   a series of distributed computer processors, or nodes, that are all interconnected via a computer network.

Dead time   the elapsed time between the initiation of an input (measured, controlled, or manipulated variable), change, or other stimulus and the point at which the resulting response can be measured or observed.

Decanting   a process for the separation of mixtures of immiscible liquids or of a liquid and a solid mixture such as a suspension.

Defeat   *See* bypass.

Deluge system   an emergency shutdown (ESD) system component that, when a leak is detected, sprays the area of leakage with large quantities of water droplets to dilute the concentration of vapors so they are below the toxic limit for personnel or the lower explosive limit (LEL), thereby reducing danger to personnel and the potential for fire.

Demultiplexer (DEMUX)   a device that separates single stream, multiplexed signals back into their original multistream constituents.

Density   mass per unit volume.

Depth   the distance from the surface of a liquid down to a zero reference point.

Derivative action   a controller output response that is proportional to the rate at which the controlled variable deviates from the set point.

Deviation   the difference between the set point and the controlled variable under dynamic (changing) conditions.

Deviation alarm   an alarm that is activated when the difference between two variables exceeds a set limit (e.g., in a particular catalyst bed, the difference or deviation between the outlet temperature and the inlet temperature should not exceed 50 degrees Fahrenheit); in this type of alarm, the actual temperature value is not important, what is important is the difference between the two values).

Differential pressure (Δp, D/P)   the difference between two related pressure measurements; usually used in measurements of pressure, temperature, level, and flow.

Differential pressure cell (D/P cell)   an instrument that measures the difference between two related pressure points and produces a corresponding output signal that is proportional to the process being measured.

Digital signal   data that is represented as coded information in the form of binary numbers; used to transmit data to and from field transmitters on a twisted pair of wires; may also be between computers and computer components.

Digital   a transmission method that employs discrete electrical signals as opposed to continuous signals or *waves*.

Digital to analog (D/A)   the conversion of a digital signal to an analog signal so that it can be interpreted and reported on an analog gauge or instrument (e.g., using an I/P converter to convert a digital signal to a pneumatic signal that triggers the opening or closing of a control valve); also referred to as "D to A."

Digital transmission   a signal transmission method that uses digital devices to create a binary signal.

Direct acting controller   a controller whose output signal value increases as the controller input signal value increases or decreases as the controller input value decreases.

Direct current (DC)   electrical current that flows in a single direction through a conductor.

Direct level measurement   measures the process variable directly in terms of itself (e.g., using a sight glass or dipstick).

Discrete controllers   digital controllers that can accept and produce only digital input and output.

Discrete sensing element   a sensing element that can stand alone or is individually distinct; connected to the transmitter by sensor wires.

Displacer   a sealed cylindrically shaped tube that uses the principle of buoyancy to displace an amount of fluid equal to the container level.

Distributed control system (DCS)   a computer-based system consisting of a series of controllers used to monitor and control a process.

Drift   the change in the accuracy of an instrument signal or indication over time; it can be attributed to factors such as temperature and aging.

Duty cycle   the period of time it takes for a signal to complete an on-and-off cycle.

E/I conversion   the conversion of a signal from voltage to current.

E/P conversion   the conversion of a signal from voltage to analog pneumatic.

Electrical transmission   a signal transmission method that uses electrical devices to create a signal.

Electrolyte   a nonmetallic substance that is an ionic conductor; the greater a substance conducts an electrical charge, the more ionic the substance becomes.

Electromagnetic flow meters (magmeters)   magnetic flow meters designed to determine volumetric flow of electrically conductive liquids, slurries, and corrosive and/or abrasive materials.

Electronic signal   analog or digital signal; current or voltage signal.

Electronic   powered by electricity.

Emergency shutdown (ESD) system   a system consisting of interlocks, breakers, sensors, and other equipment responsible for automatically shutting down equipment in an extremely abnormal situation in order to protect people and equipment and reduce the risk of catastrophe.

Fahrenheit   scale of measurement to determine temperature; freezing point of water is 32 degrees Fahrenheit (F), and the boiling point of water is at 212 degrees F.

Fail last (fail in place)   the position of the valve seat upon loss of instrument air or power failure.

Feedback control   a closed loop control strategy where the difference in the set point and measurement drives the output of the controller to the final control element.

Feedback loop   the most common type of control loop where the change caused by the output of the controller is fed back to the process, providing a self-regulating action.

Feedforward control   an open loop control strategy in which a disturbance in the process is anticipated, and the controller output is adjusted to compensate for its effect.

Filled thermal system   a temperature sensing bulb filled with a liquid, vapor, or gas and connected by means of a capillary tube to a pressure measuring element.

Final control element   the last active device in an instrument control loop; it directly controls the manipulated variable; usually a control valve, a louver, or an electric motor.

Flat-glass level gauge   there are three types of flat-glass level gauges: reflex, transparent, and welded pad; the heavy, thick-bodied glass promotes safety in high temperature and high-pressure hydrocarbon or steam service.

Float gauge   a local measuring device that uses cables, pulleys, levers, or any other mechanism to convey the position of a float to a liquid level.

Flow   the movement of fluids.

Flow measurement   flow rate (an instantaneous flow measurement) and total flow (a summation of instantaneous flow rates over a time interval or an accumulation of counts provided by a positive displacement device); measured in volume or mass units without respect to time.

Flow nozzle   a device similar to a Venturi tube but with an extended tapered inlet commonly installed in a short piece of pipe called a spool piece.

Flow rate   the specific amount of fluid moving past a given point per unit of time, usually expressed in volume units per unit of time, such as gallons per minute (GPM) or cubic feet per minute (CFM).

Fluid   substances, usually liquids or vapor, that can be made to flow.

Force   energy that causes a change in the motion of an object, involving strength or direction of push or pull.

Gain   change in output multiplied by the change in input.

Gas chromatograph (GC) analyzer   an analyzer that provides the necessary means to accomplish the chromatographic separation and analysis.

Gas chromatography   an analytical method that can provide a molecular separation of one or more individual components in a sample.

Generator   a device that converts mechanical energy into electrical energy.

Glass stem thermometer   a temperature measuring device constructed of a glass bulb and a capillary (small) tube made of glass with a numbered scale; the bulb and the capillary tube contain a liquid such as mercury or colored alcohol that expands, allowing a reading to be made from the etched scale on the length of the tube; usually utilized in laboratory settings for calibration.

Handwheel the mechanism that raises and lowers a valve stem to control the flow of fluid through a valve.

Heat added energy that causes an increase in the temperature of a material (sensible heat) or a phase change (latent heat); the transfer of kinetic energy to increase the temperature of a substance via the process of conduction, convection, or radiation.

Height a distance from a zero reference point to the surface.

High (H) alarm the first alarms triggered when a process variable (e.g., fluids in a tank) rises above a predetermined high level; the purpose of this alarm is simply to notify an operator that the level is abnormal.

High high (HH) alarm alarm triggered when a process variable (e.g., fluids in a tank) continues to rise above a predetermined maximum level after a high (H) alarm has already been triggered. The purpose of this alarm is to notify a process technician that the alarm condition of a process variable is becoming increasingly abnormal and to initiate a shutdown or other corrective action regardless of any action by the process technician.

Hunting a condition that exists when a control loop is improperly designed, installed, or calibrated, or when the controller itself is not properly aligned; overcorrection above and/or below the set point.

I/E conversion the conversion of a signal from current to voltage.

I/P conversion the conversion of an analog signal from current to pneumatic.

I/P transducer (current-to-pneumatic transducer) a device that converts a milliampere signal into a pneumatic pressure signal.

If/then a conditional logic statement used to control a process in which one condition leads to the next.

Impulse tubing a tube that is usually made of stainless steel and allows the process variable to be sensed by the sensor located in the transmitter.

Inches of mercury (in. Hg) most common measurement scale for a manometer; measures in pounds per square inch; 1 inch Hg = 13.0687 inches H_2O.

Inches of water (in. H_2O) a very small measurement of pressure equal to 0.036 pounds per square inch at 4 degrees Celsius (39.2 degrees Fahrenheit).

Inches of water column a pressure measurement scale that can be used for liquid level in low pressure containers, generally open or vented to the atmosphere; accuracy is dependent on the liquid's specific gravity; the measurement can be converted to pounds per square inch (PSI).

Indicating the showing of a current process variable condition via a readout (e.g., digital or analog).

Indirect flow measurement measuring one variable of a process to infer another.

Indirect level measurement measures another process variable (e.g., head pressure or weight) in order to determine level.

Infrared thermometer a device that infers temperature using thermal radiation emitted by the substance being measured; also called *pyrometer*.

Inline analyzer an analyzer (either continuous or contiguous) installed directly in a process line that may be either a sample system or of the newer probe type.

Innage the measurement from the bottom of a tank to the surface of the product.

Input the process of entering data or information into a computer system or program; the data entered into a computer program or process.

Instrument scale a range of ordered marks that indicate the numerical values of the process variable.

Insulation any substance that prevents the passage of heat, light, electricity, or sound from one medium to another.

Integral action (reset) a controller output response that is proportional to the length of time the controlled variable has been away from the set point.

Integrally mounted sensing element a sensing element that is a physical part of the transmitter.

Interface level the surface at which two immiscible fluids meet; the interface can be between two liquids, a liquid and a slurry, or a liquid and a foam.

Interleave to mix (two or more digital signals) by alternating between them.

Interlock a system for connecting mutually independent equipment that can stop upstream equipment when downstream equipment is shut down.

Inverter a device used to change direct current (DC) into alternating current (AC).

ISA the International Society of Automation (originally known as the Instrument Society of America); a global, nonprofit technical society that develops standards for automation, instrumentation, control, and measurement.

Jumper *See* bypass.

Kelvin scale of measurement that is sometimes called an *absolute scale* because 0 Kelvin (K) is the point at which no heat exists; freezing point of water is 273 K and boiling point of water is 373 K.

Ladder diagrams diagrams that guide electricians in the fabrication process.

Lag time a measure of the elapsed time between two events, states, or processes.

Laminar flow a condition in which fluid flow is smooth and unbroken; viewed as a series of laminations or thin cylinders of fluid slipping past one another inside a tube.

Lead/lag control the manner in which a process reacts to a disturbance or other manipulating conditions.

Legend a section of a drawing that explains or defines the information or symbols contained within the drawing.

Level a position of height or depth along a vertical axis; in the process industries, level usually means the height of the surface of a material compared to a zero reference point.

Level measurement the act of establishing the height of a liquid surface in reference to a zero point.

Limit switch an electromechanical device used to limit travel of a component beyond a predetermined point.

Line symbols connectors between the basic pieces of equipment without which process streams could not be moved.

Linear scaling a linear relationship between two scales (input versus output).

Liquid (hydrostatic) head pressure the pressure exerted by the height of a column of liquid; the most common indirect level measurement.

Live zero a standard bias added to the instrument signal (e.g., pneumatic 3 to 15 PSIG or electronic 4 to 20 mA); instead of reading zero, 0 percent of scale would have a reading of 3 PSIG (pneumatic) or 4 mA (electronic).

Load cell a level measuring device consisting of a transducer that measures force or weight (called a strain gauge), which is bonded to a robust support column called a force beam; generally built into the supporting structure of a vessel.

Load change a change in any variable in the process that affects the value or state of the controlled variable.

Local related to instrumentation in close physical proximity to the process, generally found outside the control room.

Local automatic control the act of controlling an instrument loop by automated means within the processing area.

Local controller when the controller is physically mounted in the processing area near the other instruments in the loop.

Local indicators instruments placed on equipment in the field only to be read in the field; may be used for comparison with transmitted instrumentation readings or used on noncritical processes to indicate process values.

Local manual control the act of controlling a process variable by hand within the processing area.

Loop error the accumulated error of each device in the loop; calculated as the square root of the sum of the squares of individual device accuracy.

Louvers (dampers) devices used to control airflow.

Low (L) alarm the first alarm triggered when a process variable (e.g., fluids in a tank) drops below a predetermined low level; the purpose of this alarm is to notify a process technician that the level is abnormal.

Low low (LL) alarm alarm triggered when a process variable (e.g., fluids in a tank) continues to drop below a predetermined minimum level after a low (L) alarm has already been triggered. The purpose of this alarm is to notify a process technician that the alarm condition of a process variable is becoming increasingly abnormal and to initiate a shutdown or some other corrective action (e.g., shutting off power to a pump to prevent cavitations, which could result in pump damage).

Lower range value (LRV) the number at the bottom of the scale.

Manipulating element the final control element (e.g., control valve) manipulated by the corrective response of the controller output to maintain the process variable at the correct set point value.

Mass the amount of matter in a body or object measured by its resistance to a change in motion.

Mass flow meter a meter that eliminates the need to compensate for typical process variations such as temperature, pressure, density, and even viscosity; the most common type of true mass flow meter is the Coriolis.

Mass flow rate amount of mass passing through a plane per unit of time.

Mass flow units a unit of measure for the weight being passed through a certain location per unit of time; usually expressed in pounds per unit of time, as in pounds per minute.

Mass spectrometer (MS) a device capable of separating a gaseous stream into a spectrum according to mass and charge.

Measured variable a process variable that is measured.

Mechanical link a way of mechanically transmitting the motion of a primary sensor to a controlling mechanism; conveys linear or rotary motion by using a pivoting crank.

Mechanical transmission a signal transmission method that uses mechanical methods to create a signal.

Meniscus the curved upper surface of a column of liquid having either a convex or concave shape.

Molecular speed the rate of motion of molecules in a substance; it may change as a result of factors such as heat or collision with other molecules (e.g., the walls of a vessel).

Motor control center (MCC) hard-wired relays, switches, or contacts, housed in large metal cabinets that control the starting, stopping, and sequencing of motors and other devices.

Multiplexer (MUX) a device that merges or interleaves multiple signals into one stream or output signal in such a way that each individual signal can be recovered or separated out using a demultiplexer.

Multiport pitot tube multiport tube having four impact points spaced across the pipe and facing the flow, with another tube sensing static pressure; measures the average pressure produced by the four impact points; common brand name Annubar®.

Multivariable input industrial processes that involve an interaction between two or more process variables.

Node a single computer terminal located on a computer network or data highway.

Noise any unwanted or excessive sound; disturbances that affect a signal and may distort the information carried by the signal.

Noisy signal a signal that fluctuates dramatically and that is most likely the result of a loose sensor connection.

Nuclear device uses a tightly controlled gamma radiation source with a detector to inversely infer a tank's level; used when other technologies are unsuccessful; located on the outside of a tank and impervious to the effects of adverse process conditions; common gamma radiation sources are the radioisotopes cobalt 60 and cesium 137.

Offset the difference between the set point and the controlled variable under steady-state (not changing) conditions.

On/off a conditional logic statement used to control a process so that the process is either fully engaged or not engaged at all.

On/off control a controller that simply drives the manipulated variable from fully closed to fully open, depending on the position of the controlled variable relative to the set point.

Opacity analyzers optical analyzers used to determine how much particulate matter is in a gas sample.

Open control loop when a control loop does NOT have feedback (e.g., controller is in manual mode).

Optical measurement a measurement that uses reflection, refraction, or absorption properties of light to measure the chemical or physical properties of a sample.

Optical meters any or all of the following: turbidity meters, opacity meters, and color meters.

Optical transmission a signal transmission method that uses light frequencies to create a signal.

Orifice flanges flanges with holes drilled in them through to the pipe to allow pressures upstream and downstream of an orifice plate to be measured.

Orifice plate a piece of ⅛ in. to ½ in. thick metal with a calibrated hole drilled (or cut) through it; types of orifice plates include the concentric plate, eccentric bore, and segmental plate.

ORP analyzer an electrochemical analyzer that measures a small electromotive force (EMF) across a hydrogen-ion-sensitive glass bulb; an ORP (oxidation reduction potential) meter also measures a small voltage (EMF) across its electrode and has the capability of detecting all oxidizing and reducing ions in the solution.

Output the number or value that comes out from a computer program or process.

Overrange a condition that exists when the value of a measured variable exceeds the upper limit of an instrument.

Override controller a device or program that remains inactive until a specific constraint (highest or lowest permitted extreme) on the measured variable is about to be reached.

Overspeed a dangerous condition that can occur in a turbine or other type of equipment that moves too fast.

Oxidation reduction potential (ORP) a measure of redox potential created by the ratio of reducing agents to oxidizing agents present in the sample.

P/I conversion the conversion of an analog signal from pneumatic to current.

Parallel data communication one wire per bit or 64+ wires for a 64-bit binary word; used primarily in short distances (a few feet).

Percent level measurement a measurement of level based upon percentage, with 0 percent level being the lowest measuring point and 100 percent level being the maximum measuring point.

Percentage flow rate a common way to indicate a flowing process, with 100 percent equating to an actual quantity such as specified gallons per minute (GPM).

Permissives a set of conditions designed to ensure safe operations that must be met before a piece of equipment can be turned on.

pH a measurement of the hydrogen ion concentration in a solution that can react and indicate whether a substance is an acid or a base.

pH meter an electrochemical instrument that measures the acidity and alkalinity of a solution; an analyzer that measures hydrogen concentration in a solution.

Photometers color analyzers that operate in the visible light spectrum (400–800 μm).

Photometry system an automated system used to determine the color of a sample.

Piping and instrumentation diagram (P&ID) detailed illustration that graphically represents the relationship of equipment, piping, instrumentation, and flows contained within a process in the facility.

Pitot tube an L-shaped tube that is inserted into a pipe with its open end facing the flow and another tube sensing static pressure.

Plug the only movable component in the valve; it is actuated to open or close the flow path through the valve. May also be called a *disc* or *ball*.

Pneumatic powered by a gas.

Pneumatic signal an instrument communication with a range of 3 to 15 PSIG; must have an instrument air supply; has a lag time associated with the signal; relatively short transmission distances.

Pneumatic transmission an analog signal transmission method that uses compressed air/gas to create a signal.

Point level measurement level is measured at one distinct point in a tank.

Positive displacement flow measurement measurement of flow in absolute volumes where the flowing material is admitted into a chamber of known volume and then transferred to a discharge point; a counter registers the number of times the chamber fills and discharges.

Positive displacement meters piston, oval gear, nutating disk, and rotary vane types of positive displacement flow meters.

Pounds per square inch (PSI) a pressure measurement scale often used in liquid level measurement, using the liquid's hydrostatic head pressure; also expressed as inches of water column.

Precision how close repeated measurements are versus the action; reproducibility; the closeness of repeated measurements of the same quantity.

Pressure the amount of force a substance or object exerts over a specific area.

Pressure gauge the most common local instrument type for measuring pressure; the three primary types of pressure gauge measuring scales are absolute, gauge, and vacuum.

Pressure-reducing regulator a device used to regulate and/or control the pressure of a process fluid downstream of the device location.

Primary controller perceives the secondary loop as a separate entity; responds to one process variable (the controlled variable) while the secondary controller responds to another (the manipulated variable).

Process analyzers unattended analytical instruments capable of continuously monitoring a process stream.

Process control the act of regulating one or more process variables so that a product of a desired quality can be produced.

Process equilibrium the condition that exists when there is a balance of material and energy within a given system or process; also referred to as "steady state."

Process error the difference between set point and process variable (SP − PV).

Process flow diagram (PFD) basic illustration that uses symbols and direction arrows to show the primary flow path of material through a process.

Process variables (PVs) measurements of the chemical or physical properties of a process stream, generally relating to operating parameters or product specifications.

Process variable switch a type of switch that actuates when a predetermined value of a process variable (e.g., pressure, temperature, level, flow, or analytics) is present.

Programmable logic controller (PLC) a computer-based controller that uses inputs to monitor processes and outputs to control processes.

Properties characteristics of a substance, such as pH, ORP, conductivity, optical measurement, composition, and/or combustion capability.

Proportional action a controller output response that is proportional to the amount of deviation of the controlled variable from the set point.

Proportional band (proportional gain) the amount of deviation of the controlled variable from the set point required to move the output of the controller through its entire range (expressed as a percent of span).

Proportional, integral, derivative (PID) control tuning parameters utilized in a controller to ensure appropriate response to process variable changes in comparison to set point.

Proximity switch a type of switch operation requiring the presence, or closeness, of an object or device to facilitate its operation.

PSIA pounds per square inch absolute; zero on the absolute scale would represent the absence of any pressure.

PSIG pounds per square inch gauge; pressure measurement that references 14.7 PSIA as its zero point.

Pyrometer see infrared thermometer.

Qualitative a measurement of the properties of a substance.

Quantitative a measurement of the amount of properties within a substance.

Radiation the transfer of heat energy through electromagnetic waves.

Rankine an absolute scale of measurement to determine temperature; freezing point of water is 492 degrees Rankine (R) and boiling point of water is 672 degrees R.

Rate action the faster the rate of change of the process variable, the greater the output response (derivative action).

Ratio the proportion of two separate elements, such as the proportion of flow between two separate streams entering a mixing point.

Ratio control loop control loop designed to mix two or more flowing streams together while maintaining a quantitative ratio between them.

Ratio control control used in blending raw materials to make a final product and in regulating the correct proportion of flow rates of reactants feeding into a reactor; also a term used to describe maintaining the proper fuel-air ratio feeding into a furnace.

Ratio controller controllers that are designed to ratio (or proportion) flow rate between two separate flows entering a mixing point; designed so that its output represents the exact flow rate needed by the controlled flow loop to remain in alignment with the desired ratio to the uncontrolled flow; may also be capable of receiving two separate flow inputs and ratioing its output to the control valve located in the control line.

Recording the keeping of historical data by making a documented record.

Rectilinear speed linear speed expressed in distance per unit of time (e.g., feet per second).

Redundancy a design feature that provides more than one operation for accomplishing a given task, so the failure of one operation does not impair the system's ability to operate.

Reflex level gauge a local measuring device that has a single flat-glass panel capable of refracting light off a prism like backside, creating a silvery contrast above the liquid level that allows it to be seen from a distance.

Regulator a self-contained and self-actuating controlling device used to manipulate variables such as pressure, flow, level, and temperature in a process.

Relay an electromechanical device that boosts, maintains, or controls the electrical flow of a signal through a separate circuit.

Remote relating to instrumentation located away from the process (e.g., in a control room) and controlled by the distributed control system (DCS) or other panel board mounted controller.

Remote automatic control the act of controlling an instrument loop remotely from a control room.

Remote controller any controller that is not located in the processing area with the transmitter and control valve.

Remote manual control the act of controlling a valve manually from a remote location, such as a control room.

Remote processing unit (RPU) devices that control remote nodes in a DCS.

Remote set point (RSP) a set point received from an external source.

Representative sample a sample that contains portions of the process stream combined together over time or distance.

Resistance temperature detector (RTD) primary element that measures temperature changes in terms of electrical resistance.

Reverse acting controller a controller whose output signal value decreases as the controller input signal value increases, and vice versa.

Reynolds number a mathematical computation describing the quality of the flow of fluids numerically.

Rotameter a direct read variable area (tapered) flow tube in which fluid enters through the bottom, then flows upward, lifting a free floating indicator plummet (float); the position of the float against the calibrated marks on the glass tube indicates flow rate.

Rotational speed number of revolutions per unit of time (e.g., revolutions per minute, or rpms).

Sample system the various components of a sampling apparatus that obtains, transports, and returns the sample with the analyzer itself conditioning and analyzing the sample.

Scaling the act of equating the numerical value of one scale to its mathematically proportional value on another scale.

Seat the stationary part of the valve trim, connected to the body, that comes in contact with the valve plug; when the valve plug is fully seated, the flow through the valve ceases.

Secondary controller a special type of controller, sometimes called a cascade controller or remote set point controller; a standard controller with the added capability of choosing to receive a remote set point from an external source or a local (internal) set point.

Sensing the act of detecting.

Sensing element the control loop component that detects, or senses, the process variable.

Sensor (also known as a *primary element*) the control loop component that is in direct contact with the process variable to be measured; it senses or detects change and provides a signal to the transmitter.

Serial data communication two data wires; the most common means of communication used between plant equipment.

Set point the desired RPM indication where a controller is set for optimum operation.

Set point control the mechanism by which a technician can manipulate the set point.

Shelf position the contact position of an electrical device when de-energized (e.g., a "normally open" switch is normally open when it is de-energized).

Shielding a technique used to control external electromagnetic interference (EMI) by preventing transmission of noise or static signals from the source to the receiver; consists of foil, mesh, or woven wire.

Shutdown switch a switch used to actuate a circuit that shuts down a process system or equipment unit.

Signal converter (transducer) part of a transmitter that converts the process variable into a standard instrument signal; a device that converts one energy form into another.

Smart instruments instruments that have one or more microprocessors or "smart chips" included in their electronic circuitry so that they may be programmed and have diagnostic capability.

Span a number that expresses the difference between the upper and lower range values (URV and LRV) on a scale; sometimes referred to as an *operating range*.

Spanning the process of setting or calibrating the lower and upper range values of a device; also the process of testing a device or system by traversing up and down the scale through the entire calibrated or operational range.

Specific gravity the ratio of the density of a liquid or solid to the density of pure water *or* the density of a gas to the density of air at standard temperature and pressure (STP).

Spectrometer analyzer an analyzer used to detect chemical components in a process sample by measuring variations in transmittance (or absorption) of a spectrum of light passed through the sample; may use visible light (VIS), ultraviolet light (UV), or infrared light (IR and NIR) as a source and detection measurement.

Speed the distance traveled per unit of time irrespective of direction (e.g., feet per second).

Speed monitor a device that measures speed; comprises a speed sensor and a readout/receiving device.

Split-range controller where the output signal is divided between two final control elements; if there are more than two valves, it is called a dual split-range controller.

Spring the device that provides the energy to move a valve in the opposite direction of the diaphragm loading motion so that the valve can be opened and closed proportionate to the instrument signal; also provides energy to return the valve back to its fail-safe condition.

Standard instrument signal any kind of detectable quantity used to communicate information in a standard format for control.

Standard signal the language that instruments use to communicate with one another (4 to 20 mA, 3 to 15 PSIG, or digital).

Steady state a trend condition in which the process is in equilibrium and running smoothly, producing a relatively straight line on a graph or chart.

Stem the pushing rod that transfers the motion of the actuator to the valve plug.

Strain gauge transducer a pressure sensing and measuring device consisting of a group of wires that stretch when pressure is applied, creating resistance, thereby changing the process pressure into an electronic signal.

Switch an electrical device used to start, stop, or otherwise reconfigure the flow of electricity in a circuit.

Symbology various graphical representations used to identify equipment, lines, instrumentation, or process configurations.

Tape a narrow strip of calibrated ribbon (steel) used to measure length.

Tape gauge a local level-measuring device consisting of a metal tape that has one end attached to an indicator and the other end attached to a float residing on top of the liquid level.

Temperature the average kinetic energy of a material; the indication of heat energy available to flow between bodies of differing temperatures; the measure of the thermal energy of a substance.

Temperature differential (deltaT or ΔT) difference between two related temperatures or between temperatures measured at two different points.

Temperature gauge an independent analog device with a sensing element such as a bimetallic strip, bourdon tube, or bellows that is linked to a pointer displaying the temperature on a calibrated face.

Thermistor a type of resistor used to measure temperature changes, relying on the change in its resistance with changing temperature.

Thermocouple primary element consisting of two wires of dissimilar metals connected at one end; when it is exposed to heat, it generates a voltage proportional to the change in temperature.

Thermowell a thick walled, typically stainless steel device shaped like a tube, which is inserted into a hole in piping or equipment; it is specifically prepared to house a temperature sensing and measuring element.

Total carbon analyzers analyzers used to determine how much carbon is in a sample; used to detect carbon based contaminants in steam condensate and wastewater.

Total flow the quantity of fluid moving past a given point per day (e.g., 4 GPM × 60 min/hr × 24 hr = 240 × 24 = 5760 gal/day).

Transducer a device that receives information in one form of instrument signal and changes it into another form of instrument signal.

Transmitter the control loop component that receives a signal from the sensor, measures that signal, and sends it forward to the controller or indicator.

Transmitting communicating via electronic or pneumatic signal from one control loop component to another.

Transparent level gauge a local measuring device that has two transparent flat-glass panels, front and back. These flat-glass panels, in conjunction with the metal sides, form a vertical chamber where the process level can be seen and measured.

Trend the plotting of a process variable over time.

Trending the analysis of a change in measured data over at least three data measurement intervals.

Troubleshooting the systematic search for the source of a problem so that it can be solved.

Tubular type sight glass a local measuring device that has a reflex or clear glass tube open into a vessel on top and bottom.

Tuning adjusting the control action settings so that they produce an appropriate dynamic response to the process resulting in good control.

Turbidity analyzer an optical analyzer used to determine the cloudiness of a liquid.

Turbidity and opacity meters optical meters; measure the transmittance or absorption of light passing through a sample.

Turbine meter a flow tube containing a free spinning turbine (fan) wheel where the revolutions per minute (rpm) are proportional to flow.

Turbulent flow a condition in which the fluid flow pattern is disturbed, so there is considerable mixing.

U-tube manometer a gravity balanced pressure measuring device with two fluid chamber tube gauges connected by a U-shaped tube so fluid (a liquid or mercury) flows freely between the chambers.

Ullage (outage) the measurement from the surface of the product to a reference point at the top of the tank.

Ultrasonic and radar device an accurate distance measuring instrument usually inserted into the top of a vessel; emits a pulse of energy that is reflected off the surface of the material back to the receiver; may be either a single unit (transponder) or two separate units (transmitter and receiver).

Uncontrolled flow flow in a process line where there is no control valve.

Uninterruptible power supply (UPS) a backup power unit, usually consisting of large batteries, rectifier, inverter, battery charger, and transfer switch that provides continuous auxiliary power when the normal power supply is interrupted.

Upper range value (URV) the number at the top of the scale.

UPS alarm an audible or a visual signal used to draw attention to problem situations.

Vacuum a condition in which the pressure measured is less than atmospheric pressure; usually measured in inches of mercury (in. Hg).

Valve positioner a device that ensures a valve is positioned properly according to the signal received from the controller.

Velocity the distance traveled over time or change in position over time.

Venturi tube a primary element used in pipelines to create a differential pressure (D/P) such that when converted to flow units (e.g., GPM), the flow in the line is measured.

Vibration the periodic motion of an object.

Vibration alarm an alarm designed to trigger if the level of vibration increases above an acceptable level in order to protect vibrating equipment.

Vibration meter a device used to measure displacement, velocity, or acceleration due to vibration; consists of a pickup device, an electronic amplification circuit, and an output meter.

Vibration sensor a device used to determine the velocity, acceleration, displacement, or any combination of these characteristics for the purpose of predicting wear or impending failure. High vibration would activate a switch for an alarm or equipment shutdown.

Vibration sensors or monitors devices used to sense the effects of vibration and send a signal to a meter or monitor, or to shut down a device if operating limits are exceeded.

Visual color analyzers compare a physical sample with a standard by shining visible light through a sample; the color comparison is made by the human eye/brain.

Volumetric flow units a unit of measure for the volume being passed through a certain location in a pipe or channel per unit of time; usually measured in gallons per minute and cubic feet per minute.

Voting logic   computer logic that analyzes the signals from several devices, all of which are monitoring the same condition; it will initiate shutdown if a majority or a single high priority variable of the monitoring devices signals a dangerous condition.

Welded pad gauge   a local measuring device generally made of flat glass and integrally mounted to the vessel by either a welded or flanged connection with threaded pipe; extremely rugged to prevent vessel drainout.

Windup   the saturation of the controller output signal to its minimum or maximum value if the process variable does not return to the set point.

Zeroing out   adjusting a measuring instrument to the proper output value for a zero measurement signal.

Index

A

Absolute pressure, 27, 29
   measuring, 29–34
Accuracy, of DCS, 312
Accuracy vs. precision, 184
Actuator, 228, 230–31
Actuator symbols, 144–45
Air supply pressure gauge, 233
Alarm systems, 159–61
   alarm switches, 153
   in emergency shutdown (ESD) systems, 333–36
   uninterruptible power supply (UPS) alarms, 322–23
Algorithms, 185
Alternating current (AC). *See* Power supply to process plant
Amplification of signal voltage, 168
Analog instruments, 12
   signals, 12, 168–69, 187–88, 290
Analytical instruments
   of DCS, 313–14
   elements, malfunction of, 351–52
   purpose of, 111–12
   role of process technician, 124
   sensing or measurement instruments, 111–21
   types of instruments and terms, 109–11
And/or logic, 301
Atmospheric pressure, 28
Automatic/manual switch, 208
Automatic process control
   history of, 3–5, 184
   local, 268–69
Automatic to manual and manual to automatic switching procedure, 211
Automatic transfer switch, (ATS), 322
Autostart switch, 154, 155

B

Back-pressure regulators, 239
Backup
   copies of plant configuration, 312
   feature of DCS, 311
Badge monitors, 120
Ball control valve, 237–38
Balloon, 137
Bar/millibar, 28
Barometer, 28
Basic equipment symbols, 136
Battery power. *See* Power supply to process plant
Bernoulli's principle, 93
Bimetallic strip, 51–52
Binary data
   conversion of signals to, 169, 290–91
   and data highway, 309
   and error correction, 168

Block flow diagram (BFD), 130
Body, of valve, 227
Boiling point of water, 47–48
   for Celsius and Fahrenheit, 45
   effects of pressure and contaminants on, 47
   and relationship to pressure, 17, 24
Bonnet, of valve, 227
Boolean logic, 301
Bourdon tube, 31
Breakers. *See* Emergency shutdown (ESD) systems; Power supply to process plant
British thermal unit (BTU), 44
Bubblers, 78–79
Bump, 211
Bumpless transfer, 211
   of DCS, 313
Buoyancy, measuring, 80–81
Butterfly control valves, 236–37
Bypass switches, 154, 155

C

Calibration, in troubleshooting, 345–46, 363
Calorie, 44
Capacitance transducers, 34
Carbon analyzers, 119–20
Cascade control loop, 209
Cascade/remote set point (RSP) controllers, 214–15
Cascading
   control, 274–78
   of remote control, 270–71
Categories
   of instruments, 6–13
   of level measurement, 67–8
Celsius
   Anders, 45
   scale, 47
Centigrade. *See* Celsius
Changes in variables, effect of, on measurement, 18
Chromatography, 111
   gas chromatography (GC), 117–18, 119
Closed-loop control (feedback), 180, 181–82, 248, 267
Coefficient of thermal expansion, 51
Cold or reference junction, 53–54
Color/optical analyzers, 114
Column, defined, 111
Combined control action, 261
Combustion, 111
Communications network. *See* Distributed control systems (DCSs)
Compound vacuum/pressure gauge, 32
Computers. *See* Microprocessors
Conditioning, signal, 168
Conduction, 44, 110
   conductivity meters, 113–14

383

Configuration station, 312
Continuous environmental monitoring systems (CEMS), 111
Continuous level measurement
  defined, 67
  displacers as, 81
  floats as, 75
  nuclear devices as, 82
Controllers, 10. *See also* Control loops; Digital control systems
  applications, 212–16
  automatic to manual mode, 212
  bumpless transfer, 211
  characteristics, 208–11
  combined control action, 261
  comparator part of, 185
  control actions or modes, 247–52, 268–71
  control loops, 207–8
  control schemes, 266–68, 274–78, 282–85
  defined, 247–48
  malfunction of, 352–53
  manual to automatic mode, 212
  process recording, 252–54
  with split-range valves, 213–14
  tuning modes, 255–61
Control loops, 9, 10. *See also* Controllers; Signal transmission
  automatic, history of, 4–5
  closed-loop control or feedback loop, 181–82, 248, 267
  controller, 207–8 (*see also* Controllers)
  final control element, 216–19
  functions and signals, 182–86
  observation of, 201
  open-loop control, 182, 249, 268
  ratio control loops, 278–82
  response to instrumentation failure, 352–54
Control mode, 210
Control systems. *See* Control loops
Control valves, 207, 217–18, 227
  actuators, 230–32
  for control configurations, 235–38
  failure conditions, 229–30
  instrument air regulator, 238–40
  key components, 227–28
  output signals, 234–35
  positioner, 232–34
  sliding stem valves, actions for, 230–31
Convection, 44
Conversions
  analog and digital, 293, 294
  pressure measurement tables, 28
  pressure scales, 34–37
  of signals, 169–70
  temperature scales, 57–58
  transducer, role of in, 185–86
  and transmitting in control loop, 183–84
Cooling water flow transmitter, 210
Coriolis meter, 98
Cost-effectiveness, of DCS, 313
Crosshatched lines, 146
Current to pneumatic converter (I/P), 185
  as relay, 159
Current-to-pneumatic transducer, 228
Cybersecurity, 314
Cycling, 262

D
Dampers (louvers), 218
Data highway, 305, 308–10
Data management improvement, of DCS, 310–13

Datum line, 68
Dead time, 248
Delta. *See* Differential pressure
Deluge systems, 336, 339
Demultiplexers (DEMUX), 292
Density
  and hydrostatic head pressure, 72–73
  and relationship to pressure, 25–26
Derivative action (rate), 211, 252, 259–60
Desired value. *See* Set point
Deviation, 253
  fixed, in troubleshooting, 347
Differential pressure. *See also* Pressure
  D/P cell as sensor, 32–33, 77–78, 79, 196, 351
  gauge, 32
  as measurement, 26–27
  as process variable, 6–7
  transmitters, 98–99, 197–98, 351
Digital control systems
  versus analog, 290–91
  components of, 289, 291–92
  history of, 290
  signals, 169, 188, 292–94
Direct acting/reverse acting (DIR and REV), 209
Direct control. *See* Digital control systems
Direct current. *See* Power supply to process plant
Direct flow measurement, 91
Direct level measurement, 67
Direct or reverse acting controllers, 249, 250
Discrete controllers, 292
Displacers, 80–81
Distance-measuring devices, 81–82
Distributed control systems (DCSs), 290
  advantages of, 312–13
  architecture of, 307–12
  history of, 305–6
  operation of, 306–307
  processes, equipment and safety, 313–14
Dosimeters, 120
Duty cycle, 219
Dynamic equilibrium, 245

E
Electrical power supply. *See* Power supply to process plant
Electrical transmission of signals, 169
Electric eye sensors, 333
Electric motor, 218–19
Electrolyte, 114
Electromagnetic interference (EMI), and signal interference, 166–67
Electromagnetic meters (magmeters), 96–97
Electromotive force (EMF), 53
Electronic instruments
  analog signals, 187–88
  digital signals, 188, 201
Emergency shutdown (ESD) systems
  alarms, 333–36
  avoiding shutdowns, 336–37
  breaker switches, 332
  bypassing, 340
  electric eye sensors, 333
  interlocks, 330, 331–32
  permissives, 329–30
  resetting and testing, 338–40
Emissivity, 50
Environmental monitoring and reporting, 111
Environmental Protection Agency (EPA), 111
Equilibrium, process, 245

Equipment
　damage, during testing, 339
　mechanical integrity of fixed, 112
　permissives, 329–30
Equivalency, unit, 29
Equivalent units method, 36
Error
　in control loop, 189
　correction and signal transmission, 167–68
　process, 182
Ethernet, 309

F
Fahrenheit
　Gabriel, 45
　scales, 47
Fail closed, control valve, 229
Fail in place, control valve, 230
Fail last (fail in place), 217
Fail open, control valve, 229
Fail-safe situations
　and distributed control systems (DCSs), 305
　and shutdown switches, 153
Faraday's law of induction, 96–97
Feedback loop. *See* Control loops
Feedforward loop. *See* Control loops
Fiberoptic transmission of signals, 169
Filled thermal system, 56
Final control element, in control loop, 216
　control valves, 217–18
　electric motor, 218–19
　louvers (dampers), 218
Final control elements, 144
Flat-glass level gauges, 75
Floats and float gauges, 75–76, 95, 96
Flow, 143
Flow/flow rate
　measurement, types of, 92–101
　orifice meter malfunction, 350–51
　as process variable, 90
　Reynolds number, 90–91
　sensing and measurement instruments, 92–101
　signal conversion, 172
Flow nozzles, 94
Force. *See* Pressure
Free electrons, 110, 113
Freezing point of water, 47–48
　for Celsius and Fahrenheit, 45
Front panel controllers, 208–9

G
Gain (proportional band), 210, 251, 255–58, 260
Gas detection alarms, 336
Gas sample analyzers. *See* Chromatography
Gauge or sight glass, to measure level, 74–75
Gauge pressure, 27
　measuring, 29–34
General formula for converting pressure units, 36–37
Generator, emergency, 322
Glass-stem thermometer, 49–50
Globe control valves, 235
Grounding and shielding, in signal transmission, 166–67

H
Hand-drawn clouds, 145
Hand-Off-Auto (HOA) switches, 153, 158
Handwheel, 228

Head pressure
　liquid (hydrostatic), 70, 72–73
　measuring in closed environment, 78
Heat energy
　and molecular movement, 44
　transfer of, 45
Height of liquid, calculating, 70–71
Hunting, 254
Hydrostatic head pressure. *See* Head pressure

I
Ideal Gas Law in physics (PV=nRT), 45
If/then logic, 301
Impulse tubing, 196
Inches of mercury, 28, 35
Inches of water, 28, 35
Indicating instruments, 30
　pressure gauges, 31
　temperature gauges, 56
Indirect flow measurement, 91
Indirect level measurement, 68
　load cells, 83
Infrared thermometer, 50–51
Inline analyzer, 100
Innage, 66
Input/output (I/O) devices, 307–8
Instrument air regulator, 238–40
Instrumentation process, of components, 9–13
Instrumentation symbols
　basic equipment symbols, 136
　interpretation, 137–46
　line symbols, 136–37
Instrument balloon symbols, 19
　examples, 139–40
Instrument input pressure gauge, 233
Instrument malfunction. *See also* Troubleshooting
　failure modes, 349–52
　identifying, 344–49
　loop response to, 352–54
Insulation, 165, 167
Integral action (reset), 210–11
Integral (reset) controller response, 251, 258, 259
Integral orifice flow meter, 100–1
Interface level, 68–69, 81
Interlocks. *See* Emergency shutdown (ESD) systems
International Organization for Standardization (ISO), 133
International Society of Automation (ISA), 188
Internet, operator access to, 312
Internet of Things, 314
Internet Protocol, 173–74, 309
Intrinsically safe devices, 314
I/P converter, 185
I/P transducer, 228
ISA instrument tag number, 134–36
ISA S5.1, 133

K
Kelvin temperature scale, 46–47
K-factor, 98

L
Ladder diagrams, 298, 299–300
Lag time, 246, 276
Laminar flow, 89
Lead/lag control, 266–67
LEL. *See* Lower explosive limit (LEL)

Level, 142
  in relation to temperature, density, and volume, 71–73
    sensing and measurement instruments, 73–73
    terms and definitions, 65–71
Limit switches, 155, 156
Line symbols, 136
Live zero, 189, 201
Load cells, 83
Load change, 245, 246
Local area network (LAN), 312
Local controllers, 212
Local or field instruments
  controllers, 268–69
  pressure gauges, 31
  temperature gauges, 56
Local/remote (L/R) switch, 209
Logic
  in programmable logic controllers, 301
  voting, as redundancy in alarm systems, 337
Louvers (dampers), 218
Lower explosive limit (LEL), 117
Lower range value (LRV), 171, 172

M

Magnetic flowmeter (magmeter), 96–97
Malfunction of equipment. *See* Instrument malfunction; Troubleshooting
Manipulating process stream in control loop, 186
Manometers, 30–31
Mass and relationship to pressure, 25
Mass flow meters, 98
Mass flow units, 91–92
Mass spectrometers, 118–19
Measured junction, 53
Measurement, of temperature, 46–47
Measuring devices
  for pressure, 29–34
  for temperature, 46–47
Mechanical integrity of fixed equipment, 112
Mechanical link in control loop, 188–89
Mechanical transmission of signals, 168
Meniscus, 30, 69–70
Mercury
  inches of, 28, 35
  and temperature, 44
Microprocessors. *See also* Programmable logic controllers (PLCs)
  in controllers, 285
  role in development of digital control, 294
  in smart transmitters, 201
  use of, in gas chromatography, 118
Millibar, 28
Molecular speed, 25
Monitoring process variables, 5–6
Monitors, 120–21
Motor control centers (MCCs), 298, 318–19, 322
Multiplexers (MUX), 292
Multiport pitot tube, 95
Multivariable control, 283–85

N

Networks. *See* Distributed control systems (DCSs)
Nodes. *See* Distributed control systems (DCSs)
Noisy signals, 347, 348
Normally open (N.O.)/normally closed (N.C.) contacts, 152, 300
Nuclear devices, 82
  radiation monitors, 120

O

Offset, 253, 262
Online product analysis, 112
On/off control, 250, 251, 266, 300
Opacity analyzers, 114–17
Open internet protocol transmission/addressing, 173–74, 309
Open-loop control, 180, 182, 249, 268
Operator or work stations, 310–11
Optical analyzers
  color, turbidity, opacity, 114–17
  measurement of electromagnetic waves, 111
Optical transmission of signals, 169
Orifice plates, 26–27, 92–93, 100–101
  meter malfunction, 350–51
Outage, 66
Output pressure gauge, 233
Over range, 246, 247, 358
Override controllers, 285
Oxidation reduction potential (ORP), 110
  meter, 112
Oxygen analyzer, 117

P

Parallel data communication, 188
Percentage flow rate, 91
Percent D/P, conversion of, 172
Permissives. *See* Emergency shutdown (ESD) systems
Personnel monitors, 120
pH, 110
  meters, 112–13
  scale, 113
Piezoelectric transducers, 34
Piping and instrumentation diagram (P&ID), 131–33
Piston-type actuators, 231–32
Pitot tubes, 94–95
Plug, of valve, 227
Plummet, 96
Pneumatically driven actuators, 231–32
Pneumatic instruments
  pneumatic (analog) signal, 168, 187, 201
Pneumatic valve positioners, 233–34
Point level measurement
  defined, 67
  displacers as, 81
  floats as, 75
  nuclear devices as, 82
Positive displacement flow measurement, 91
Pounds per square inch scales, 27
Power supply to process plant, 318–21
  uninterruptible power supply (UPS), 319, 321–23
Preact mode, 252, 259–60
Precision *vs.* accuracy, 184
Pressure, 141. *See also* Differential pressure
  boiling point of water, 17
  conversions, 34–37
  defined, 23–24
  elements malfunction, 351
  force exerted by molecules, 24–26
  gauges, 29, 31–32
  height of liquid effect, 15–16
  measurement of, 26–29
  pressure-sensing and measurement instruments, 29–34
  temperature change effect, 14–15
  units, 35–36
Pressure-reducing regulators, 239–40
Primary controller, 214
Printers and other accessory devices, 312

Process analyzers, 109
Process control, 179–80. *See also* Control loops
Process diagrams
   block flow diagram, 130
   piping and instrumentation diagram, 131–33
   process flow diagram, 131
Process equilibrium, 245
Process error in feedback control loop, 182
Process flow diagram (PFD), 131
Process technician, role of, with analytical instruments, 124
Process variable (PV), 6
   analytics, 8–9
   flow, 7
   level, 8
   pressure, 6–7
   relationships, 14–18
   switches for, 156
   temperature, 7
Programmable logic controllers (PLCs), 292, 300
   historical background of, 298–99
   logic in, 301
   programming, 301–2
Properties, measured, 180
Proportional band, 210
Proportional gain, 210, 251, 255–58, 260
Proportional, integral, and derivative (PID) settings, 251–52, 255, 260–61
Protection well, 55
Protocol, open Internet (intranet), 173–74, 309
Proximity switches, 155

Q
Qualitative and quantitative analyses, 110
Quality assurance, role of analyzers in, 112, 119

R
Radar distance-measuring device, 81
Radiation, 44
Radio frequency interference (RFI), and signal interference, 166–67
Range, 171, 172
Rankine temperature scale, 46–47
Rate action. *See* Derivative action (rate)
Rate mode, 252, 259–60
Ratio control, 278–82
Ratio controllers, 215–16
Recording process, 252–54
Reduction-oxidation (redox). *See* Oxidation reduction potential (ORP)
Redundancy feature
   of DCS, 311–12
   of UPS, 337
Reference or cold junction, 53–54
Reflex level gauges, 74
Regulator, 228
Relays, 158–59
   applications, 159
   types and functions, 159
Reliability, of DCS, 313
Remote controllers, 213
Remote instruments
   controllers, 269–70
   remote processing units (RPUs), 291
Remote set point (RSP), 209, 274
Repeatability, of DCS, 312
Reset (integral) controller response, 251, 258, 259
Resetting and testing. *See* Emergency shutdown (ESD) systems

Resistance temperature device (RTD), 52–53
   in control loop, 183, 196
   failure modes of, 349–50
Reverse or direct acting controllers, 249, 250
Reversible electric motor, 219
Reynolds number, 90–91
Rotameters, 95–96
Rotary valves. *See* Butterfly control valves

S
Safety concerns in signal transmission, 168. *See also* Emergency shutdown (ESD) systems
Safety, of DCS, 313
Sample system, 109
Scales
   pressure measurement, 27
   temperature measurement, 46–47
   and transmitters, 198–200
Scaling calculations for signal conversion, 170
Seat, 227
Secondary controller, 214
Seebeck effect, 53
Segmented ball control valves, 237–38
Selection relays, 159
Sensors, 195–97
   purpose and operation, 194–95
   sensing element in control loop, 183
Set point (SP), 180, 185, 209
   remote (RSP), 274
   track, 274
Shelf position, 300
Shielding, 165, 166–67
Shutdown
   circuits, 330
   switches, 153, 154
Side panel, 209–11
Sight glass or gauge, to measure level, 74–75
Signal transmission, 186–89
   common types of, 168–69
   conversion, 169–70, 292–94
   noisy, 347, 348
   and selection relays, 159
   standard signal, 197
   by transducers, 200–1
   by transmitter, 198–200
Single point measurement, 46
"Smart" digital devices. *See* Transmitters
Smart instruments, 364
Space efficiency, of DCS, 313
Span/spanning, 171, 172
Specific gravity, 26
   in formulas for head pressure, 66–69
Spectrometers, 119
Speed and relationship to pressure, 25
Split-range control, 282–83
Split-range valves, 213–14
Spring, 228
Spring and diaphragm actuators, 231
Stand-alone digital controllers, 294
Standard bias (live zero), 189
Standard cell, 114
Standard line symbols, 137
Steady state, 245
Stem, 228
Strain gauge transducers, 33–34

Streamlined flow, 89
Switches
    categories, 151–52
    nomenclature and identification, 152–57
    symbols for, 158
Symbology, 133
    ISA instrument tag number, 134–36
    ISA S5.1, 133
Symbols for switches, 158

T

Tapes and tape gauges, 76–77
Target flow meter, 100
Temperature, 141–42
    coefficient of resistance, 53
    conversion, 57–58
    defined, 43
    differential (delta), 46
    gauge, 56–57
    heat energy and molecular movement, 44–45
    heat transfer, 45
    measurement, 45–48, 46f
    and molecular speed, 25
    sensing and measurement instruments, 49–47
    standard temperature, 26
Testing. *See* Emergency shutdown (ESD) systems
Thermistors, 54–55
Thermocouples, 53–54, 195
    failure modes of, 349–50
Thermometers, 49–51
Thermowell, 55–56
Three-way control valves, 235–36
Toggle switches, 153
Total carbon analyzers, 119–20
Transducers
    capacitance, 34
    in control loops, 195
    and conversion of signals, 169–70
    load cells, 83
    measuring, 185–86
    signal transmission, 200–1
    strain gauge, 33–34, 83
Transmission Control Protocol/Internet Protocol (TCP/IP), 173–74, 309
Transmitters, 53, 80–81, 197–98. *See also* Differential pressure; Signal transmission
    in control loops, 195
    and converting-to-measurement process, 183–84
    malfunction of, 352
    scaling, 198–200
    and signal transmission, 165
    smart, 201, 294, 307
Transparent level gauges, 75
Transponder, 81
Trend, 253, 254
    of DCS, 312–14
Troubleshooting. *See also* Instrument malfunction
    calibration, 363
    chart recorders as tool for, 252
    communication during, 362
    controller tuning errors, 262
    equipment, 362–64
    methods, 358–62
    safety and environmental issues, 364
    smart instruments, 364
    typical malfunctions, 358
    of uninterruptible power supply (UPS), 323
Tubular type sight glass, 74
Tuning control loop, 209, 210–11, 255–62
Turbidity meters, 115–16
Turbine meter, 97–98
Turbulent flow, 90

U

Ullage (outage), 66
Ultrasonic
    detection methods, 120
    distance-measuring device, 81–82
Ultrasonic flow meter, 101
Uncontrolled flow, 215
Uninterruptible power supply (UPS). *See* Power supply to process plant
Unit equivalency, 29
Upper explosive limit (UEL), 117
Upper range value (URV), 171, 172
U-tubes, 28–29

V

Vacuum pressure, 27, 28, 29
    measuring, 29–34
Valve positioner, 228
Variable-speed electric motors, 219
Venturi tubes, 93
Vibration
    alarm and shutdown, 335–36
    sensors, 156
Volumetric flow units, 91
Vortex flow meter, 100
Voting logic system, 337

W

Water
    boiling and freezing, 47–48
    inches of, 28, 35
    and temperature, 43
    water column, inches of, 70
Welded pad gauges, 75
Wet leg instruments
    level transmitter, 78
    zeroing out, 170–71
Windup, 251
Wireless technologies, 173
Work or operator stations, 310–11

Z

Zeroing out, 171
Zero, live, 189, 201